LONDON MATHEMATICAL SOCIETY LECTURE NOTE SERIES

Managing Editor: Professor N.J. Hitchin, Mathematical Institute, 24–29 St. Giles, Oxford OX1 3DP, UK

All the titles listed below can be obtained from good booksellers or from Cambridge University Press. For a complete series listing visit http://publishing.cambridge.org/stm/mathematics/lmsn/

LONDON MATHEMATICAL SOCIETY LECTURE NOTE SERIES. 328

Fundamentals of Hyperbolic Geometry:
Selected Expositions

Edited by

RICHARD D. CANARY
University of Michigan

DAVID EPSTEIN
University of Warwick

ALBERT MARDEN
University of Minnesota

CAMBRIDGE
UNIVERSITY PRESS

CAMBRIDGE
UNIVERSITY PRESS

University Printing House, Cambridge CB2 8BS, United Kingdom

One Liberty Plaza, 20th Floor, New York, NY 10006, USA

477 Williamstown Road, Port Melbourne, VIC 3207, Australia

314-321, 3rd Floor, Plot 3, Splendor Forum, Jasola District Centre, New Delhi - 110025, India

103 Penang Road, #05-06/07, Visioncrest Commercial, Singapore 238467

Cambridge University Press is part of the University of Cambridge.

It furthers the University's mission by disseminating knowledge in the pursuit of
education, learning and research at the highest international levels of excellence.

www.cambridge.org
Information on this title: www.cambridge.org/9780521615587

First published 2006

A catalogue record for this publication is available from the British Library

ISBN 978-0-521-61558-7 Paperback

....

Contents

Preface

During the academic year 1983/84, the Science and Engineering Research Council of the United Kingdom gave generous financial support for two symposia, at the Universities of Warwick and Durham, on hyperbolic geometry, Kleinian groups and 3-dimensional topology. The symposium at Durham was also sponsored by the London Mathematical Society. I would like to express my thanks to both the SERC and the LMS for their help and support. It is a pleasure to acknowledge the help of my co-organizer at Durham, Peter Scott, who was also an unofficial co-organizer at Warwick. He made an essential contribution to the great success of the symposia.

The world's foremost contributors to this very active area were all invited, and nearly all of them came. The activity centred on the University of Warwick, and climaxed with a 2-week long intensive meeting at the University of Durham during the first 2-weeks of July 1984. There was earlier a period of intense activity during the Easter vacation of 1984, when a number of short introductory lectures were given. The text of the most important of these series of lectures, by S.J. Patterson, is published in these Proceedings.

The papers published here are the result of an invitation to all those attending the two Symposia to submit papers. Not all the papers submitted were the subject of talks given during the Symposia – the contents of the Proceedings are based on their relevance to the subject, and not on their accuracy as documents recording the events of the Symposia. Also, a number of important contributions to the Symposia are not published here, having been previously promised elsewhere.

One of the few advantages of being an editor is that one can confer certain rights and privileges on oneself. I have taken the opportunity of accepting as suitable for publication several rather large papers of which I was the author or co-author, and which have a substantial element of exposition. This is a field which is expanding very quickly, mainly under Thurston's influence, and more material of an expository nature is sorely needed, as many of those attempting to penetrate the area will testify. I hope that my own efforts in this direction will be of some help.

Due to the amount of material submitted, it has been necessary to publish two separate books. The division was done on the basis of first sorting into five different fairly narrowly defined subject areas, and then trying to balance the sizes of the two books. The books are entitled "Analytical and Geometric Aspects of Hyperbolic Space" and "Low-Dimensional Topology and Kleinian Groups".

At an early stage I made the decision to set the books by computer, with the advantage that the entire process would be under my own control. This has been an interesting experience. I will content myself with the comment that computer typesetting is not the joy and wonder that I once thought it would be; the considerable delay in publication has been largely due to the unforseen difficulties encountered in this process.

My particular thanks go to Russell Quin, without whose help the typesetting difficulties would never have been overcome. I am grateful to Kay Dekker, who has spent many hours creating fonts of special characters needed for this work. I would like to thank the University of Warwick Computer Unit for the use of their facilities for the printing.

Finally I must thank the contributors for their patience and forbearing during the long delay before publication.

D.B.A. Epstein
Mathematics Institute,
University of Warwick,
Coventry, CV4 7AL,
ENGLAND.

12 July 1986

Preface 2005

In 1987 Cambridge University Press published the proceedings of meetings in Warwick and Durham as Lecture Notes 111 and 112. The original preface is reprinted above. Taking account of demand, CUP suggested that certain of the articles in the original two volumes be reprinted. We have chosen four of them to comprise a single new volume. We believe they will continue to be helpful to those learning about the field. The authors have kindly given permission to reissue their work.

Part I, "Notes on Notes ..." (N&N), has a new foreword by Canary. Particular topics in Chapters 8 and 9 of Thurston's original lecture notes had formed the basis for much of N&N. Still, there is other important material in those chapters that has not been widely digested by the mathematical community at large, yet has turned out to play key roles in later developments. The purpose of the foreword is to provide a guide to the recent literature where explanations of this additional material are now available. Also included are brief accounts of the recently announced solutions of the tameness and ending lamination conjectures, direct generalizations of topics in those fateful chapters.

Likewise, in Part II, the centrality of the convex core in studying hyperbolic manifolds has become even more apparent in the intervening years as important additional details of its structure have been worked out. Accordingly, the careful account of this topic given in the original article on convex hulls and Sullivan's theorem has been brought up to date with a new addendum by Epstein and Marden.

Part III in this volume is Thurston's famous paper that develops the notion of "earthquake" and proves the Earthquake Theorem. Earthquake maps are to hyperbolic geometry much like quasi-conformal maps are to complex analysis.

Part IV consists of five lectures by S.J. Patterson that develop from basic principles the theory of measures on limit sets of Kleinian groups. The theory developed applies to discrete groups acting on any dimension hyperbolic space. In particular the exponent of convergence is discussed in reference to the geodesic flow and to the Hausdorff dimension of the limit set. The Sullivan–Patterson measures play an increasing role, in differential geometry, in parts of ergodic theory and in geometric group theory.

The mistakes (that we are aware of) have been corrected. We are grateful to Curt McMullen and to Caroline Series for bringing to our attention a number of infelicities.

Looking back over the 18 years since the original volumes were written, it is amazing how much has been accomplished in the subject. This period could well serve as a model example of collective efforts of the many researchers in a field resulting in a very dramatic increase in the knowledge and depth of understanding of it.

We thank David Tranah for suggesting the reprinting, and for his continuing patient support and encouragement along the way. It was a great relief when he was able to arrange for retyping the original articles into LaTeX. It turns out that automated conversion from Troff to LaTeX only works well with relatively simple files.

Richard D. Canary
David B.A. Epstein
Albert Marden

February 2005

PART I

Notes on Notes of Thurston

R.D. Canary, Department of Mathematics, University of Michigan,
Ann Arbor, MI 48109, USA

D.B.A. Epstein, Mathematics Institute, University of Warwick,
Conventry CV4 7AL, UK

P.L. Green, Barton Peveril College, Eastleigh, Hampshire SO5 5LX, UK

A New Foreword

*Richard D. Canary**

The article "Notes on Notes of Thurston" was intended as an exposition of some portions of Thurston's lecture notes *The Geometry and Topology of Three-Manifolds*. The work described in Thurston's lecture notes revolutionized the study of Kleinian groups and hyperbolic manifolds, and formed the foundation for parts of Thurston's proof of his Geometrization theorem. At the time, much of the material in those Notes was unavailable in a published form. In this foreword, we point the reader to some more recent publications where detailed explanations of the material in Thurston's original lecture notes are available. We will place a special emphasis on Thurston's Chapters 8 and 9. This material was the basis for much of our original article and it still represents the part least well-digested by the mathematical community. This is also the material which has been closest to the author's subsequent interests, so the selection will, by necessity, reflect some of his personal biases.

We hope this foreword will be useful to students and working mathematicians who are attempting to come to grips with the very beautiful, but also sparingly described, mathematics in Thurston's notes. No attempt has been made to make this foreword self-contained. It is simply a rough-and-ready guide to some of the relevant literature. In particular, we will not have space to define all the mathematical terms used, but we hope the reader will make use of the many references to sort these out. In particular, we will assume that the reader has a copy of Thurston's notes on hand. We would also like to suggest that it would be valuable for a publisher to make available Thurston's lecture notes, in their original form. The author would like to apologize, in advance, to the mathematicians whose relevant articles have been omitted due to the author's ignorance.

* partially supported by grants from the National Science Foundation

In a final section, we describe some recent progress on the issues dealt with in Chapters 8 and 9 of Thurston's notes.

We would like to thank Francis Bonahon, Al Marden and Yair Minsky for their helpful comments on earlier versions of this foreword.

1. General references

Before focusing in a more detailed manner on the material in Chapters 8 and 9, we will discuss some of the more general references which have appeared since the publication of our article. In order to conserve space, we will be especially telegraphic in this section.

Thurston [153] recently published volume I of a new version of his lecture notes under the title *Three-dimensional Geometry and Topology*. This new volume contains much of the material in Chapter 1, 2, and 3 of the original book, as well as material which comes from Sections 5.3 and 5.10. However, the most exciting and novel portions of his original notes have been left for future volumes. A number of other books on Kleinian groups and hyperbolic manifolds have been published in the last 15 years, including books by Apanasov [14], Benedetti and Petronio [23], Buser [50], Kapovich [83], Katok [84], Maskit [104], Matsuzaki and Taniguchi [105], Ohshika [128] and Ratcliffe [137].

There are now several complete published proofs of Thurston's Geometrization Theorem for Haken 3-manifolds available. McMullen [107] used his proof of Kra's Theta Conjecture to outline a proof of the Geometrization Theorem for Haken 3-manifolds that do not fiber over the circle. A more complete version of this approach is given by Otal [132], who also incorporates work of Barrett and Diller [15]. Kapovich [83] has recently published a book on the proof of the Geometrization Theorem. His approach to the main portion of the proof is based on work of Rips (see [21]) on the actions of groups on **R**-trees. (Morgan and Shalen [120, 121, 122] first used the theory of **R**-trees to prove key portions of the Geometrization Theorem. See Bestvina [20] or Paulin [133] for a more geometric viewpoint on how actions of groups on **R**-trees arise as limits of divergent sequences of discrete faithful representations.) An outline of Thurston's original proof of the main portion of the Geometrization Theorem was given by Morgan in [119]. Portions of this proof are available in Thurston's article [152] and preprint [155]. Thurston's original proof develops much more structural theory of Kleinian groups than the later proofs.

Otal [131] also published a proof of the Geometrization Theorem in the case where the 3-manifold fibers over the circle. Otal's proof makes use of the theory of **R**-trees, and in particular uses a deep theorem of Skora [144] which characterizes certain types of actions of surface groups on **R**-trees. (Kleineidam and Souto [90] used some of Otal's techniques to prove a spectacular generalization of Thurston's Double Limit Theorem to the setting of hyperbolic structures on compression bodies.) Thurston's original proof of the Geometrization Theorem for 3-manifolds which fibre over the circle is available at [154]. (A survey of this proof is given by Sullivan [146]; see also McMullen [108].)

We will now briefly indicate where one might look for details on some of the material in Thurston's notes which is not in Chapters 8 and 9. The material in Sections 4.1–4.7 of Thurston's notes is discussed in Chapter E of Benedetti–Petronio [23]. The material in Sections 4.8 and 4.9 was further developed by Epstein in [65]. The results in Sections 4.10 and 4.11 were generalized in Floyd–Hatcher [73] and Hatcher–Thurston [76].

The material in Section 5.1 is the subject matter of Sections 1.5–1.7 of [57]. In Sections 5.2, 5.5 and 5.6, Thurston develops a useful estimate for the dimension of the representation variety, which was proven carefully by Culler and Shalen in Section 3 of [62]. Thurston's Hyperbolic Dehn Surgery theorem is established in Section 5.8, using the dimension count established in the previous sections and the theory developed in Section 5.1. This version of the proof is discussed in Hodgson–Kerckhoff [77] and, in more detail, in Bromberg [46]. Bromberg also develops generalizations of Thurston's Hyperbolic Dehn Surgery theorem to the infinite volume setting, see also Bonahon–Otal [31] and Comar [61]. A complete proof of the Hyperbolic Dehn Surgery theorem using ideal triangulations is given by Petronio and Porti [135]. The proof of the Mostow–Prasad rigidity theorem given in Section 5.9 follows the same outline as Mostow's original proof [123], see also Marden [98], Mostow [124] and Prasad [136]. In Sections 5.11 and 5.12, Thurston proves Jørgensen's theorem that given a bound C, there exists a finite collection of manifolds, such that every hyperbolic 3-manifold of volume at most C is obtained from one of the manifolds in the collection by Dehn Filling; see also Chapter E in [23].

In Sections 6.1–6.5, Thurston gives Gromov's proof of the Mostow–Prasad rigidity theorem and develops Gromov's theory of simplicial volume; see Gromov [74] and Chapter C of Benedetti and Petronio [23]. In Section 6.6, Thurston proves that the set of volumes of hyperbolic 3-manifolds is well-ordered, again see Chapter E of Benedetti and Petronio [23]. Dunbar and Meyerhoff [64] generalized Thurston's arguments to show that the set of volumes of hyperbolic 3-orbifolds is well-ordered.

Chapter 7 of the original notes, concerning volumes of hyperbolic manifolds, was written by John Milnor and much of the work in this chapter appears in appendices to [110] and [111]. Portions of the material in the incomplete Chapter 11 appear in Appendix B of McMullen [108]. Chapter 13 begins with the theory of orbifolds, see for example Scott [142] and Kapovich [83]. Scott [142] also discusses the orbifold viewpoint on Seifert fibered spaces and the geometrization of Seifert fibered spaces. The remainder of Chapter 13 concerns Andreev's theorem and its generalizations. Andreev's original work appeared in [11] and [12]. Andreev's theorem has been generalized by Rivin–Hodgson [138] and Rivin [139].

2. Chapter 8 of Thurston's notes

Sections 8.1 and 8.2 largely deal with basic properties of the domain of discontinuity and the limit set of a Kleinian group. Variations on this material can be found in any text on Kleinian groups, for example [104] or [105].

2.1. Geometrically finite hyperbolic 3-manifolds

In Section 8.3, Thurston offers a new viewpoint on two of the main results in Marden's seminal paper "The geometry of finitely generated Kleinian groups." Marden's Stability theorem (Proposition 9.1 in [98]) asserts that any small deformation of a convex cocompact Kleinian group is itself convex cocompact and is quasiconformally conjugate to the original group. Thurston's version of this theorem (Proposition 8.3.3 in his notes) appears as Proposition 2.5.1 in our article [57]. Marden's Stability theorem also includes a relative version of this result, which asserts that any small deformation of a geometrically finite Kleinian groups that preserves parabolicity, is itself geometrically finite and is quasiconformally conjugate to the original manifold.

Marden's Isomorphism Theorem (Theorem 8.1 in [98]) asserts that any homotopy equivalence between two geometrically finite hyperbolic 3-manifolds which extends to a homeomorphism of their conformal boundaries, is homotopic to a homeomorphism which lifts (and extends) to a quasiconformal homeomorphism of $\mathbf{H}^3 \cup S^2_\infty$. Thurston's Proposition 8.3.4 is a variation on Marden's Isomorphism theorem.

Proposition 8.3.4: *Let $N_1 = \mathbf{H}^n/\Gamma_1$ and $N_2 = \mathbf{H}^n/\Gamma_2$ be two convex cocompact hyperbolic n-manifolds and let M_1 and M_2 be strictly convex submanifolds of N_1 and N_2. If $\phi: M_1 \to M_2$ is a homotopy equivalence which is a homeomorphism from ∂M_1 to ∂M_2, then there exists a map $f: \mathbf{H}^n \cup S^{n-1}_\infty \to \mathbf{H}^n \cup S^{n-1}_\infty$ such that the restriction $\hat{f}$ of f to S^{n-1}_∞ is quasiconformal, $\hat{f}\Gamma_1\hat{f}^{-1} = \Gamma_2$, and the restriction of f to $\mathbf{H}^n$ is a quasi-isometry.*

In Section 8.4, Thurston continues his study of geometrically finite hyperbolic 3-manifolds. Theorem 8.4.2 is Ahlfors' result, see [3], that the limit set

$\Lambda(\Gamma)$ of a geometrically finite hyperbolic manifold $N = \mathbf{H}^n/\Gamma$ either has measure zero or is all of the sphere at infinity S_∞^{n-1} and Γ acts ergodically on S_∞^{n-1}. Ahlfors' Measure Conjecture asserts that this is the case for all finitely generated Kleinian groups. In Section 8.12, Thurston proves Ahlfors' conjecture for freely indecomposable geometrically tame Kleinian groups. Proposition 8.4.3 discusses three equivalent definitions of geometric finiteness. The various definitions of geometric finiteness are treated thoroughly by Bowditch [33].

2.2. Measured laminations and the boundary of the convex core

In Section 8.5, Thurston introduces geodesic laminations and observes that the intrinsic metric on the boundary of the convex core is hyperbolic. This result is established in Chapter 1 of Epstein–Marden [66] and by Rourke [140]. Later, Thurston will observe that the boundary of the convex core is an uncrumpled surface. Uncrumpled surfaces are now known as pleated surfaces. Geodesic laminations are treated in Chapter 4 of our article [57] and in Chapter 4 of Casson-Bleiler [60].

In Section 8.6, Thurston introduces transverse measures on geodesic laminations. In particular, he develops the bending measure on the bending locus of the boundary of the convex core. The bending measure is discussed in Section 1.11 of Epstein–Marden [66]. Measured laminations are discussed by Hatcher [75] and Penner–Harer [134]. The parallel theory of measured foliations is developed in great detail in the book by Fathi, Laudenbach and Poenaru [70]. The connection between measured laminations and measured foliations is made explicit by Levitt [97]. Hubbard and Masur [80] showed that measured foliations can themselves be naturally linked to the theory of quadratic differentials, see also Marden–Strebel [101]. One of the most spectacular applications of the theory of measured laminations was Kerckhoff's proof [87] of the Neilsen Realization Theorem.

Bonahon developed the theory of geodesic currents, which are a generalization of measured laminations, in [25] and [26]. This theory provides a beautiful and flexible conceptual framework for the theory of measured laminations and was put to central use in Bonahon's proof [25] that finitely generated, freely indecomposable Kleinian groups are geometrically tame. Bonahon [26] also used geodesic currents to give a beautiful treatment of Thurston's compactification of Teichmüller space. Bonahon is currently preparing a research monograph [29] which covers geodesic laminations, measured laminations, train tracks and geodesic currents. It also describes Bonahon's more recent work on transverse cocycles and transverse Hölder distributions for geodesic laminations which provide powerful new tools for the study of deformation spaces of hyperbolic manifolds. As one application of these techniques Bonahon [27] has computed the derivative of the function with domain a deformation space of geometrically finite hyperbolic 3-manifolds given by considering the volumes of the convex cores. His formula is a generalization of Schläfli's formula for the variation of volumes of hyperbolic polyhedra. Bonahon's briefer survey paper [28] covers some of the same material; both the research monograph and the survey paper are highly recommended.

2.3. Quasifuchsian groups and bending

In Section 8.7, Thurston begins his study of quasifuchsian groups. A finitely generated, torsion-free Kleinian group is said to *quasifuchsian* if its limit set is a Jordan curve and both components of its domain of discontinuity are invariant under the entire group. Thurston's definition of a quasifuchsian group is incomplete as it leaves out the condition on the domain of discontinuity. His definition allows Kleinian groups which uniformize twisted I-bundles over surfaces, as well as those which uniformize product I-bundles. Proposition 8.7.2 offers several equivalent definitions of quasifuchsian groups. We give a corrected version of Thurston Proposition 8.7.2 below:

Proposition 8.7.2. (Maskit [102]) *If* Γ *is a finitely generated, torsion-free Kleinian group, then the following conditions are equivalent:*

(1) *Γ is quasifuchsian.*
(2) *The domain of discontinuity* $\Omega(\Gamma)$ *of* Γ *has exactly two components, each of which is invariant under the entire group.*
(3) *Γ is quasiconformally conjugate to a Fuchsian group, i.e. there exists a Fuchsian group* $\Theta \subset \mathrm{PSL}_2(\mathbf{R})$ *(such that its limit set* $\Lambda(\Theta) = \mathbf{R} \cup \infty$*) and a quasiconformal map* $\phi \colon \hat{\mathbf{C}} \to \hat{\mathbf{C}}$ *such that* $\Gamma = \phi\Theta\phi^{-1}$.

This characterization is originally due to Maskit, see Theorem 2 in [102], although Thurston follows the alternative proof given by Marden in Section 3 of [98].

Example 8.7.3 is the famous Mickey Mouse example, which is produced using the bending construction. Bending has been studied extensively by Apanasov [14] and Tetenov [13], Johnson and Millson [81], Kourouniotis [91] and others. Universal bounds on the bending lamination of a quasifuchsian group and hence on the bending deformation, are obtained by Bridgeman [34, 35] (and generalized to other settings by Bridgeman–Canary [37]). These bounds are discussed in more detail in the addendum to Epstein–Marden [66] in this volume.

After the Mickey Mouse example, Thurston discusses simplicial hyperbolic surfaces, although he does not give them a name. A simplicial hyperbolic surface is a map, not necessarily an embedding, of a triangulated surface into a 3-manifold such that each face is mapped totally geodesically and the total angle around each vertex is at least 2π. The restriction on the vertices guarantees that the induced metric, usually singular, on the surface has curvature ≤ -1 in the sense of Alexandrov. Simplicial hyperbolic surfaces were used extensively by Bonahon [25] in his proof that freely indecomposable Kleinian groups are geometrically tame and they are discussed in detail in Section 1.3 of [25].

Proposition 8.7.7 asserts that every complete geodesic lamination is realizable in a quasifuchsian hyperbolic 3-manifold. This statement is included in Theorem 5.3.11 in [57]. We will discuss realizability of laminations more fully when we come to Sections 8.10 and 9.7.

2.4. Pleated surfaces and realizability of laminations

Section 8.8 of Thurston's notes concerns pleated surfaces, which are called uncrumpled surfaces in the notes. The results in this section form the basis of Section 5 of our original article [57]. Pleated surfaces are also discussed in Thurston's articles on the Geometrization theorem [152, 154, 155].

Section 8.9 of Thurston's notes develops the theory of train tracks. Proposition 8.9.2 and Corollary 8.9.3 assert that any geodesic lamination on a surface may be well-approximated by a train track. Three-dimensional versions of these results play a key role in Bonahon's work and Section 5 of his paper [25] discusses train track approximations to geodesic laminations in great detail. The general theory of train tracks is developed by Penner and Harer in [134].

In Section 8.10, Thurston turns to the issue of realizability of laminations in 3-manifolds. We discuss this issue in detail in Section 5.3 of [57]. One begins with an incompressible, type-preserving map $f: S \to N$ of a finite area hyperbolic surface S into a hyperbolic 3-manifold N. (An incompressible map $f: S \to N$ is said to be type-preserving if $f_*(g)$ is parabolic if and only if $g \in \pi_1(S)$ is parabolic where $f_*: \pi_1(S) \to \pi_1(N)$ is regarded as a map between the associated groups of covering transformations.) One says that a geodesic lamination λ on S is *realizable* if there is a pleated surface $h: S \to N$ which maps λ into N in a totally geodesic manner. If a realization exists then the image of λ is unique (Proposition 8.10.2 in Thurston and Lemma 5.3.5 in [57].) The map from the space of pleated surfaces (which are homotopic to f) into the space of geodesic laminations (with the Thurston topology) given by taking a pleated surface to its pleating locus is continuous (Proposition 8.10.4 in Thurston and Lemma 5.3.2 in [57].) Propositions 8.10.5, 8.10.6 and 8.10.7 develop more basic properties of geodesic measured laminations, see the earlier references for details. Theorem 8.10.8 in Thurston's notes asserts that the set $\mathcal{R}_f$ of realizable laminations is open and dense in the set $GL(S)$ of all geodesic laminations on S, see Theorem 5.3.10 in [57] for details. Thurston's Corollary 8.10.9, which asserts that if N is geometrically finite then $\mathcal{R}_f = GL(S)$ unless N virtually fibres over the circle with fibre $f(S)$, is stated as Corollary 5.3.12 in [57]. We note that Thurston's Conjecture 8.10.10, which asserts that $f_*(\pi_1(S))$ is quasifuchsian if and only if $\mathcal{R}_f = GL(S)$, is a consequence of Bonahon's work [25], see the discussion after Proposition 9.7.1 and the discussion of Bonahon's work in Section 4.

In related work, Brock [38] proved that the length function is continuous on the space of realizable laminations in $AH(S) \times ML(S)$ and extends to a continuous function on all of $AH(S) \times ML(S)$. Thurston claimed this result and used it in his proof [154] of the Geometrization theorem for 3-manifolds which fiber over the circle.

2.5. Relative compact cores and ends of hyperbolic 3-manifolds

It will be convenient to formalize the material in Section 8.11 in the language of relative compact cores. If N is a hyperbolic 3-manifold, and we choose ε less than the Margulis constant (see Section 4.5 of Thurston [153] or Chapter D in Benedetti–Petronio [23] for example) we can define the ε-thin part of N to be the portion of N with injectivity radius at most ε. Each compact part of the ε-thin part will be a solid torus neighborhood of a geodesic, while each non-compact component will be the quotient of a horoball by a group of parabolic isometries (isomorphic to either $\mathbf{Z}$ or $\mathbf{Z} \oplus \mathbf{Z}$). We obtain N^0 from N by removing its "cusps", i.e. the non-compact components of its thin part. A relative compact core M for N is a compact 3-dimensional submanifold of N^0 whose inclusion into N is a homotopy equivalence which intersects each toroidal component of ∂N^0 in the entire torus and intersects each annular component of ∂N^0 in a single incompressible annulus. Bonahon [24], McCullough [106] and Kulkarni–Shalen [93] proved that every hyperbolic 3-manifold with finitely generated fundamental group admits a relative compact core.

Feighn–McCullough [71] and Kulkarni–Shalen [93] (see also Abikoff [1]) have used the relative compact core to give topological proofs of Bers' area inequality, which asserts that the area of the conformal boundary is bounded by the number of generators (see Bers [16]) and Sullivan's Finiteness Theorem, which asserts that the number of conjugacy classes of maximal parabolic subgroups of a Kleinian group is bounded by the number of generators (see Sullivan [148]). See Section 7 of Marden [98] for a similar treatment of Bers' area inequalities in the setting of geometrically finite groups where Marden constructs an analogue of the relative compact core.

Much of the remainder of Sections 8 and 9 are taken up with understanding the geometry and topology of ends of hyperbolic 3-manifolds. Ends of N^0 are in a one-to-one correspondence with components of $\partial M - \partial N^0$, see Proposition 1.3 in [25], where M is a relative compact core for N. An end of N^0 is *geometrically finite* if it has a neighborhood which does not intersect the convex core. At the end of Section 8.11, Thurston introduces the crucial notion of a simply degenerate end of a hyperbolic 3-manifold. If M is a relative compact core for N, then an end E of N^0 which has a neighborhood bounded by an incompressible component S of $\partial M - \partial N^0$ is said to be *simply degenerate* if there exists a sequence $\{\gamma_i\}$ of non-trivial simple closed curves on S whose geodesic representatives in N all lie in the component of $N^0 - M$ bounded by S and leave every compact subset of N. (Here we have given Bonahon's version of Thurston's definition, which is equivalent to Thurston's.) A hyperbolic 3-manifold in which each component of $\partial M - \partial N^0$ is incompressible is said to be *geometrically tame* if each of its ends is either geometrically finite or simply degenerate.

We will say that the relative compact core M has *relatively incompressible* boundary if each component of $M - \partial N^0$ is incompressible. Thurston works almost entirely in the setting of hyperbolic 3-manifolds whose relative compact core has relatively incompressible boundary. If N has no cusps, the relative compact core has incompressible boundary if and only if $\pi_1(N)$ is freely indecomposable. In general, the relative compact core has relatively incompressible boundary if and only if there does not exist a non-trivial free decomposition of $\pi_1(N)$ such that every parabolic element is conjugate into one of the factors, see Proposition 1.2 in Bonahon or Lemma 5.2.1 in Canary-McCullough [58].

In Section 4.1, we will explain how the definition of geometric tameness is extended to all hyperbolic 3-manifolds with finitely generated fundamental group.

2.6. Analytic consequences of tameness

In Section 8.12, Thurston proves a minimum principle for positive superharmonic functions on geometrically tame hyperbolic 3-manifolds.

Theorem 8.12.3:　*If N is a geometrically tame hyperbolic 3-manifold (whose compact core has relatively incompressible boundary), then for every nonconstant positive superharmonic (i.e. $\Delta h \leq 0$) function h on N,*

$$\inf\nolimits_{C(N)} h = \inf\nolimits_{\partial C(N)} h$$

*where $C(N)$ denotes the convex core of N. In particular, if $C(N) = N$ (i.e. $L_\Gamma = S^2$)
then there are no positive non-constant superharmonic functions on N.*

As a corollary, he shows that Ahlfors' measure conjecture holds for
geometrically tame hyperbolic 3-manifolds.

Corollary 8.12.4:   *If $N = \mathbf{H}^3/\Gamma$ is a geometrically tame 3-manifold (whose
compact core has relatively incompressible boundary), then either L_Γ is all
of S^2_∞ or it has measure zero. Moreover, if $L_\Gamma = S^2_\infty$ then Γ acts ergodically
on S^2_∞.*

He also notes that one may combine his minimum principle with work of
Sullivan [145] to show that the geodesic flow of a geometrically tame hyperbolic
3-manifold (whose compact core has relatively incompressible boundary) is
ergodic if and only if its limit set is the entire Riemann sphere. These arguments
are generalized to the setting of analytically tame hyperbolic 3-manifolds in
[53], see also Section 7 of Culler–Shalen [63] and Sullivan [147]. (A hyperbolic
3-manifold is *analytically tame* if its convex core may be exhausted by a nested
sequence $\{C_i\}$ of compact submanifolds C_i such that there exists K and L such
that ∂C_i has area $\leq K$ and the neighborhood of radius one of ∂C_i has volume $\leq L$.)
In particular, Ahlfors' Measure conjecture is established for analytically tame
hyperbolic 3-manifolds.

Sullivan [149] and Tukia [156] showed that limit sets of geometrically finite
Kleinian groups have Hausdorff dimension less than 2, unless their quotient has
finite volume in which case the limit set is the entire sphere at infinity. Sullivan
[147] provided the first examples of finitely generated Kleinian groups whose
limit sets have measure zero but Hausdorff dimension 2 (see also Canary [52]).
Bishop and Jones [22] later proved that the limit set of every finitely generated,
geometrically infinite Kleinian group has Hausdorff dimension 2.

3. Chapter 9 of Thurston's notes

Chapter 9 is largely devoted to the study of limits of hyperbolic 3-manifolds.
In Section 9.1, the notion of geometric convergence of a sequence of Kleinian
groups is discussed. In Section 3 of [57] we prove the equivalence of several
different notions of geometric convergence. In particular Thurston's Corollary
9.1.7 appears as Corollary 3.1.7 in [57].

3.1. Algebraic and geometric limits

Many of the most interesting results in Chapter 9 concern the interplay
between algebraic and geometric convergence of Kleinian groups. If a sequence

$\{\rho_i\colon G \to \mathbf{PSL}_2(\mathbf{C})\}$ of discrete, faithful representations converges, in the compact-open topology, to $\rho\colon G \to \mathbf{PSL}_2(\mathbf{C})$, then we say that ρ is the *algebraic limit* of $\{\rho_i\}$. If G is not virtually abelian, then $\{\rho_i\}$ has a subsequence $\{\rho_j\}$ such that $\{\rho_j(G)\}$ converges geometrically to a Kleinian group $\hat{\Gamma}$ which is called the *geometric limit* of $\{\rho_j(G)\}$ (Corollary 9.1.8 in Thurston and Proposition 3.8 in Jorgensen–Marden [82]). If ρ is the algebraic limit of $\{\rho_i\}$ and $\rho(G)$ is the geometric limit of $\{\rho_i(G)\}$, then we say that $\{\rho_i\}$ *converges strongly* to ρ.

Example 9.1.4, which is due to Jorgensen (see Section 5 of Jorgensen-Marden [82]), is the most basic example of a sequence which converges algebraically but not strongly. In this example, the algebraic limit is an infinite cyclic group, while the geometric limit is a free abelian group of rank two. More complicated examples which contain this same phenomenon can be found in Marden [100], Kerckhoff–Thurston [88], Ohshika [126] and Thurston [154]. Brock [40] exhibited a sequence where the algebraic limit differs from the geometric limit, yet the geometric limit does not contain a free abelian subgroup of rank two. Anderson and Canary [7] exhibited examples where the (quotient of the) algebraic limit is topologically tame, but is not homeomorphic to any of its approximates. The most comprehensive reference on the foundations of the relationship between the algebraic and the geometric limit is Jorgensen-Marden [82].

3.2. Limits of quasifuchsian groups

In Section 9.2, Thurston begins to study limits of quasifuchsian groups. This study was crucial in his original proof of the Geometrization theorem. If S is a compact surface (with negative Euler characteristic), a discrete, faithful representation $\tau\colon \pi_1(S) \to \mathbf{PSL}_2(\mathbf{C})$ is said to be *quasifuchsian* if $\tau(\pi_1(S))$ is quasifuchsian and $\tau(g)$ is parabolic if and only if g is a peripheral element of $\pi_1(S)$, i.e. if the curve representing g is freely homotopic into a component of the boundary. Thurston's Theorem 9.2 asserts that any type-preserving algebraic limit of quasifuchsian representations is geometrically tame and is in fact a strong limit.

Theorem 9.2: *Let S be a compact surface with negative Euler characteristic. Suppose that $\{\rho_i\colon \pi_1(S) \to \mathbf{PSL}_2(\mathbf{C})\}$ is a sequence of quasifuchsian representations converging to ρ and that $\rho \circ \rho_i^{-1}\colon \rho_i(\pi_1(S)) \to \rho(\pi_1(S))$ is type-preserving for all i. Then, $N_\rho = \mathbf{H}^3/\rho(\pi_1(S))$ is geometrically tame and $\{\rho_i\}$ converges strongly to ρ.*

This result divides naturally into two pieces. Ohshika established that, in this setting, the convergence is strong in Corollary 6.1 of [130]. Evans established a generalization of this strong convergence result for surface groups as part of [67]. Generalizations of the fact that a strong, type-preserving limit of quasifuchsian groups is geometrically tame are obtained by Canary–Minsky [59] and by Ohshika [129] in the case that there are no parabolics and in the general situation by Evans [68].

3.3. The covering theorem

In the proof of Theorem 9.2, Thurston develops the covering theorem which is a very important tool in the study of algebraic and geometric limits. It asserts that, with the exception of the cover of 3-manifold which fibres over the circle associated to the fibre, a simply degenerate end can only cover finite-to-one.

Theorem 9.2.2:   *Let $\hat{N}$ be a hyperbolic 3-manifold which covers another hyperbolic 3-manifold N by a local isometry $p\colon \hat{N} \to N$. If $\hat{E}$ is a simply degenerate end of $\hat{N}^0$ then either*

(a) $\hat{E}$ has a neighborhood $\hat{U}$ such that p is finite-to-one on $\hat{U}$, or
(b) N has finite volume and has a finite cover N' which fibers over the circle such that if N_S denotes the cover of N' associated to the fiber subgroup then $\hat{N}$ is finitely covered by N_S. Moreover, if $\hat{N} \neq N_S$, the $\hat{N}$ is homeomorphic to the interior of a twisted I-bundle which is doubly covered by N_S.

In Thurston's covering theorem, the end $\hat{E}$ is required to be associated to an incompressible surface in $\hat{N}^0$ (as this is the setting in which Thurston has defined simply degenerate ends.) Canary generalized Thurston's covering theorem to the setting of simply degenerate ends of topologically tame hyperbolic 3-manifolds in [55], see also Lemma 2.2 in Ohshika [126]. For a survey of some of the remaining issues related to the covering theorem see [54].

3.4. An intersection number lemma

The key result in Section 9.3 is Theorem 9.3.5, which is an intersection number lemma for geodesic laminations. Bonahon proves a version of this result as Proposition 3.4 in [25].

Theorem 9.3.5:   (Bonahon [25]) *Let N be a hyperbolic 3-manifold with finitely generated fundamental group and let S be a properly embedded incompressible surface in N^0. There exists a constant $K \geq 0$ such that if α_1^* and α_2^* are two closed geodesics in N^0 of distance $\geq D$ from S, homotopic to 2 curves α_1 and α_2 in S by two homotopies which meet S only in α_i and arrive on the same side,* *and each geodesic is disjoint from the thin part of N, or is itself the core of a Margulis tube, then*

$$i(\alpha_1, \alpha_2) \leq K e^{-D} l(\alpha_1) l(\alpha_2) + 2$$

where i is intersection number in S, and l is length measured on S.

Thurston's version allows α_1^* and α_2^* to be measured geodesic laminations in N, but Thurston neglects to include the restriction on the geodesics.

One consequence of Theorem 9.3.5 is that a simply degenerate end admits a well-defined geodesic lamination, called the ending lamination. Let E be a simply degenerate end with a neighborhood bounded by an incompressible subsurface S of the boundary of a relative compact core for N^0. If $\{\alpha_i\}$ is a sequence of simple closed curves on S whose geodesic representatives $\{\alpha_i^*\}$ in N exit E, then the *ending lamination* $\varepsilon(E)$ of E is the limit of $\{\alpha_i\}$ in $GL(S)$. (More formally, to ensure uniqueness, we must define $\varepsilon(E)$ to be the maximal sublamination of $\lim \alpha_i$ which supports a measure). Thurston uses Theorem 9.3.5 to show that $\varepsilon(E)$ is well-defined, that every leaf of $\varepsilon(E)$ is dense in $\varepsilon(E)$ and that every simple closed curve in the complement of $\varepsilon(E)$ is peripheral. These results are also established in Bonahon [24]. This discussion is generalized to the setting of topologically tame hyperbolic 3-manifolds in [53]. For a more thorough discussion of ending laminations and the Ending Lamination Conjecture see Minsky [114].

3.5 Topological tameness

Sections 9.4 and 9.5 of Thurston are devoted to proving that a geometrically tame hyperbolic 3-manifold is *topologically tame*, i.e. homeomorphic to the interior of a compact 3-manifold.

Theorem 9.4.1: *If N is a geometrically tame hyperbolic 3-manifold (whose relative compact core has relatively incompressible boundary), then N is topologically tame.*

Bonahon offers a simpler proof of Theorem 9.4.1 in [24], based on a result of Freedman, Hass and Scott [72]. An outline of this argument is also given in [25]. Canary [53] generalized the notion of a geometrically tame hyperbolic 3-manifold to the setting where the relative compact core need not have relatively incompressible boundary and proved that topologically tame hyperbolic 3-manifolds are geometrically tame. We will discuss this further in the next section.

Thurston's method of proof is based on a scheme for interpolating between any two pleated surfaces in a simply degenerate end with a family of negatively curved surfaces. This approach is discussed in Section 5 of Ohshika's paper [130]. As part of this discussion, Thurston proves that the space *ML(S)* of measured laminations has a piecewise integral linear structure and that the space *PL(S)* of projective measured laminations has a piecewise integral projective structure, see Theorem 3.1.4 in Penner–Harer [134] for details.

In remarks at the end of Section 9.5, Thurston describes two alternative approaches to this interpolation. The first approach, which makes use of simplicial hyperbolic surfaces and elementary moves on triangulations, is carried out by Canary in [55] and has also been used by Fan [69] and Evans [68]. The second approach makes use of the theory of harmonic maps. This approach was carried out by Minsky in [112] and was utilized in his proof of the ending lamination conjecture for geometrically tame hyperbolic 3-manifolds with freely indecomposable fundamental group and a lower bound on their injectivity radius [113].

3.6. Strong convergence

In Section 9.6, Thurston generalizes Theorem 9.2 to the setting of hyperbolic manifolds whose relative compact cores have relatively incompressible boundary. Let M be a compact 3-manifold and let P be a collection of incompressible annuli and tori in ∂M. (One often explicitly requires that (M, P) be a pared 3-manifold, see Section 4 of Morgan [119].) We say that $\rho\colon \pi_1(M) \to \mathbf{PSL}_2(\mathbf{C})$ *uniformizes the pair* (M, P) if there exists a relative compact core Q for $(N_\rho)^0$ and a homeomorphism of pairs $h\colon (M, P) \to (Q, Q \cap \partial(N_\rho)^0)$ such that $h_* = \rho$.

Theorem 9.6.1: *Let M be a compact 3-manifold and let P be a collection of incompressible annuli and tori in ∂M such that each component of $\partial M - P$ is incompressible. Suppose that $\{\rho_i\}$ is a sequence of geometrically tame uniformizations of (M, P) which converge to $\rho\colon \pi_1(M) \to \mathbf{PSL}_2(\mathbf{C})$ and that $\rho \circ \rho_i^{-1}$ is type-preserving for all i. Then,*

(1) $\{\rho_i\}$ converges strongly to ρ, and

(2) ρ is a geometrically tame uniformization of (M, P).

Thurston only sketches the proof of Theorem 9.6.1 in the case that there is no essential annulus in M with one boundary component in P. A generalization of the result in part (1), that $\{\rho_i\}$ converges strongly, to the setting where M is allowed to have compressible boundary, is given by Anderson and Canary [8, 9]. A generalization of part (1) which allows the sequence to be only weakly type-preserving is given by Evans [67]. (Kleineidam [89] has given a quite nice characterization of strong convergence from a different viewpoint.) A generalization of part (2), is given in the case where P is empty by Canary–Minsky [59] and Ohshika [129], and in the general setting by Evans [67].

3.7. Ending laminations

The first result in Section 9.7 sums up what has been learned about ending laminations. It can be derived from the work of Bonahon [24, 25] but is not explicitly stated in any of his papers. We say that an isomorphism $\tau \colon \Gamma \to \Theta$ is *weakly type-preserving* if whenever $\gamma \in \Gamma$ is parabolic, then $\tau(\gamma)$ is parabolic.

Proposition 9.7.1: *Let S be a finite area hyperbolic surface and let $\rho \colon \pi_1(S) \to$* **PSL$_2$(C)** *be a discrete, faithful, geometrically tame, weakly type-preserving representation. There exist two geodesic laminations λ_+ and λ_- on S, such that $\alpha \in GL\,(S)$ is realizable if and only if α contains no component of λ_+ or λ_-.*

In Proposition 9.7.1, one first constructs a relative compact core M for $(N_\rho)^0$. Then M is homeomorphic to $S \times [0, 1]$ and $P = \partial M \cap \partial(N_\rho)^0$ is a collection of annuli which includes $\partial S \times [0, 1]$. We let γ_+ and γ_- denote the core curves of the annuli in P which lie in $S \times \{0\}$ and $S \times \{1\}$. One obtains λ_+ by appending to γ_+ the ending laminations of any simply degenerate ends bounded by components of $S \times \{0\} - P$. One forms λ_- similarly.

The remainder of Section 9.7 concerns train track coordinates for the space of measured laminations. We refer the reader again to Sections 3.1 and 3.2 of Penner–Harer [134] for details. In particular, Thurston proves that $PL(S)$ is a sphere, see Theorem 3.1.4 in Penner–Harer [134] or Proposition 1.5 of Hatcher [75].

In Section 9.9, Thurston surveys Sullivan's work [145] on ergodicity of geodesic flows on hyperbolic manifolds. Sullivan's work is also discussed in Ahlfors' book [4].

4. Selected generalizations

In this section, we will briefly discuss a few of the most direct generalizations of the material in Chapters 8 and 9 of Thurston's notes. The past few years has been a period of intense activity in the field and we will not have space to mention many important results. Other recent surveys of related material include Anderson [6], Brock–Bromberg [41], Canary [56] and Minsky [117, 118].

4.1. Tameness

Marden [98] conjectured that all hyperbolic 3-manifolds with finitely generated fundamental group are topologically tame. Marden's Tameness Conjecture has developed into a central goal of the field.

In a tour de force, Bonahon proved that every hyperbolic 3-manifold whose relative compact core has relatively incompressible boundary is geometrically tame, and hence topologically tame.

Theorem: (Bonahon [25]) *Suppose that N is a hyperbolic 3-manifold with finitely generated fundamental group and that M is a relative compact core for N^0. If each component of $\partial M - \partial N^0$ is incompressible, then N is geometrically tame.*

An immediate consequence of this result is that Ahlfors' Measure Conjecture is valid for all hyperbolic 3-manifolds whose relative compact cores have relatively incompressible boundary.

Subsequently, Canary [53] used Bonahon's work to prove that all topologically tame hyperbolic 3-manifolds are geometrically tame. In order to make sense of this result, one must first define geometric tameness for hyperbolic 3-manifolds whose relative compact core may not be relatively incompressible. For simplicity, we will assume that N has no cusps, for the full definition see [53]. We say that an end E of N is *simply degenerate* if it has a neighbourhood U which is homeomorphic to $S \times [0, \infty)$ (for some compact surface S) and there exists a sequence of simplicial hyperbolic surfaces $\{f_i \colon S \to U\}$, each of which is homotopic within U to $S \times \{0\}$, which leave every compact subset of U. A hyperbolic 3-manifold with finitely generated fundamental group is said to *geometrically tame* if all its ends are either geometrically finite or simply degenerate.

Theorem: (Canary [53]) *A hyperbolic 3-manifold with finitely generated fundamental group is topologically tame if and only if it is geometrically tame.*

Canary's result implies that topologically tame hyperbolic 3-manifolds are analytically tame so Thurston's Theorem 8.12.3 holds for topologically tame hyperbolic 3-manifolds. In particular, Ahlfors' Measure Conjecture holds.

Corollary: *If $N = \mathbf{H}^3 / \Gamma$ is a topologically tame, 3-manifold, then either L_Γ is all of S_∞^2 or it has measure zero. Moreover, if $L_\Gamma = S_\infty^2$ then Γ acts ergodically on S_∞^2.*

There has been a steady progression in our understanding of tameness properties of limits of geometrically finite hyperbolic 3-manifolds due to many authors, including Canary–Minsky [59], Ohshika [129], and Evans [68]. The best results about limits combine work of Brock–Bromberg–Evans–Souto [43] and Brock–Souto [45].

Theorem: (Brock–Bromberg–Evans–Souto [43] and Brock–Souto [45]) *Any algebraic limit of geometrically finite hyperbolic 3-manifolds is topologically tame.*

Agol [2] and Calegari–Gabai [51] have recently given complete proofs of Marden's Tameness Conjecture.

4.2. Spaces of geometrically finite hyperbolic 3-manifolds

The parameterization of the space of geometrically finite hyperbolic structures on a fixed compact 3-manifold has been well-understood since the 1970s. Roughly, a geometrically finite hyperbolic 3-manifold is known to be determined by its topological type and the conformal structure on its conformal boundary. This parameterization combines work of Ahlfors [5], Bers [18], Kra [92], Marden [98] and Maskit [103]. For complete discussions of this parameterization see Bers [19], Section 6 of Marden [99], or Section 7 of Canary-McCullouch [58].

One might hope that one could also parameterize geometrically finite hyperbolic 3-manifolds by internal geometric data, e.g. the bending lamination. Bonahon and Otal [32] characterize exactly which measured laminations arise as the bending lamination of the convex core of a geometrically finite hyperbolic 3-manifold whose relative compact core has relatively incompressible boundary. Lecuire [95] has extended their result to the general case. Keen and Series, see [85] for example, have done an extensive analysis of the "pleating rays" (i.e. lines where the support of the bending lamination is constant) in a number of concrete situations. It is conjectured that the topological type of the convex core and the bending lamination determines a geometrically finite hyperbolic 3-manifold, but Bonahon–Otal [32] and Lecuire [95] only establish this for finite-leaved laminations. Series [143] has recently established this conjecture for punctured torus groups. For more general surfaces, Bonahon [30] proved the conjecture for quasifuchsian groups lying in a neighborhood of the set of Fuchsian groups.

Similarly, it is conjectured that the topological type of the convex core and the conformal structure on the boundary of the convex core determine a geometrically finite hyperbolic 3-manifold. It is known that every possible conformal structure arises, but the uniqueness remains unknown. If the boundary of the convex core has incompressible boundary, then the existence follows immediately from the continuity of the structure on the conformal boundary, see Keen–Series [86], and Sullivan's theorem, see Epstein–Marden [66]. If the boundary of the convex core is compressible, one may replace the use of

Sullivan's theorem with the results of Bridgeman–Canary [36]. See Labourie [94] for a generalization of the existence result.

Scannell and Wolf [141] established that a quasifuchsian hyperbolic 3-manifold is determined by the conformal structure on one boundary component and the bending lamination on the associated component of the boundary of the convex core. McMullen [109] had previously established this fact for quasifuchsian once-punctured torus groups.

Bers [17], Sullivan [149] and Thurston [151] conjectured that every hyperbolic 3-manifold with finitely generated fundamental group arises as the (algebraic) limit of a sequence of geometrically finite hyperbolic 3-manifolds. Bromberg [47] and Brock–Bromberg [42] proved that all hyperbolic 3-manifolds with freely indecomposable fundamental group and no cusps arise as limits of geometrically finite hyperbolic 3-manifolds.

4.3. Thurston's Ending Lamination Conjecture

Another subject which is hinted at, although not addressed directly, in Chapters 8 and 9, is Thurston's Ending Lamination Conjecture, which Thurston first explicitly states in [151]. Thurston conjectured that a hyperbolic 3-manifold is determined by the topological type of its relative compact core, the conformal structure at infinity of each of its geometrically finite ends and the ending laminations of its simply degenerate ends. For a discussion of the background of this conjecture see Minsky [114, 118].

Minsky [113] established Thurston's Ending Lamination Conjecture for hyperbolic 3-manifolds with a lower bound on their injectivity radius and freely indecomposable fundamental group. Ohshika [127] generalized Minsky's proof to the setting of topologically tame hyperbolic 3-manifolds with a lower bound on their injectivity radius. Minsky [115] subsequently established Thurston's conjecture for punctured torus groups, i.e. weakly type-preserving representations of the fundamental group of a finite area punctured torus, see Minsky [115]. Ohshika [125] used results of Thurston [154, 155] to give a complete characterization of which laminations can arise as the ending invariants of a hyperbolic 3-manifold whose relative compact core has relatively incompressible boundary.

Brock, Canary and Minsky [116, 44] established Thurston's Ending Lamination Conjecture for hyperbolic 3-manifolds with freely indecomposable fundamental group (and more generally for hyperbolic 3-manifolds whose relative compact core has relatively incompressible boundary.) In combination with work of Ohshika [125], this establishes the Bers–Sullivan–Thurston Density Conjecture for hyperbolic 3-manifolds whose relative compact core has relatively incompressible boundary. See Minsky [118] for a survey of this work.

Brock, Canary and Minsky have announced a proof of Thurston's Ending Lamination Conjecture for topologically tame hyperbolic 3-manifolds. In combination with the recent resolution of Marden's Tameness Conjecture, see Agol [2] and Calegari–Gabai [51], this gives a complete resolution of Thurston's Ending Lamination Conjecture. One may combine work of Ohshika [125], Kleineidam–Souto [90] and Lecuire [96] with the resolution of Thurston's Ending Lamination Conjecture to give a full proof of the Bers–Sullivan–Thurston Density Conjecture.

The resolution of Thurston's Ending Lamination Conjecture gives a complete classification of hyperbolic 3-manifolds with finitely generated fundamental group. One might hope that it would also give a topological parameterization of the space $AH(M)$ of hyperbolic 3-manifolds homotopy equivalent to a fixed compact 3-manifold M. However, both the topological type (see Anderson–Canary [7]) and the ending invariants themselves (see Brock [39]) vary discontinuously, so one does not immediately obtain such a parameterization. Moreover, Holt [78, 79] showed that there are points in the closures of arbitrarily many components of the interior of $AH(M)$, i.e. arbitrarily many components can "bump" at a single point. McMullen [109] and Bromberg–Holt [49] showed that individual components of the interior of $AH(M)$ often "self-bump," i.e. there is a point in the closure of the component such that the intersection of any small enough neighborhood of the point with the component is disconnected. Most recently, Bromberg [48] has shown that the space of punctured torus groups is not locally connected. Thus, any parameterization of $AH(M)$ must be rather complicated.

To finish on a positive note, the work of Anderson, Canary and McCullough [10] may be combined with the proof of Thurston's Ending Lamination Conjecture to give a complete enumeration of the components of $AH(M)$ whenever M has incompressible boundary. This enumeration can be expressed entirely in terms of topological data.

References

[1] W. Abikoff, "The Euler characteristic and inequalities for Kleinian groups," *Proc. A.M.S.* **97**(1986), 593–601.

[2] I. Agol, "Tameness of hyperbolic 3-manifolds," preprint available at: http://front.math.ucdavis.edu/math.GT/0405568

[3] L.V. Ahlfors, "Fundamental polyhedrons and limit sets of Kleinian groups," *Proc. Nat. Acad. Sci. USA* **55**(1966), 251–4.

[4]   L.V. Ahlfors, *Möbius Transformations in Several Dimensions*, Univ. of Minnesota Lecture Notes, 1981.

[5]   L. Ahlfors and L. Bers, "Riemann's mapping theorem for variable metrics," *Annals of Math.* **72**(1960), 385–404.

[6]   J.W. Anderson, "A brief survey of the deformation theory of Kleinian groups," in *Geometry and Topology Monographs volume 1: The Epstein Birthday Schrift*(1998), 23–49.

[7]   J.W. Anderson and R.D. Canary, "Algebraic limits of Kleinian groups which rearrange the pages of a book," *Invent. Math.* **126**(1996), 205–214.

[8]   J.W. Anderson and R.D. Canary, "Cores of hyperbolic 3-manifolds and limits of Kleinian groups," *Amer. J. Math*, **118**(1996), 745–779.

[9]   J.W. Anderson and R.D. Canary, "Cores of hyperbolic 3-manifolds and limits of Kleinian groups II," *J. of the L.M.S.*, **61**(2000), 489–505.

[10] J.W. Anderson, R.D. Canary and D. McCullough, "On the topology of deformation spaces of Kleinian groups," *Annals of Math.* **152**(2000), 693–741.

[11] E.M. Andreev, "Convex polyhedra in Lobacevskii space," *Math. USSR-Sb.* **10**(1970), 413–440.

[12] E.M. Andreev, "Convex polyhedra of finite volume in Lobacevskii space," *Math. USSR-Sb.* **12**(1970), 255–259.

[13] B.N. Apanasov and A. Tetenov, "On existence of nontrivial quasiconformal deformations of Kleinian groups in space," *Dokl. Sov. Acad. Sci.* **239**(1978), 14–17.

[14] B.N. Apanasov, *Discrete groups in space and uniformization problems*, Kluwer Academic Publishers Group, 1991.

[15] D. Barrett and J. Diller, "Contraction properties of the Poincaré series operator," *Mich. Math. J.* **43**(1996), 519–538.

[16] L. Bers, "Inequalities for finitely generated Kleinian groups," *J. d'Anal. Math.* **18**(1967), 23–41.

[17] L. Bers, "On boundaries of Teichmüller spaces and on Kleinian groups. I," *Annals of Math.* **91**(1970), 570–600.

[18] L. Bers, "Spaces of Kleinian groups", in *Maryland Conference in Several Complex Variables I.* Springer-Verlag Lecture Notes in Math, vol. **155**(1970), 9–34.

[19] L. Bers, "On moduli of Kleinian groups," *Russian Math Surveys* **29**(1974), 88–102.

[20] M. Bestvina, "Degenerations of the hyperbolic space," *Duke Math. J.* **56**(1988), 143–161.

[21] M. Bestvina and M. Feighn, "Stable actions of groups on real trees," *Invent. Math.* **121**(1995), 287–321.

[22] C.J. Bishop and P.W. Jones, "Hausdorff dimension and Kleinian groups," *Acta Math.* **179**(1997), 1–39.

[23] R. Benedetti and C. Petronio, Lectures on Hyperbolic Geometry, Springer-Verlag Universitext, 1992.

[24] F. Bonahon, "Bouts des variétés hyperboliques de dimension 3," Prépublicationes d'Orsay 85T08, 1985.

[25] F. Bonahon, "Bouts des variétés hyperboliques de dimension 3," *Annals of Math.* **124**(1986), 71–158.

[26] F. Bonahon, "The geometry of Teichmüller space via geodesic currents," *Invent. Math.* **92**(1988), 139–62.

[27] F. Bonahon, "A Schläfli-type formula for convex cores of hyperbolic 3-manifolds," *J. Diff. Geom.* **50**(1998), 24–58.

[28] F. Bonahon, "Geodesic laminations on surfaces," in *Laminations and foliations in dynamics, geometry and topology (Stony Brook, NY, 1998)*, Contemp. Math. **269**, American Mathematical Society, 2001, 1–37.

[29] F. Bonahon, "Closed curves on surfaces," preliminary version available at: http://www-rcf.usc.edu/fbonahon/Research/Preprints/Preprints.html

[30] F. Bonahon, "Kleinian groups which are almost Fuchsian," *J. Reine. Angew. Math.*, to appear, preliminary version available at: http://front.math.ucdavis.edu/math.DG/0210233

[31] F. Bonahon and J.P. Otal, "Variétés hyperboliques à géodésiques arbitrairement courtes," *Bull. L.M.S.* **20**(1988), 255–261.

[32] F. Bonahon and J.P. Otal, "Laminations mesurées de plissage des variétés hyperboliques de dimension 3," *Ann. of Math.*, **160**(2004), 1013–1055.

[33] B. Bowditch, "Geometrical finiteness for hyperbolic groups," *J. Funct. Anal.* **113**(1993), 245–317.

[34] M. Bridgeman, "Average bending of convex pleated planes in hyperbolic three-space," *Invent. Math.* **132**(1998), 381–391.

[35] M. Bridgeman, "Bounds on the average bending of the convex hull of a Kleinian group," *Mich. Math J.* **51**(2003), 363–378.

[36] M. Bridgeman and R.D. Canary, "From the boundary of the convex core to the conformal boundary," *Geom. Ded.* **96**(2003), 211–240.

[37] M. Bridgeman and R.D. Canary, "Bounding the bending of a hyperbolic 3-manifold," *Pac. J. Math.*, **218**(2005), 299–314.

[38] J. Brock, "Continuity of Thurston's length function," *G.A.F.A.* **10**(2000), 741–797.

[39] J. Brock, "Boundaries of Teichmüller spaces and end-invariants for hyperbolic 3-manifolds," *Duke Math. J.* **106**(2000), 527–552.

[40] J. Brock, "Iteration of mapping classes and limits of hyperbolic 3-manifolds," *Invent. Math.* **143**(2001), 523–570.

[41] J. Brock and K. Bromberg, "Cone-manifolds and the density conjecture," in *Kleinian Groups and Hyperbolic 3-manifolds*, Cambridge University Press, 2003, 75–93.

[42] J. Brock and K. Bromberg, "On the density of geometrically finite Kleinian groups," *Acta Math.* **192**(2004), 33–93.

[43] J. Brock, K. Bromberg, R. Evans and J. Souto, "Tameness on the boundary and Ahlfors' Measure Conjecture," *Publ. Math. I.H.E.S.* **98**(2003), 145–166.

[44] J. Brock, R.D. Canary and Y.N. Minsky, "The classification of Kleinian surface groups II: the ending lamination conjecture," preprint available at: http://front.math.ucdavis.edu/math.GT/0412006

[45] J. Brock and J. Souto, "Algebraic limits of geometrically finite manifolds are tame," *G.A.F.A.*, to appear, preprint available at: http://www.math.brown.edu/brock/pub.html

[46] K. Bromberg, "Hyperbolic Dehn surgery on geometrically infinite 3-manifolds," preprint available at: http://front.math.ucdavis.edu/math.GT/0009150

[47] K. Bromberg, "Projective structures with degenerate holonomy and the Bers density conjecture," preprint available at: http://front.math.ucdavis.edu/math.GT/0211402

[48] K. Bromberg, "The topology of the space of punctured torus groups," in preparation.

[49] K. Bromberg and J. Holt, "Self-bumping of deformation spaces of hyperbolic 3-manifolds," *J. Diff. Geom.* **57**(2001), 47–65.

[50] P. Buser, *Geometry and Spectra of Compact Riemann Surfaces*, Birkhäuser, 1992.

[51] D. Calegari and D. Gabai, "Shrinkwrapping and the taming of hyperbolic 3-manifolds," preprint available at: http://front.math.ucdavis.edu/math.GT/0407161

[52] R.D. Canary, "On the Laplacian and geometry of hyperbolic 3-manifolds," *J. Diff. Geom.* **36**(1992), 349–367.

[53] R.D. Canary, "Ends of hyperbolic 3-manifolds," *J.A.M.S.* **6**(1993), 1–35.

[54] R.D. Canary, "Covering theorems for hyperbolic 3-manifolds," in *Proceedings of Low-Dimensional Topology*, International Press, 1994, 21–30.

[55] R.D. Canary, "A covering theorem for hyperbolic 3-manifolds and its applications," *Topology* **35**(1996), 751–778.

[56] R.D. Canary, "Pushing the boundary," in *In the Tradition of Ahlfors and Bers, III*, Contemporary Mathematics **355**(2004), American Mathematical Society, 109–121.

[57] R.D. Canary, D.B.A. Epstein, and P. Green, "Notes on notes of Thurston," in *Analytical and Geometrical Aspects of Hyperbolic Spaces*, Cambridge University Press, 1987, 3–92.

[58] R.D. Canary and D. McCullough, "Homotopy equivalences of 3-manifolds and deformation theory of Kleinian groups," *Mem. A.M.S.*, **172**(2004).

[59] R.D. Canary and Y.N. Minsky, "On limits of tame hyperbolic 3-manifolds," *J. Diff. Geom.* **43**(1996), 1–41.

[60] A.J. Casson and S.A. Bleiler, *Automorphisms of surfaces after Nielsen and Thurston*, London Mathematical Society Student Texts 9, Cambridge University Press, 1988.

[61] T.D. Comar, *Hyperbolic Dehn surgery and convergence of Kleinian groups*, Ph.D. thesis, University of Michigan, 1996.

[62] M. Culler and P.B. Shalen, "Varieties of group representations and splittings of 3-manifolds," *Annals of Math.* **117**(1983), 109–146.

[63] M. Culler and P.B. Shalen, "Paradoxical decompositions, 2-generator Kleinian groups, and volumes of hyperbolic 3-manifolds," *J.A.M.S.* **5**(1992), 231–288.

[64] W. Dunbar and R. Meyerhoff, "Volumes of hyperbolic 3-orbifolds," *Ind. Math. J.* **43**(1994), 611–637.

[65] D.B.A. Epstein, "Transversely hyperbolic 1-dimensional foliations," *Astérisque* **116**(1984), 53–69.

[66] D.B.A. Epstein and A. Marden, "Convex hulls in hyperbolic space, a Theorem of Sullivan, and measured pleated surfaces," in *Analytical and Geometrical Aspects of Hyperbolic Space*, Cambridge University Press, 1987, 113–253.

[67] R. Evans, "Weakly type-preserving sequences and strong convergence," *Geom. Ded.* **108**(2004), 71–92.

[68] R. Evans, "Tameness persists in weakly type-preserving strong limits," *Amer. J. Math.* **126**(2004), 713–737.

[69] C. Fan, "Injectivity radius bounds in hyperbolic I-bundle convex cores," preprint available at: http://front.math.ucdavis.edu/math.GT/9907052

[70] A. Fathi, F. Laudenbach and V. Poenaru, *Travaux de Thurston sur les surfaces, Astérisque* **66–67**(1979).

[71] M. Feighn and D. McCullough, "Finiteness conditions for 3-manifolds with boundary", *Amer. J. Math.* **109**(1987), 1155–69.

[72] M. Freedman, J. Hass, and P. Scott, "Least area incompressible surfaces in 3-manifolds," *Invent. Math.* **71**(1983), 609–642.

[73] W. Floyd and A. Hatcher, "Incompressible surfaces in punctured-torus bundles," *Top. Appl.* **13**(1982), 263–282.

[74] M. Gromov, "Volume and bounded cohomology," *Publ. I.H.E.S.* **56**(1982), 5–99.

[75] A.E. Hatcher, "Measured lamination spaces for surfaces, from the topological viewpoint," *Top. Appl.* **30**(1988), 63–88.

[76] A. Hatcher and W.P. Thurston, "Incompressible surfaces in 2-bridge knot complements," *Invent. Math.* **79**(1985), 225–246.

[77] C.D. Hodgson and S.P. Kerckhoff, "Rigidity of hyperbolic cone-manifolds and hyperbolic Dehn surgery," *J. Diff. Geom.* **48**(1998), 1–59.

[78] J. Holt, "Some new behaviour in the deformation theory of Kleinian groups," *Comm. Anal. Geom* **9**(2001), 757–775.

[79] J. Holt, "Multiple bumping of components of deformation spaces of hyperbolic 3-manifolds," *Amer. J. Math.*, **125**(2003), 691–736.

[80] J. Hubbard and H. Masur, "Quadratic differentials and foliations," *Acta Math.* **142**(1979), 221–274.

[81] D. Johnson and J. Millson, "Deformation spaces associated to compact hyperbolic manifolds," in *Discrete groups in geometry and analysis (New Haven, Conn., 1984)*, Progr. Math. 67, Birkhäuser, 1987, 48–106.

[82] T. Jørgensen and A. Marden, "Algebraic and geometric convergence of Kleinian groups," *Math. Scand.* **66**(1990), 47–72.

[83] M. Kapovich, *Hyperbolic Manifolds and Discrete groups: Lectures on Thurston's Hyperbolization*, Birkhauser, 2000.

[84] S. Katok, *Fuchsian groups*, University of Chicago Press, 1992.

[85] L. Keen and C. Series, "Pleating coordinates for the Maskit embedding of the Teichmüller space of punctured tori," *Topology* **32**(1993), 719–749.

[86] L. Keen and C. Series, "Continuity of convex hull boundaries," *Pac. J. Math.* **168**(1995), 183–206.

[87] S. Kerckhoff, "The Nielsen realization problem," *Annals of Math.* **117**(1983), 235–265.

[88] S. Kerckhoff and W.P. Thurston, "Non-continuity of the action of the mapping class group at Bers' boundary of Teichmuller space," *Invent. Math.* **100**(1990), 25–47.

[89] G. Kleineidam, "Strong convergence of Kleinian groups," G.A.F.A. **15**(2005), 416–452.

[90] G. Kleineidam and J. Souto, "Algebraic Convergence of Function Groups," *Comm. Math. Helv.* **77**(2002), 244–269.

[91]   C. Kourouniotis, "Deformations of hyperbolic structures," *Math. Proc. Cambr. Phil. Soc.* **98**(1985), 247–261.

[92]   I. Kra, "On spaces of Kleinian groups," *Comm. Math. Helv.* **47**(1972), 53–69.

[93]   R.S. Kulkarni and P.B. Shalen, "On Ahlfors' finiteness theorem," *Adv. in Math.* **76**(1989),155–169.

[94]   F. Labourie, "Métriques prescrites sur le bord des variétés hyperboliques de dimension 3," *J. Diff. Geom.* **35**(1992), 609–626.

[95]   C. Lecuire, "Plissage des variétés hyperboliques de dimension 3," preprint.

[96]   C. Lecuire, "An extension of Masur domain," preprint.

[97]   G. Levitt, "Foliations and laminations on hyperbolic surfaces," *Topology* **22**(1983), 119–135.

[98]   A. Marden, "The geometry of finitely generated Kleinian groups," *Annals of Math.* **99**(1974), 383–462.

[99]   A. Marden, "Geometrically finite Kleinian groups and their deformation spaces," in *Discrete Groups and Automorphic Functions*, Academic Press, 1977, 259–293.

[100] A. Marden, "Geometric relations between homeomorphic Riemann surfaces," *Bull. A.M.S.* **3**(1980), 1001–1017.

[101] A. Marden and K. Strebel, "The heights theorem for quadratic differentials on Riemann surfaces," *Acta Math.* **153**(1984), 153–211.

[102] B. Maskit, "On boundaries of Teichmüller spaces and on Kleinian groups II", *Annals of Math.* **91**(1970), 607–639.

[103] B. Maskit, "Self-maps of Kleinian groups," *Amer. J. Math.* **93**(1971), 840–56.

[104] B. Maskit, *Kleinian groups*, Springer-Verlag, 1988.

[105] K. Matsuzaki and M. Taniguchi, *Hyperbolic manifolds and Kleinian groups*, Oxford University Press, 1998.

[106] D. McCullough, "Compact submanifolds of 3-manifolds with boundary," *Quart. J. Math. Oxford* **37**(1986), 299–307.

[107] C.T. McMullen, "Riemann surfaces and the geometrization of 3-manifolds," *Bull. A.M.S.* **27**(1992), 207–216.

[108] C.T. McMullen, *Renormalization and 3-manifolds which Fiber over the Circle*, Annals of Mathematics Studies 142, Princeton University Press, 1996.

[109] C.T. McMullen, "Complex earthquakes and Teichmüller theory," *J.A.M.S.* **11**(1998), 283–320.

[110] J. Milnor, "Hyperbolic geometry: the first 150 years," *Bull. A.M.S.* **6**(1982), 9–24.

[111] J. Milnor, "On polylogarithms, Hurwitz zeta functions, and the Kubert identities," *Enseign. Math.* **29**(1983), 281–322.

[112] Y.N. Minsky, "Harmonic maps into hyperbolic 3-manifolds," *Trans. A.M.S.* **332**(1992), 607–632.

[113] Y.N. Minsky, "On rigidity, limit sets and end invariants of hyperbolic 3-manifolds," *J.A.M.S.* **7**(1994), 539–588.

[114] Y.N. Minsky, "On Thurston's ending lamination conjecture," Low-dimensional topology (Knoxville, TN, 1992), ed. by K. H. Johannson, International Press, 1994, 109–122.

[115] Y.N. Minsky, "The classification of punctured torus groups," *Annals of Math.*, **149**(1999), 559–626.

[116] Y.N. Minsky, "The classification of Kleinian surface groups I: models and bounds," preprint, available at: http://front.math.ucdavis.edu/math.GT/0302208

[117] Y.N. Minsky, "Combinatorial and geometrical aspects of hyperbolic 3-manifolds," in *Kleinian groups and hyperbolic 3-manifolds (Warwick, 2001)*, London Math. Soc. Lecture Note Ser. vol. 299, Cambridge Univ. Press, 2003, 3–40.

[118] Y.N. Minsky, "End invariants and the classification of hyperbolic 3-manifolds," in *Current developments in mathematics, 2002*, International Press, 2003, 181–217.

[119] J.W. Morgan, "On Thurston's uniformization theorem for three-dimensional manifolds," in *The Smith Conjecture*, ed. by J. Morgan and H. Bass, Academic Press, 1984, 37–125.

[120] J.W. Morgan and P. Shalen, "Valuations, trees and degenerations of hyperbolic structures I," *Annals of Math.* **120**(1984), 401–476.

[121] J.W. Morgan and P. Shalen, "Degeneration of hyperbolic structures II: Measured laminations in 3-manifolds," *Annals of Math.* **127**(1988), 403–456.

[122] J.W. Morgan and P. Shalen, "Degeneration of hyperbolic structures III: Actions of 3-manifold groups on trees and Thurston's compactification theorem," *Annals of Math.* **127**(1988), 457–519.

[123] G.D. Mostow, "Quasiconformal mappings in n-space and the rigidity of hyperbolic space forms," *Publ. I.H.E.S.* **34**(1968), 53–104.

[124] G.D. Mostow, *Strong rigidity of locally symmetric spaces*, Annals of Mathematics Studies 78, Princeton University Press, 1973.

[125] K. Ohshika, "Ending laminations and boundaries for deformation spaces of Kleinian groups," *Jour. L.M.S.* **42**(1990), 111–121.

[126] K. Ohshika, "Geometric behaviour of Kleinian groups on boundaries for deformation spaces," *Quart. J. Math. Oxford* **43**(1992), 97–111.

[127] K. Ohshika, "Topologically conjugate Kleinian groups," *Proc. A.M.S.* **124**(1996), 739–743.

[128] K. Ohshika, *Discrete groups*, Translations of Mathematical Monographs **207**, American Mathematical Society, 2002.

[129] K. Ohshika, "Kleinian groups which are limits of geometrically finite groups," *Mem. Amer. Math. Soc.* **177**(2005).

[130] K. Ohshika, "Divergent sequences of Kleinian groups," in *Geometry and Topology Monographs volume 1: The Epstein Birthday Schrift* (1998), 419–450.

[131] J.P. Otal, *Le théorème d'hyperbolisation pour les variétés fibrées de dimension 3,"* *Astérisque* **235**(1996).

[132] J.P. Otal, "Thurston's hyperbolization of Haken manifolds," in *Surveys in differential geometry, Vol. III (Cambridge, MA, 1996)*, International Press, 1998, 77–194.

[133] F. Paulin, "Topologie de Gromov équivariante, structures hyperboliques et arbres réels," *Invent. Math.* **94**(1988), 53–80.

[134] R.C. Penner with J.L. Harer, *Combinatorics of train tracks*, Annals of Mathematics Studies, 125, Princeton University Press, 1992.

[135] C. Petronio and J. Porti, "Negatively oriented ideal triangulations and a proof of Thurston's hyperbolic Dehn filling theorem," *Expo. Math.* **18**(2000), 1–35.

[136] G. Prasad, "Strong rigidity of Q-rank 1 lattices," *Invent. Math.* **21**(1973), 255–286.

[137] J. Ratcliffe, *Foundations of hyperbolic manifolds*, Graduate Texts in Mathematics, vol. 149, Springer-Verlag, 1994.

[138] I. Rivin and C.D. Hodgson, "A characterization of compact convex polyhedra in hyperbolic 3-space," *Invent. Math.* **111**(1993), 77–111.

[139] I. Rivin, "A characterization of ideal polyhedra in hyperbolic 3-space," *Annals of Math.* **143**(1996), 51–70.

[140] C.P. Rourke, "Convex ruled surfaces," in *Analytical and Geometrical Aspects of Hyperbolic Spaces*, Cambridge University Press, 1987, 255–272.

[141] K. Scannell and M. Wolf, "The Grafting Map of Teichmüller Space," *J.A.M.S.* **15**(2002), 893–927.

[142] G.P. Scott, "The geometries of 3-manifolds," *Bull. L.M.S.* **15**(1983), 401–487.

[143] C. Series, "Thurston's bending measure conjecture for once punctured torus groups," preprint available at: http://front.math.ucdavis.edu/math.GT/0406056

[144] R. Skora, "Splittings of surfaces," *J.A.M.S.* **9**(1996), 605–616.

[145] D.P. Sullivan, "On the ergodic theory at infinity of an arbitrary discrete group of hyperbolic motions," in *Riemann Surfaces and Related Topics Proceedings of the 1978 Stony Brook Conference*, Annals of Math. Studies **97**(1980), 465–96.

[146] D.P. Sullivan, "Travaux de Thurston sur les groupes quasi-fuchsiens et les variétés hyperboliques de dimension 3 fibrées sur S^1," in *Bourbaki Seminar, Vol. 1979/80*, Lecture Notes in Math., 842, Springer-Verlag, 1981, 196–214.

[147] D.P. Sullivan, "Growth of positive harmonic functions and Kleinian group limit sets of zero planar measure and Hausdorff dimension 2," in *Geometry Symposium Utrecht 1980*, Lecture Notes in Mathematics 894, Springer-Verlag, 1981, 127–144.

[148] D.P. Sullivan, "A finiteness theorem for cusps," *Acta Math.* **147**(1981), 289–299.

[149] D. Sullivan, "Entropy, Hausdorff measures old and new, and limit sets of geometrically finite Kleinian groups," *Acta Math.* **153**(1984), 259–277.

[150] W.P. Thurston, *The geometry and topology of 3-manifolds*, lecture notes, Princeton University, 1980, currently available at: http://www.msri.org/publications/books/gt3m/

[151] W.P. Thurston, "Three-dimensional manifolds, Kleinian groups, and hyperbolic geometry," *Bull. A.M.S.* **6**(1982), 357–381.

[152] W.P. Thurston, "Hyperbolic structures on 3-manifolds I: Deformation of acylindrical 3-manifolds," *Annals of Math.* **124**(1986), 203–246.

[153] W.P. Thurston, *Three-dimensional Geometry and Topology: Volume 1*, Princeton University Press, 1997.

[154] W.P. Thurston, "Hyperbolic structures on 3-manifolds II: Surface groups and 3-manifolds which fibre over the circle," available at: http://front.math. ucdavis.edu/math.GT/9801045

[155] W.P. Thurston, "Hyperbolic structures on 3-manifolds III: Deformations of 3-manifolds with incompressible boundary," available at: http://front.math. ucdavis.edu/math.GT/9801058

[156] P. Tukia, "On the dimension of limit sets of geometrically finite Möbius groups," *Ann. Acad. Sci. Fenn.* **19**(1994), 11–24.

Introduction to Part I

This part is based on our study of Bill Thurston's notes (Thurston, 1979), which consist of mimeographed notes produced by Princeton University Mathematics Department as a result of the course given by Thurston in 1978/79. We shall refer to these notes as Thurston (1979). Thurston plans to expand parts of his notes into a book (Thurston, 1979). There is very little overlap between the projected book and this part, whose basis was the joint M.Sc. dissertation written by two of us and supervised by the third. Thanks are due to Thurston who gave us help and encouragement, and also to Francis Bonahon for additional help.

A useful reference for background information on hyperbolic geometry is Epstein (1983) or Beardon (1983).

Our work should be regarded as exposition of results of Thurston. There is not much genuinely original material. Nevertheless the effort of production has been considerable and we hope that readers will find it helpful. One way to use this part would be to read it at the same time as reading Thurston's notes. Certainly Thurston's notes cover ground we do not cover, even in those areas to which we pay particular attention. There is some overlap between our work and that contained in Lok (1984). Two good expositions of related work of Thurston are Morgan-Bass (1984) and Scott (1983).

(Editors' comments. In the intervening years, [T] has become Thurston (1979). This projected book has materialized as Thurston (1997).)

Chapter I.1
(G, X)-structures

I.1.1. (G, X)-structures on a manifold

The material of this section is discussed in Chapter 3 of Thurston (1979).

Let X be a real analytic manifold and let G be a Lie group acting on X faithfully and analytically. Let N be a compact manifold, possibly with boundary, having the same dimension as X.

I.1.1.1. Definition. A (G, X)-*atlas* for N is a collection of charts $\{\phi_\lambda : U_\lambda \to X\}_{\lambda \in \Lambda}$ satisfying the following conditions:

(1) The $\{U_\lambda\}$ form an open covering of N.
(2) Each ϕ_λ is a homeomorphism onto its image. The image of the boundary, $\phi_\lambda(U_\lambda \cap \partial N)$, is locally flat in X.
(3) For each $x \in U_\lambda \cap U_\mu$, there is a neighbourhood $N(x)$ of x in $U_\lambda \cap U_\mu$ and an element $g \in G$, such that

$$\phi_\lambda | N(x) = g \circ \phi_\mu | N(x).$$

We call g a *transition function*.

The last condition gives us a locally constant map

$$g_{\lambda\mu} = g \colon U_\lambda \cap U_\mu \to G.$$

Notice that $g_{\lambda\mu}$ is determined by $x \in U_\lambda \cap U_\mu$, λ, and μ. To see this, note that $N(x)$ has a non-empty intersection with the interior of N. This means that g is equal to $\phi_\lambda \phi_\mu^{-1}$ on some open subset of X. Since the action of G is faithful, g is determined. Notice also that it does not work to insist that $g_{\lambda\mu}$ should be constant on all of $U_\lambda \cap U_\mu$. An example is given by taking $S^1 = \mathbb{R}/\mathbb{Z}$, with the standard $(\mathbb{R}, \mathbb{R})$-structure, where $\mathbb{R}$ acts on $\mathbb{R}$ by addition. If we cover S^1 by two open intervals U_1 and U_2, then $U_1 \cap U_2$ is the disjoint union of two open intervals, and g_{12} is not a constant element of $\mathbb{R}$.

Any (G,X)-atlas determines a unique maximal (G,X)-atlas.

I.1.1.2. Definition. We define a (G,X)-structure on N to be a maximal (G,X)-atlas.

We usually think in terms of atlases which are not maximal. The above definition refers to a C^0-structure, though the differentiability class only really depends on what happens at the boundary. If N is a C^r−manifold $(r \geq 1)$, we can insist that each ϕ_λ is a C^r-embedding. In that case it is automatic that the boundary is locally flat.

Given any open covering of N by coordinate charts $\{U_\lambda\}$, one may choose a refinement $\{V_j\}$, such that the same element $g \in G$ can be used as a transition function throughout the intersection of any two coordinate charts (see Godement, 1958, p. 158). We shall assume from now on that each $g_{\lambda\mu}$ is a transition function which works throughout $U_\lambda \cap U_\mu$. (As we have already pointed out, this is not possible if we insist on a maximal atlas.)

When X is a complete Riemannian manifold, G is the group of isometries of X, and N is a manifold without boundary, we say that N has a Riemannian (G,X)-structure. Under these circumstances, N has an induced Riemannian metric. In this paper, we shall mainly be interested in the case where X is $\mathbb{H}^n$ and G is the group of all isometries of X.

I.1.2. Developing map and holonomy

The material of this section is discussed in Section 3.5 of Thurston (1979).

Let M be a manifold with a (G, X)-structure and $\gamma \colon I \to M$ a path (here I is the unit interval $[0, 1]$). Holonomy results from the attempt to define a single chart in a coherent way, over the whole of γ and is a measure of the failure of that attempt. We shall give a short sketch of the construction. For more details we refer the reader to Epstein (1984).

It works like this: cover γ with a finite number of charts $\{U_i\}_{i=1}^k$ where $U_i \cap U_{i+1} \neq \emptyset$ and each transition function is constant as opposed to locally constant. Consider $U_1 \cap U_2$. There exists some $g_1 \in G$ such that $\phi_1 | U_1 \cap U_2 = g_1 \circ \phi_2 | U_1 \cap U_2$. If we replace ϕ_2 with $g_1 \circ \phi_2$ we still have a chart, which "extends" ϕ_1 to U_2. Similarly, associated with $U_2 \cap U_3$, is some $g_2 \in G$. Replacing ϕ_3 by $g_2 \circ \phi_3$ extends ϕ_2. So, replacing ϕ_3 by $g_1 \circ g_2 \circ \phi_3$ extends ϕ_1. Continuing along γ we arrive at $g_1 \circ g_2 \circ \ldots \circ g_{k-1} \circ \phi_k$ which replaces ϕ_k.

I.1.2.1. Definition. We now define the *holonomy* of γ, denoted by $H(\gamma)$, to be $g_1 \circ g_2 \circ \ldots \circ g_{k-1}$. It can be shown that the holonomy depends only on the homotopy class of γ, keeping the endpoints fixed, and on the germs of the endpoints of γ. If we choose some basepoint x_0 and a germ about this point, then, by just considering closed loops from x_0, the holonomy gives a map $H: \pi_1(M, x_0) \to G$ which is a homomorphism. Changing the germ conjugates the image by some $g \in G$. In order that H be a homomorphism, rather than an anti-homomorphism, we need to take the correct definition of multiplication in $\pi_1(M, x_0)$. Here $g_1 g_2$ means traversing first g_1 and then g_2.

Now we are in a position to define the developing map. This map can be thought of as the result of analytically continuing the germ of some chart at the basepoint along all possible paths in M. Let $\tilde{M}$ be the universal cover of M. Fix a germ of a chart at x_0, say ϕ_0; then the *developing map* D: $\tilde{M} \to X$ is defined as follows. Take some point $[\omega]$ in $\tilde{M}$, that is $[\omega]$ is a homotopy class of paths represented by ω: $I \to M$ with $\omega(0) = x_0$; let ϕ_1 be a chart at $\omega(1)$. Then we define $D([\omega]) = H(\omega) \circ \phi_1(\omega(1))$. D is independent of ϕ_1 because ϕ_1 is used in the definition of $H(\omega)$. D does depend on the germ of ϕ_0, but changing ϕ_0 merely composes D in X with some $g \in G$.

I.1.2.2. Theorem: Equivalent definitions of completeness. *Let M be a manifold with a Riemannian (G, X)-structure. Then if M has no boundary the following statements are equivalent:*

(1) *The developing map $D: \tilde{M} \to X$ is the universal cover of X.*
(2) *M is metrically complete.*
(3) *For each $r \in \mathbb{R}^+$, and $m \in M$ the closed ball $B(m, r)$ of radius r about m is compact.*
(4) *There exists a family of non-empty compact subsets $\{S_t\}_{t \in \mathbb{R}^+}$ with $N_a(S_t) \subset S_{t+a}$ (where $N_a(S_t) = \{x \in M : d(x, S_t) < a\}$).*

For M with boundary, (2),(3), and (4) are all equivalent.

This result is proved in Section 3.6 of Thurston (1979).

If M is a manifold with boundary, we cannot expect $D: \tilde{M} \to X$ to be a covering map onto its image. Take, for example, the structure on the closed disk D^2 determined by an immersion into the plane as seen in Figure I.1.2.3. Since D^2 is simply connected the developing map is simply this immersion and is not a covering of its image.

I.1.2.3.

Figure I.1.2.3

I.1.3. Convexity

See Section 8.3 of Thurston (1979).

I.1.3.1. Definition.   Let M be a Riemannian manifold possibly with boundary. Then a *geodesic* in M is a path satisfying the geodesic differential equation.

For example let M be the annulus, as shown in Figure I.1.3.4, with a metric structure induced from its embedding in $\mathbb{R}^2$. Then γ as drawn is not a geodesic, even though it minimizes the arc length between its endpoints.

I.1.3.2. Definition. M is said to be *convex*, if, given any two points of M, each homotopy class of paths between them contains a geodesic arc. We say M is *strictly convex* if the interior of this arc is contained in the interior of M. A *rectifiable* path in M is a path "whose length makes sense". That is, let $\rho: I \to M$ be a path. For a subdivision $\tau = \{0 = t_0 < t_1 < \cdots < t_n = 1\}$ (where n is an arbitrary integer), define $l_\tau(\rho)$ by

$$l_\tau(\rho) = \sum_{i=0}^{n-1} d(\rho(t_i), \, \rho(t_{i+1})).$$

If $l(\rho) = \sup\{l_\tau(\rho)\}$ (over all subdivisions τ) exists (i.e. if this set of real numbers is bounded), then we say ρ is rectifiable and that its length is l.

I.1.3.3. Definition. A map $f: M_1 \rightarrow M_2$ between Riemannian manifolds is an *isometric map* if it takes rectifiable paths to rectifiable paths of the same length. A *path space* is a path connected metric space in which the metric is determined by path length.

I.1.3.4.

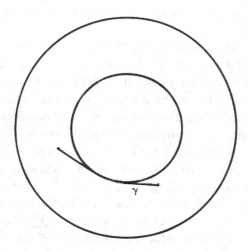

Figure I.1.3.4

In a path space, there is not necessarily a path between a and b whose length is $d(a, b)$. For example the punctured disk $D^2 \setminus \{0\}$ is a path space, in which there exist such a and b. The annulus (see the above figure), with the subspace metric induced from $\mathbb{R}^2$, is not a path space.

It is usual to consider only rectifiable paths parametrized proportionally to arc length. We leave to the reader the simple task of seeing how any rectifiable path γ can be converted, in a canonical way, into a path parametrized proportionally to arc length, having the same length as γ.

We shall develop two useful criteria for convexity (see Section 8.3 of Thurston (1979)). First we prove a generalization of a theorem which is well known for Riemannian manifolds without boundary (See Gromov, 1981b).

I.1.3.5. Theorem: Hopf Rinow. *Let (X, d) be a complete, locally compact path space. Then*:

(1) $B(x, L) = \{y: d(x, y) \le L\}$, *the ball of radius L about x, is compact for all $x \in X$ and any $L > 0$.*

(2) *Given any two points, there is a distance minimizing path between them.*

Proof. (1) Suppose there exists $x \in X$ and $L > 0$ such that $B(x, L)$ is non-compact. Then we claim that $B(x_1, L/2)$ is non-compact for some x_1 in $B(x, L)$. To prove this claim we argue by contradiction. So suppose not. Let M be the least upper bound of those $r \leq L$ for which $B(x, r)$ is compact. Then $M \geq L/2$ and $B(x, M - L/16)$ is compact. We may cover $B(x, M - L/16)$ by a finite number of balls of radius $L/8$ with centres $y_1, \ldots, y_n$ for some integer n. Enlarge each ball to an $L/2$-closed ball (each of which is compact by assumption). Then $\cup_{i=1}^{i=n} B(y_i, L/2)$ is compact.

Let y be an arbitrary point in $B(x, M + L/16)$. We choose a path from x to y of length $\leq M + L/8$. Let z be the point which is a distance $M - L/16$ along this path from x. Since $z \in B(x, M - L/16)$, we have $z \in B(y_j, L/8)$ for some j. Since $d(z, y) \leq L/8$, we have $y \in B(y_j, L/2)$. This argument shows that $B(x, M + L/16) \subset \cup_{i=1}^{i=n} B(y_i, L/2)$ is compact. Since M is maximal, this implies that $M = L$ and that $B(x, L)$ is compact. The resulting contradiction establishes our claim that $B(x_1, L/2)$ is non-compact for some $x_1 \in B(x, L)$.

Continue this process inductively to obtain a sequence $\{x_n\}$ such that $B(x_n, L/2^n)$ is non-compact and $x_n \in B(x_{n-1}, L/2^{n-1})$. Since this sequence is Cauchy, x_n converges to some $v \in X$. Since X is locally compact, there exists $\delta > 0$ such that $B(v, \delta)$ is compact. Choose N such that $\sum_{n=N}^{n=\infty} L/2^n < \delta/2$. Then $B(x, L/2^N)$ would be a non-compact closed subset of $B(v, \delta)$, which is a contradiction. This completes the proof that $B(x, L)$ is compact for each $x \in X$ and each $L > 0$.

To prove (2), note that we may choose a sequence $\{\omega_i\} \colon I \to X$ such that $\omega_i(0) = x$ and $\omega_i(1) = y$, and

(1) ω_i is rectifiable for all i;
(2) $L(\omega_i)$ converges to $d(x, y)$ as i tends to infinity;
(3) ω_i is parametrized proportional to arc length for all i.

Let $F = \{\omega_i\}$ be considered as a subset of $C(I, X)$ equipped with the compact-open topology. There exists L such that $L(\omega_i) \leq L$ for all i. F is equicontinuous since $|s - t| < \delta$ implies that $d(\omega_i(s), \omega_i(t)) < \delta L$. Since $B(x, L)$ is compact, we can apply Ascoli's theorem and deduce that ω_i converges (after taking a subsequence) to $\omega \colon I \to X$ in the compact-open topology on $C(I, X)$.

We claim that ω is a distance minimizing path. To prove this, first note that $\omega(0) = x$ and $\omega(1) = y$. For any partition $0 = t_0 < \cdots < t_k = 1$, we have

$$\sum_{i=0}^{i=k} d(\omega(t_i), \omega(t_{i+1})) = \lim_{j \to \infty} \sum_{i=0}^{i=k} d(\omega_j(t_i), \omega_j(t_{i+1}))$$

$$\leq \lim_{j \to \infty} L(\omega_j) = d(x, y).$$

This shows that ω is rectifiable and that $L(\omega) \le d(x,y)$. Hence $L(\omega) = d(x,y)$. This completes the proof of the claim.

$\square$

I.1.3.6. Definition. A metrically complete Riemannian manifold with boundary is *locally convex* if every point has a convex neighbourhood.

Remark: Whitehead's theorem (see Kobayashi–Nomizu, 1963), tells us that any point in the interior of M has a convex neighbourhood, so local convexity is really only a boundary condition. Furthermore, even for points x on the boundary, if y is near x and there is a geodesic from x to y, then this is a distance minimizing path.

I.1.3.7. Corollary: Local convexity implies convexity. *If M is a complete, locally convex, Riemannian manifold with boundary, then M is convex. In particular, any complete Riemannian manifold without boundary is convex.*

Proof. We may assume that M is simply connected, since we may always work in the universal cover.

By Theorem I.1.3.5 (*Hopf Rinow*), there is a distance minimizing path between any two points in M, which we may parametrize proportional to arc length; denote this path by $\omega\colon I \to M$. We claim that ω is a geodesic. If not, there exists $t \in [0,1]$ such that $\omega[t - \eta, t + \eta]$ is not a geodesic for any $\eta > 0$. Choose η such that $\omega[t - \eta, t + \eta]$ is contained in a convex neighbourhood of $\omega(t)$. Now replace $\omega[t - \eta, t + \eta]$ by the geodesic from $\omega(t - \eta)$ to $\omega(t + \eta)$ to obtain $\bar{\omega}\colon I \to M$ with $L(\bar{\omega}) < L(\omega) = d(x,y)$, which is a contradiction.

$\square$

Remarks: One may similarly prove that local strict convexity (i.e. every point on the boundary has a strictly convex neighbourhood) implies strict convexity.

I.1.4. The developing map and convexity

See Section 8.3 of Thurston (1979).

In this section we shall require M to be a (G, X)-manifold possibly with boundary, where G acts on X, a simply connected Riemannian manifold, as a group of isometries. We shall assume that if $\pi\colon \tilde{M} \to M$ is a covering map, then π is a local isometry and the covering translations are isometries. The following lemma is an immediate consequence of the covering homotopy property.

I.1.4.1. Lemma: Universal cover convex. *M is convex if and only if $\tilde{M}$ is.*

The next result is a natural generalization of Theorem I.1.2.2 (*Equivalent definitions of completeness*) and Proposition 8.3.2. in Thurston (1979).

I.1.4.2. Proposition: Coverings and convexity. *Suppose M is a (G,X)-manifold possibly with boundary, where X is a simply connected Riemannian manifold of non-positive curvature and G is the group of isometries of X. Then M is convex and metrically complete if and only if the developing map $D: \tilde{M} \to X$ is a homeomorphism onto a convex complete submanifold of X. In this case D is an isometry onto $D\tilde{M}$.*

Proof. Suppose M is convex and metrically complete. We have seen that $\tilde{M}$ is convex and it is clear that $\tilde{M}$ is metrically complete. Since the curvature is non-positive, no geodesic in X intersects itself. Since D takes geodesics in $\tilde{M}$ to geodesics in X, D is injective on any geodesic of $\tilde{M}$. But $\tilde{M}$ is convex, so any two points of $\tilde{M}$ can be joined by a geodesic. Hence D is injective. Thus D is an isometric homeomorphism onto its image in X. Since $\tilde{M}$ is convex, so is $D\tilde{M}$.

To prove the converse, suppose D is a homeomorphism of $\tilde{M}$ onto a convex complete submanifold of X. By Lemma I.1.4.1 (*Universal cover convex*), M is convex. To show M is metrically complete, let $\{x_i\}$ be a Cauchy sequence in M. Taking a subsequence, we may assume that $d(x_i, x_{i+1}) < 2^{-i-1}$. Let $\omega: [0,1] \to M$ be a path such that $\omega(2^{-i}) = x_i$. Let $\tilde{\omega}: (0,1] \to \tilde{M}$ be a lift of ω. Since $\tilde{M}$ is complete, $\tilde{\omega}(t)$ converges to a limit as t tends to zero. Hence $\omega(t)$ has a limit as t tends to zero. This completes the proof of the proposition.

$\square$

Remark: One may similarly prove the analogous result for strictly convex manifolds with boundary.

I.1.5. The deformation space

See Chapter 5 of Thurston (1979).

I.1.5.1. Definition. We wish to consider the space of all possible (G,X)-structures on a fixed manifold N possibly with boundary. This is called the *deformation space* of N and is denoted $\Omega(N)$.

Suppose we have a fixed $M_0 \in \Omega(N)$, and a fixed covering by charts $\{\phi_i\}$: $U_i \to X$ (locally finite with one element of G acting as transition function for the whole of the intersection of any two of them) and a shrinking $\{U_i'\}$. A sub-basis for the topology of $\Omega(N)$ is given by sets of the following form

$$N_j(M_0) = \{M \in \Omega(N)|M \text{ is defined by } \{\psi_i\}: U_i' \to X \text{ and } \psi_j \subset V_j\}$$

where V_j is an open neighbourhood of ϕ_j in the compact-open topology on $C(U_j', X)$. If N is a C^r-manifold, we can restrict to C^r-charts and take V_j to be a neighbourhood of ϕ_j in the compact C^r-topology. We have chosen a shrinking $\{U_i'\}$ so that we may use $\{U_i'\}$ as coordinate charts for all "nearby" structures, that is we may deform ϕ_i a small amount in any direction without causing any self-intersections (see Figure I.1.5.2). $\Omega(N)$ is infinite-dimensional if it is non-empty.

I.1.5.2.

Figure I.1.5.2

Suppose $\{M_i\}$ converges to M in $\Omega(N)$. Then it is clear that we may choose associated developing maps $\{D_i\}$ and D such that D_i converges to D in the compact C^r-topology of $C(\tilde{N}, X)(0 \leq r \leq \infty)$. Intuitively, any compact set in $\tilde{N}$ is covered by a finite number of lifts of coordinate charts, and one may control the behaviour on each coordinate chart.

I.1.5.3. Theorem: Limit is an embedding. *Suppose M_i converges to M in $\Omega(N)$. Let D_i and D be the associated developing maps such that D_i converges to D, and suppose that $D|K$ is an embedding, for some compact subset K of $\tilde{N}$. Then $D_i|K$ is also an embedding for sufficiently large i.*

Proof. We need only prove this in the C^0-case. Suppose that the result is false. Then there exists $\{x_i\}, \{y_i\} \subset K$ such that $D_i(x_i) = D_i(y_i)$ and $x_i \neq y_i$. Since K is compact we may assume that x_i converges to some x, and y_i converges to some y, both in K. Since D_i converges to D, $D(x) = D(y)$ and so $x = y$. Let $\pi: \tilde{N} \to N$ be the universal cover of N. It follows that, for sufficiently large i, $\pi(x_i)$, $\pi(y_i)$, $\pi(x)$, and $\pi(y)$, all lie in U'_j for some fixed j. We may assume that U'_j is contractible in N, so that there exists $\alpha\colon U'_j \to \tilde{N}$, a well defined (G, X)-homeomorphism onto its image with $\pi \circ \alpha\colon U'_j \to N$ equal to the inclusion map and $\alpha(\pi(x_i)) = x_i$, and $\alpha(\pi(y_i)) = y_i$, for large i. Since ψ_i and $D_i\alpha$ are (G, X)-embeddings of U'_j in X, $D_i \circ \alpha | U'_j = g \circ \psi_i | U'_j$ for some $g \in G$. Then since $g \circ \psi_i$ is a homeomorphism onto its image and $\pi x_i \neq \pi y_i$, we have $D_i x_i \neq D_i y_i$, which is a contradiction.

$\square$

I.1.5.4. Space of developing maps.

It is often easier to regard $\Omega(N)$ as a function space. We can characterize a developing map $D\colon \tilde{N} \to X$ as a local C^r-diffeomorphism (homeomorphism if $r = 0$) such that for covering translation γ of $\tilde{N}$, there is an element $H(\gamma)$ of G with $D \circ \gamma = H(\gamma) \circ D$. We topologize the set of developing maps by means of a sub-basis consisting of sets of the following form:

(1) U where U is open in the compact C^r-topology on $C(\tilde{N}, X)$;
(2) $N(K) = \{D \mid D|K \text{ is an embedding}\}$ where K is a compact subset of $\tilde{N}$.

Then G acts continuously on the set of developing maps by composition on the left. We can identify $\Omega(N)$ with the quotient of the space of developing maps by G.

I.1.5.5. Remarks.

(1) If we restrict our attention to C^r-manifolds for some fixed $r(1 \le r \le \infty)$, this topology on the space of developing maps is simply the compact C^r-topology (i.e. sub-basis sets of the second type are unnecessary.) It is a standard fact in differential topology that the space of C^r-embeddings is open in the space of all C^r-maps.
(2) Our topology is strictly finer than the compact-open topology – see for example Figure I.1.5.6.
(3) We shall use the notation $d_M(x, y)$ to denote the distance between x and y in N as measured in the path metric induced by M. If M_i converges to M, d_{M_i} does not necessarily converge to d_M. In fact, d_M may be infinite on a compact hyperbolic manifold with a boundary which is not smooth. In other words, it is possible to have two points on the boundary of M, with no rectifiable path joining them (see Figure I.1.5.7). To be sure that d_{M_i}

converges to d_M, one needs to be working with the compact C^r-topology ($r \geq 1$).

(4) If we consider a closed surface of negative Euler characteristic, Teichmüller space (see Section I.3.1 (*The Geometric Topology*)) is a quotient of $\Omega(N)$. More precisely, $T(N) = \Omega(N)/H$, where H is the group of isotopies of N to the identity. An isotopy h acts on a developing map D by lifting h to an isotopy $\tilde{h}$, which starts at the identity, of $\tilde{N} = \mathbb{H}^2$ and taking $D \circ \tilde{h}_1$. For a general (G, X)-structure, it is also quite usual in the literature to define the space of structures as $\Omega(N)$ modulo isotopy.

I.1.5.6.

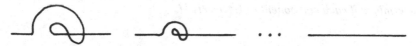

Figure I.1.5.6 *The picture shows the images of various developing maps which are immersions not embeddings. This sequence converges in the compact-open topology, but not in ours.*

I.1.5.7.

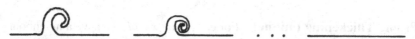

Figure I.1.5.7 d_{Mn} *does not converge to d_M in this example, as we may choose the spiral to be arbitrarily long, and distances are measured by paths. In fact d_M is infinite for some pairs of points.*

I.1.6. Thickenings

The subject we are about to discuss is capable of considerable generalization. A very general treatment is given in Haefliger (1958). Our discussion applies to a number of other situations (almost) verbatim. For example, one can obtain a proof of the Whitney–Bruhat theorem (Whitney–Bruhat 1959), which states that every real analytic manifold can be thickened to a complex analytic manifold.

Let N be a manifold of dimension n, possibly with boundary. Let M be a fixed (G, X)-structure on N.

I.1.6.1. Definition. A *thickening* of M is a (G, X)-structure on a manifold without boundary N_T, of dimension n, containing N as a submanifold, which induces the given (G, X)-structure on N.

I.1.6.2. Definition. If M_1 and M_2 are Riemannian (G, X)-manifolds and M_1 is (G, X)-embedded in M_2, we say M_2 is an ε-*thickening* of M_1 if, for each point $x \in M_1$, there is an ε-neighbourhood in M_2 which is isometric to an ε-ball in X.

I.1.6.3. Theorem: Thickenings exist. *M has a thickening, and the germ of the thickening is unique in the following sense. Let $M \subset M_1^*$ and $M \subset M_2^*$ be two thickenings. Then we can find U_1^* and U_2^* with $M \subset U_i^* \subset M_i^*$, such that each U_i^* is a thickening of M and there is a (G, X)-isomorphism between U_1^* and U_2^* which extends the identity on M. Moreover this isomorphism is uniquely determined if each component of U_1^* meets M.*

Proof. Thickenings exist:

First we prove uniqueness.

I.1.6.4. Lemma: Thickening unique. *Given two thickenings M_1^* and M_2^* of M, there exist isomorphic open neighbourhoods U_1^* of M in M_1^* and U_2^* of M in M_2^*. Moreover the isomorphism is unique if each component of U_i^* meets M.*

Proof. Thickening unique: For each point $x \in M$, we have a neighbourhood U, which is open in M_1^*, and a (G, X)-embedding $f: U \to M_2^*$, which is the identity on $U \cap M$. We may suppose that we have a family $\{f_i: U_i \to M_2^*\}$ of such embeddings, such that the $\{U_i\}$ form a locally finite family and cover. M. Let $\{V_i\}$ be a shrinking of $\{U_i\}$.

Let W be the set of $w \in \cup V_i$ such that if $w \in \overline{V}_i \cap \overline{V}_j$ then $f_i = f_j$ in some neighbourhood of w. From local finiteness, it is easy to deduce that W is open. Analytic continuation shows that $M \subset W$. We get a well defined (G, X)-immersion of W in M_2^*. By restricting to a smaller neighbourhood of M we obtain an embedding, as required.

The uniqueness of the embedding follows by analytic continuation.

> **Thickening unique**

Continuation, proof of Thickenings exist: We now prove the existence of thickenings. The clearest form of the proof is given in Figure I.1.6.5.

I.1.6.5.

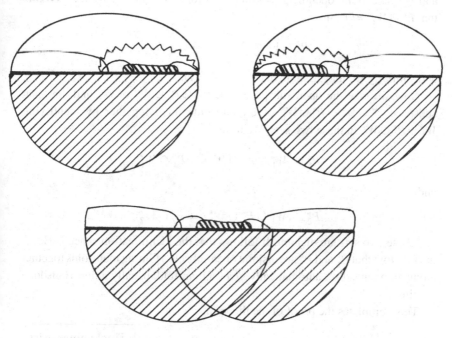

Figure I.1.6.5

Let $\{\phi_i\colon U_i \to X\}$ be a finite atlas of coordinate charts for N. First note that U_i itself may be thickened. To see this, note that we can identify a single U_i with a subspace of X. U_i is open in X if and only if $\partial U_i(=U_i \cap \partial N) = \varnothing$. By adding small open neighbourhoods in X of each point in ∂U_i, we obtain an open subset V_i of X such that U_i is closed in V_i.

We now find thickenings of $U_1 \cup \cdots \cup U_i$ by induction on i. What we have to show is that if a (G, X)-manifold M (with boundary) is the union of two open submanifolds U_1 and U_2, and if U_1 has a thickening U_1^* and U_2 has a thickening U_2^*, then M also has a thickening. To see this, we write $X_1 = M \backslash U_1$ and $X_2 = M \backslash U_2$. Then X_1 and X_2 are disjoint closed subsets of M. Let $X_1 \subset V_1$ and $X_2 \subset V_2$, where V_1 and V_2 are open in M, $\overline{V}_1 \cap \overline{V}_2 = \varnothing$, $\overline{V}_1 \subset U_1$ and $\overline{V}_2 \subset U_2$. Let $P_i^* \subset U_i^*$ be an open neighbourhood of X_i, such that $P_i^* \cap (M \backslash V_i) = \varnothing$. $M \backslash (V_1 \cup V_2) = P$ is closed subspace and $P \cap \overline{P}_i^* = \varnothing$ for $i = 1, 2$. Note that $P \subset U_1 \cap U_2$. Let U^* be a thickening of $U_1 \cap U_2$. By the Lemma I.1.6.4 (*Thickening unique*), we may assume that $U^* \subset U_1^*$ and $U^* \subset U_2^*$. Let P^* be an open neighbourhood of P in U^* such that $\overline{P}^* \cap \overline{P}_i^* = \varnothing$ for $i = 1, 2$. $\overline{V}_1 \backslash P^* \cup P_1^*$ and $\overline{V}_2 \backslash P^* \cup P_2^*$ are disjoint closed subsets of M, each contained in $U_1 \cap U_2$. Let W_1^*

and W_2^* be disjoint open neighbourhoods of these sets in U^*. We may assume that $\overline{P}_2^* \cap W_1^* = \emptyset$, since

$$\overline{P}_2^* \cap \overline{V}_1 \subset \overline{P}_2^* \cap (M \setminus V_2) = \emptyset$$

and similarly we may assume that $\overline{P}_1^* \cap W_2^* = \emptyset$.

Now we take the open subspace $P_1^* \cup W_1^* \cup P^* \cup W_2^*$ of U_1^* and the open subspace $P_2^* \cup W_2^* \cup P^* \cup W_1^*$ of U_2^* and glue them together by identifying $W_1^* \cup P^* \cup W_2^*$ in these two sets. We have

$$P_1^* \cap (W_1^* \cup P^* \cup W_2^*) = P_1^* \cap W_1^*$$

and

$$P_2^* \cap (W_1^* \cup P^* \cup W_2^*) = P_2^* \cap W_2^*$$

and these two sets are disjoint. It follows that the identification space is Hausdorff. (Note that P_1^* and P_2^* are disjoint open subsets.) Gluing manifolds together along an open subset automatically gives a manifold, usually a non-Hausdorff manifold.

This completes the proof of the theorem.

Thickenings exist

I.1.6.6 Remark. The same proof works for a non-compact manifold. Any (G,X)-manifold of dimension n with a countable basis can be covered with a finite number of charts. (These charts are not, of course, connected.) In fact $(n+1)$ charts will do.

I.1.7. Varying the structure

Let N be a compact C^r-manifold ($0 \le r \le \infty$) with boundary. Let N_{Th} be the union of N with a collar $\partial N \times I$, where $\partial N \subset N$ is identified with $\partial N \times 0 \subset \partial N \times I$. We fix a (G,X)-structure M_{Th} on N_{Th}. By fixing a basepoint in $\tilde{N}$, the universal cover of N, we can identify the group of covering translations of $\tilde{N}$ with $\pi_1(N) = \pi_1(N_{\text{Th}})$. The same group of covering translations acts on the universal cover of N_{Th}.

A developing map $D \colon \tilde{N} \to X$ induces a homomorphism from $\pi_1 N$, the group of covering translations, to G. Let $\mathcal{D}(N)$ be the space of developing maps with the topology described in Section I.1.5.4 (*Space of developing maps*). Topologize $\text{Hom}\,(\pi_1 N, G)$ with the compact-open topology.

The holonomy gives us a map

$$H: \mathcal{D}(N) \to \operatorname{Hom}(\pi_1 N, G)$$

which is easily seen to be continuous. The induced map

$$\Omega(N) \to R(\pi_1 N, G)$$

of the space of (G, X)-structures on N into the space of conjugacy classes of homomorphisms of $\pi_1 N$ into G is also continuous. In general $R(\pi_1 N, G)$ is not Hausdorff. Let M be a (G, X)-structure on N and let M_{Th} be an extension of M to N_{Th}. (From Theorem I.1.6.3 (*Thickenings exist*) and the Collaring Theorem (Connelly, 1971) we can see that such an extension exists.)

The next theorem may be viewed as one of the ways to make the discussion in Section 5.1 of Thurston (1979) more formal.

I.1.7.1. Theorem: Neighbourhood is a product. *Let* $D_{\mathrm{Th}}: \tilde{N}_{\mathrm{Th}} \to X$ *be a fixed developing map for* M_{Th}. *Then a small neighbourhood of* $D_{\mathrm{Th}} | \tilde{N}$ *in the space of developing maps of* N *is homeomorphic to* $\mathcal{A} \times \mathcal{B}$, *where* $\mathcal{A}$ *is a small neighbourhood of the obvious embedding* $N \subset N_{\mathrm{Th}}$ *in the space of locally flat embeddings, and* $\mathcal{B}$ *is small neighbourhood of the holonomy* $h_M: \pi_1 N \to G$ *in the space of all homomorphisms of* $\pi_1 N$ *into* G. *The projection of the neighbourhood of* $D_{\mathrm{Th}} | \tilde{M}$ *to* $\mathcal{B}$ *is given by the holonomy* $H: \mathcal{D}(N) \to \operatorname{Hom}(\pi_1, N, G)$ *defined above.*

The action of G *on the space of developing maps correspond to conjugation in* $\mathcal{B}$.

Proof. Neighbourhood is a product: Let N_{TT} be the union of N_{Th} with a collar (i.e. a thickening of a thickening) and let M_{TT} be a (G, X)-thickening of M_{Th} with underlying manifold N_{TT}. We now show how to map a small neighbourhood $\mathcal{B}$ of $h_M: \pi_1 N \to G$, the holonomy of M, continuously into the space of structures on N_{Th}.

I.1.7.2. Lemma: Holonomy induces structure. *There is a continuous map*

$$D: \mathcal{B} \to \mathcal{D}(N_{\mathrm{T}}).$$

Proof. Holonomy induces structure: Let $\{U_i\}_{0 \le i \le k}$ be a finite open covering of N_{TT} such that $U_0 = N_{\mathrm{TT}} \backslash N_{\mathrm{Th}}$ and such that $U_i \subset \operatorname{int} N_{\mathrm{TT}}$ for

$1 \leq i \leq k$. Let $U_i^1 = U_i$ ($1 \leq i \leq k$). For each $r > 0$ let $\{U_i^{r+1}\}_{0 \leq i \leq k}$ be a shrinking of $\{U_i^r\}_{0 \leq i \leq k}$. We may assume that U_i^r is simply connected for $i > 0$.

Let $D_M: \tilde{N}_{TT} \rightarrow X$ be a developing map for M_{TT} and let $\pi: \tilde{N}_T \rightarrow N_T$ be the universal cover.

Choose an h near h_M. We show how to construct a (G, X)-structure on N_{Th}. The method is to construct a developing map $\tilde{N}_{Th} \rightarrow X$. We define D_1 to be equal to D_M on some component U_1^* of the preimage $\{\pi^{-1}(U_1)\}$. Extend D_1 equivariantly to all of $\{\pi^{-1}(U_1)\}$ using the chosen holonomy h. Then inductively construct an equivariant local homeomorphism $D_s: \pi^{-1}U_1^s \cup \cdots \cup \pi^{-1}U_s^s \rightarrow X$ which is equal to D_{s-1} on $\pi^{-1}U_1^s \cup \cdots \cup \pi^{-1}U_{s-1}^s$.

I.1.7.3.

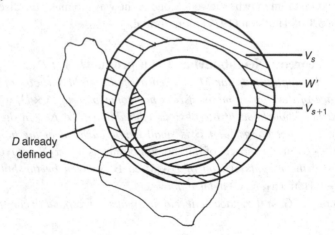

V_s

W'

V_{s+1}

D already
defined

Figure I.1.7.3

To carry out the induction step we need to define D_{s+1} on $\pi^{-1}U_{s+1}^{s+1}$. We need to do this on only one component, because we can extend by h-equivariance. Let π map $V_s \subset \tilde{N}$ homeomorphically onto U_{s+1}^s and let $V_{s+1} \subset V_s$ correspond to U_{s+1}^{s+1}. Let W be an open neighbourhood of $\overline{V}_{s+1}$ in $\tilde{N}$ whose closure is contained in V_s. We define $f: V_s \rightarrow X$ as follows. On $V_s \setminus \overline{W}, f$ is equal to D_M. On $V_{s+1} \cap (\pi^{-1}U_1^{s+1} \cup \cdots \cup \pi^{-1}U_s^{s+1}) f$ is equal to D_s. On the remainder of V_s, f is given using standard theorems about C^r-manifolds. If $r = 0$ we use a result of Edwards-Kirby (1971), in the form explained in Siebenmann (1972). If $r > 0$ one can use standard bump function techniques and the openness of the space of embeddings in the space of all maps. We define $D_{s+1}|V_{s+1}$ to be equal to $f|V_{s+1}$. Note that whether $r = 0$ or $r > 0$, the extension f depends continuously in the C^r-topology, on D_s and on h. Hence D_{s+1} depends continuously on D_s and on h.

It follows that the developing map depends continuously on h. When $s = k$, the induction is done and we have defined a (G, X)-structure on N_{Th} which depends continuously on h.

Holonomy induces structure

Continuation, proof of Neighbourhood is a product: Given an element $(i: N \rightarrow N_{\text{Th}}, h)$ of $\mathcal{A} \times \mathcal{B}$, we define an element of $\mathcal{D}$ as follows. We obtain a developing map $D(h): \tilde{N}_{\text{Th}} \rightarrow X$ from Lemma I.1.7.2 (*Holonomy induces structure*). By lifting i to the universal cover, choosing a lift $\tilde{i}$ which is near the identity, we obtain a composite developing map

$$\tilde{N} \xrightarrow{\tilde{i}} \tilde{N}_{\text{Th}} \xrightarrow{D(h)} X.$$

This composition is near $D_M | \tilde{N}$.

Conversely, given a developing map $D: \tilde{N} \rightarrow X$, near to $D_M | \tilde{N}$, we have the holonomy $h = h(D)$ which is near to h_M, and hence the developing map $D(h): \tilde{N}_{\text{Th}} \rightarrow X$ of Lemma I.1.7.2 (*Holonomy induces structure*). To complete the proof of Theorem I.1.7.1 (*Neighbourhood is a product*) we need to construct an embedding $i: N \rightarrow N_{\text{Th}}$, near the identity, such that $D(h) \circ \tilde{i} = D$. This will follow once we have proved the next lemma. (The proof that the maps between $\mathcal{A} \times \mathcal{B}$ and the neighbourhood of $D_T | \tilde{N}$ are inverse to each other is left to the reader.)

I.1.7.4. Lemma: Embedding exists. *There is a unique equivariant map $\tilde{i}: \tilde{N} \rightarrow \tilde{N}_{\text{Th}}$ such that $D(h) \circ \tilde{i} = D$.*

Proof. Embedding exists: To carry out the construction we take a covering $\{U_i\}$ of N_{Th} by simply connected coordinate charts, and a shrinking $\{U_i'\}$, also consisting of simply connected open sets. We assume that $U_1' \cup \cdots \cup U_r'$ is connected for $1 \leq r \leq k$.

Let V_i be a lifting to $\tilde{N}_{\text{Th}}$ of U_i and let $V_i' \subset V_i$ be the corresponding lift of U_i'. We choose the lifts so that $V_1' \cup \cdots \cup V_r'$ is connected for $1 \leq r \leq k$. The map $D(h) | V_i$ is a homeomorphism onto its image. We may assume $D(V_i' \cap \tilde{N}) \subset D(h)(V_i)$ since D and $D(h)$ are both near D_M. The equation $D(h) \circ \tilde{i} = D$ then determines $\tilde{i}$ uniquely on $\overline{V}_i' \cap \tilde{N}$. We shall show that i is well defined on $\tilde{N}$ and equivariant at the same time. Let $i_1: V_i' \cap \tilde{N} \rightarrow X$ satisfy $D(h)i_1 = D$ on $V_i' \cap \tilde{N}$ and let $i_2: V_j' \cap \tilde{N} \rightarrow X$ satisfy $D(h)i_2 = D$ on $V_j' \cap N$. Let

γ be a covering translation of $\tilde{N}$. Then, on $V_i' \cap \gamma^{-1}V_j'$ we have

$$D(h)i_1 = D = h(\gamma^{-1})D\gamma$$
$$= h(\gamma^{-1})D(h)_{i_2}\gamma$$
$$= D(h)\gamma^{-1}i_2\gamma.$$

Therefore $i_1 = \gamma^{-1}i_2\gamma\colon V_i' \cap \gamma^{-1}V_j' \cap \tilde{N} \to V_i \cap \gamma^{-1}V_j.$

Taking $\gamma = \mathrm{id}$ we see that $\tilde{i} = i_1 = i_2$ is well defined on $V_1' \cup \cdots \cup V_k'$. It also follows that we have a well defined and unique equivariant extension to $\tilde{N}$.

> **Embedding exists**

This completes the proof of the theorem.

> **Neighbourhood is a product**

I.1.7.5. Corollary: Epsilon thickenings exist. *If N is a manifold with a Riemannian (G,X)-structure then there is a neighbourhood of M in $\Omega(N)$ in which ε-thickenings exist for some value of $\varepsilon > 0$. (See Definition I.1.6.2 (Epsilon thickening) for the definition of an ε-thickening.)*

I.1.7.6. Weil's Theorem. In the particular case where N has no boundary, Theorem I.1.7.1 (*Neighbourhood is a product*) implies the famous theorem of Weil (1960), that, up to isotopy, deformations in holonomy corresponding to deformations of the (G,X)-structure. Our theorem gives a very precise version of that result and is also a generalization to manifolds with boundary.

Chapter I.2

Hyperbolic Structures

I.2.1. Möbius groups

The material in this section is covered in Chapter 8 of Thurston (1979).

Suppose that M is a complete hyperbolic manifold without boundary; that is, M has an $(\mathbb{H}^n, \mathrm{Isom}(\mathbb{H}^n))$ structure. Then the developing map is a homeomorphism of $\tilde{M}$ with $\mathbb{H}^n$ and we will consider $\tilde{M}$ to be equal to $\mathbb{H}^n$. The covering transformations (which form a group isomorphic to $\pi_1 M$) are a discrete subgroup Γ of $\mathrm{Isom}(\mathbb{H}^n)$ and $M = \mathbb{H}^n/\Gamma$.

I.2.1.1. Definition. We define a *Möbius group* to be a discrete subgroup of $\mathrm{Isom}(\mathbb{H}^n)$. When the group consists of orientation preserving isometries and $n = 3$, it is called a *Kleinian group* and when $n = 2$ a *Fuchsian group*. If Γ is torsion free, $\mathbb{H}^n/\Gamma$ is a complete hyperbolic manifold. From now on, Γ will denote a Möbius group, which will be assumed to be torsion free unless stated otherwise. L_Γ will denote the limit set of Γ (i.e. $\overline{\Gamma(x)} - \Gamma(x)$ where $x \in \mathbb{H}^n$ and the closure is taken in the disc $\mathbb{H}^n \cup S_\infty^{n-1}$), and $D_\Gamma = S_\infty^{n-1} \backslash L_\Gamma$ the ordinary set. (We note that Γ acts properly discontinuously on $\mathbb{H}^n \cup D_\Gamma$.)

I.2.1.2. Definition. The *convex hull* $\mathcal{C}(L_\Gamma)$ of a Möbius group Γ is defined to be the convex hull of its limit set; that is, the intersection of all closed hyperbolic half spaces of $\mathbb{H}^n \cup S_\infty^{n-1}$ containing L_Γ. We define three manifolds associated with Γ:

(1) $\mathcal{C}_\Gamma = (\mathcal{C}(L_\Gamma)\backslash L_\Gamma)/\Gamma$ the *convex core*
(2) $M_\Gamma = \mathbb{H}^n/\Gamma$ the *Kleinian manifold*
(3) $O_\Gamma = \mathbb{H}^n \cup D_\Gamma/\Gamma$ the *Kleinian manifold with boundary*.

The convex core carries all the essential information about Γ. O_Γ is often a compact manifold, even when M_Γ is non-compact.

Further information on and references for Möbius groups can be found in Beardon (1983) and Harvey (1977).

I.2.2. The thick–thin decomposition

See Section 5.10 of Thurston (1979).

I.2.2.1. Definition.   Let M be a complete hyperbolic n-manifold. We define inj(x), the *injectivity radius* at x, by

$$\text{inj}(x) = \inf_\gamma \{\text{length}(\gamma)\}/2$$

where γ varies over homotopically non-trivial loops through x. Given $\varepsilon > 0$ let $M_{[\varepsilon,\infty)} = \{x \in M \,|\, \text{inj}(x) \geq \varepsilon\}$ (this is often called the *thick* part of the manifold), and let $M_{[0,\varepsilon]} = \{x \in M \,|\, \text{inj}(x) \leq \varepsilon\}$ (known as the *thin* part). Note that given $x \in M_{(\varepsilon,\infty)}$, the ε-neighbourhood of x is isometric to an ε-ball in $\mathbb{H}^n$.

We shall need to know the structure of the thin part of a hyperbolic n-manifold. The following result is due to Margulis (Kazdan-Margulis, 1968) although the following formulation is due to Thurston (1979)

I.2.2.2. Theorem: Thick–thin decomposition.   *There is a universal constant ε (called the Margulis constant), depending only on the dimension n, such that, given any complete hyperbolic n-manifold M, the thin part $M_{(0,\varepsilon]}$ consists of a disjoint union of pieces of the following diffeomorphism types:*

(1) $N^{n-1} \times [0, \infty)$ *where N^{n-1} is an euclidean manifold. (These non-compact components of $M_{(0,\varepsilon]}$ are neighbourhoods of cusps of M.)*

(2) *A neighbourhood of a closed geodesic.*

I.2.2.3. Remarks.

(1) Neighbourhoods of cusps of M have finite volume if and only if N^{n-1} is compact. Thus M has finite volume if and only if $M_{[\varepsilon,\infty)}$ is compact.

(2) The neighbourhood of a geodesic can be non-orientable. (e.g. a Möbius band in two dimensions)

When $n = 2$ or $n = 3$ and M is orientable, we know, even more specifically, that the non-compact components of $M_{(0,\varepsilon]}$ are of the form

$$H_1/\Gamma_p \subset M_{(0,\varepsilon]} \subseteq H_2/\Gamma_p$$

where H_1 and H_2 are horoballs centred at p (a parabolic fixed point of Γ) and Γ_p is the stabilizer of p in Γ. We also have the following corollary which will be used implicitly throughout the section on geodesic laminations.

I.2.2.4. Corollary: Simple geodesics go up a cusp. *If M is a hyperbolic surface or 3-manifold, there exists an ε such that if a simple geodesic enters a non-compact component of $M_{(0,\varepsilon]}$ it must continue straight up the cusp (i.e. it must have a lift with endpoint p).*

I.2.2.5. Remarks.

(1) In the orientable case, each non compact component of $M_{(0,\varepsilon]}$ is of the form H_1/Γ_p, where H_1 is a horoball centred on p.
(2) A *uniform horoball* is a horoball whose images under the group Γ are disjoint or equal. If $n \leq 3$, then each cusp gives rise to a uniform horoball. There are examples in higher dimensions (Apanasov, 1985) where no uniform horoball exists, but the matter is still not entirely cleared up because all known counter-examples require an infinite number of generators for Γ.

The sort of decomposition we get in the case of a surface is illustrated in Figure I.2.2.6.

I.2.2.6.

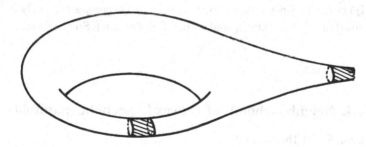

Figure I.2.2.6

I.2.3. The nearest point retraction

See Section 8.4 of Thurston (1979).

This section discusses properties of the nearest point retraction from hyperbolic space (with the sphere at infinity included) onto a convex subset. In general we will be considering it as a retraction of hyperbolic space onto the convex hull of the limit set of a Kleinian group. In this form it will induce a retraction from the Kleinian manifold with boundary onto the convex core (see also Part II).

Given a closed convex subset C of $\mathbb{H}^n \cup S_\infty^{n-1}$, there is a canonical retraction $r\colon \mathbb{H}^n \cup S_\infty^{n-1} \to C$. If $x \in C$ then $r(x) = x$, and if $x \in \mathbb{H}^n/C$ then, since C is closed, there is a ball B_h of radius h about x and disjoint from C, for some h. We increase the radius of this ball continuously until it first touches C. This point of first contact we define to be $r(x)$. If $x \in S_\infty^{n-1}/C$, we do the same construction but with horoballs centred at x. Again we define the point of first contact to be $r(x)$.

I.2.3.1. Proposition: r continuous.   *r is well defined and continuous.*
For the proof of this result see Part II.

Now let Γ be a Möbius group and let O_Γ and C_Γ be the Kleinian manifold with boundary and the convex core, respectively (see Section I.2.1 (*Möbius groups*)). Then we can use r to define a map $\bar{r}\colon O_\Gamma \to C_\Gamma$ as follows. Given $x \in O_\Gamma$ we define $\bar{r}(\pi(x)) = \pi r(x)$ where $x \in \mathbb{H}^n \cup D_\Gamma$. (Here $\pi\colon \mathbb{H}^n \cup D_\Gamma \to O_\Gamma$). Clearly $\bar{r}$ is well defined.

I.2.3.2. Proposition: r bar proper.   *$\bar{r}$ is proper.*

Proof. Note that $r\colon \mathbb{H}^n \cup S^{n-1} \to C(L_\Gamma)$ is proper. Therefore the induced map $\mathbb{H}^n \cup D_\Gamma \to C(L_\Gamma)\backslash L_\Gamma$ is proper. Let K be a compact subset of C_Γ. For each $x \in K$, we choose $\tilde{x} \in \mathbb{H}^n \cup D_\Gamma$ such that $\pi\tilde{x} = x$ and we choose a compact neighbourhood $N(\tilde{x})$ of $\tilde{x}$ in $C(L_\Gamma)$. Let $x_1, \ldots, x_k$ be chosen so that $\pi \circ N(\tilde{x}_1), \ldots,$ $\pi \circ N(\tilde{x}_k)$ cover K. Since r is proper, $L_i = r^{-1}N(\tilde{x}_i)$ is compact. Clearly $\bar{r}^{-1} K$ is contained in $\pi L_1 \cup \cdots \cup \pi L_K$ and is therefore compact. So $\bar{r}$ is proper. $\square$

I.2.4. Neighbourhoods of convex hyperbolic manifolds

See Section 8.3 of Thurston (1979).

I.2.4.1. Theorem: Embedding convex manifolds.   *Any convex complete hyperbolic manifold M with boundary can be embedded in a unique complete hyperbolic manifold with the same fundamental group as M.*

Proof. Since M is convex and complete, we can apply Proposition I.1.4.2 (*Coverings and convexity*) and think of $\tilde{M}$ as a closed convex subset C of $\mathbb{H}^n$ together with a group $\Gamma \subset \text{Isom } \mathbb{H}^n$ which acts properly discontinuously on C.

Since r is proper, Γ also acts properly discontinuously on $r^{-1}C = \mathbb{H}^n \cup D_\Gamma$. Therefore it acts properly discontinuously on $\mathbb{H}^n$.

The only thing that might stop $\mathbb{H}^n/\Gamma$ being a manifold is that Γ could have torsion. For example, take the annulus in $\mathbb{H}^2$ as shown in Figure I.2.4.2. Then the quotient by a rotation of π about the centre gives a manifold. But $\mathbb{H}^2/\Gamma$ is a cone.

I.2.4.2.

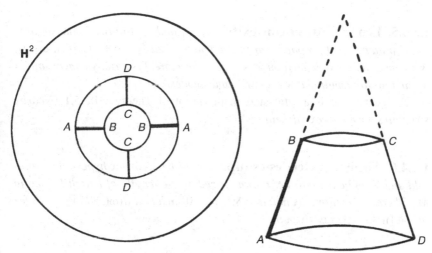

Figure I.2.4.2   *On the left we show the Poincaré disk model, with an annulus drawn. This annulus is cut out of the hyperbolic plane, and then further cut along the two arcs labelled AB. When the two edges labelled AB are glued together, we get part of a hyperbolic cone, as shown on the right.*

We shall prove by contradiction that Γ is torsion free. Suppose Γ has an element g with order n. Take any point $x \in C$ and consider the set $\{x, g(x), \ldots, g^{n-1}(x)\} \subset C$. The convex hull C_1, of these points is contained in C and also C_1 is invariant under g. But then existence of a fixed point $p \in C_1 \subset C$ for g contradicts the fact that C/Γ is a manifold. This contradiction shows that $\mathbb{H}^n/\Gamma$ is a manifold.

$\Box$

I.2.4.3. Definition.   Given a convex, metrically complete hyperbolic manifold M, with boundary, embedded in a metrically complete hyperbolic manifold, and an $\varepsilon > 0$, we define M_ε to be the ε-neighbourhood of M. ε does not need to be small. Unless otherwise stated we shall always assume that M is embedded in the manifold given by Theorem I.2.4.1 (*Embedding convex manifolds*).

I.2.4.4. Proposition: Epsilon neighbourhood smooth. *Let C be a closed, convex subset of* $\mathbb{H}^n$. *Then the distance function* $\Delta: \mathbb{H}^n \backslash C \to \mathbb{R}$ *given by* $\Delta(x) = d(x, C)$ *is* C^1.

For an elegant proof of this, due to Brian Bowditch see Part II.

We shall assume the following well-known result which is actually a consequence of negative curvature (see Douady, 1979).

I.2.4.5. Lemma: Strict convexity. *If* ω_1 *and* ω_2 *are two geodesic arcs in* $\mathbb{H}^n$ *parametrized proportional to arc length, then* $f(x, y) = d(\omega_1(x), \omega_2(y))$ *is a convex function defined on* $\mathbb{R} \times \mathbb{R}$. *Furthermore, f is strictly convex unless* ω_1 *and* ω_2 *are contained within the same geodesic.*

The next result is a vital step in the proof of Theorem I.2.5.1 (*Nearby structures have convex thickenings*).

I.2.4.6. Theorem: Geodesics dip. *Let M be a convex hyperbolic manifold and let* ω *be a geodesic, parametrized by arc length, of length* $L \geq 2\eta$ *in* M_ε. *There is a continuous function* $\delta(\varepsilon, \eta) > 0$ *such that* $d(\omega(t), M) \leq \varepsilon - \delta$ *for all* $t \in [\eta, L - \eta]$. δ *is given by*

$$\delta(\varepsilon, \eta) = \varepsilon - \text{arcsinh}\left(\frac{\sinh \varepsilon}{\cosh \eta}\right).$$

Note that $\delta(\varepsilon, \eta) < \eta$ *and that* $\delta(\varepsilon, \eta)$ *is a monotonic increasing function of* η.

Before we prove this, we prove two lemmas.

I.2.4.7. Lemma: Estimate for distance. *Let* λ *be a geodesic in* $\mathbb{H}^2$. *Let* ω *be a geodesic, and set* $u = d(\omega(0), \lambda)$. *Let the angle between* $\omega'(0)$ *and the perpendicular to* λ *through* $\omega(0)$, *oriented away from* λ, *be* θ. *Then*

$$\sinh d(\omega(t), \lambda) = \sinh u \cosh t + \sinh t \cosh u \cos \theta.$$

Proof.   Left to reader.

$\square$

I.2.4.10. Lemma: Distance function convex. *If C is a closed convex subset of* $\mathbb{H}^n$, *then* $d(\omega(t), C)$ *is convex.*

I.2.4.8.

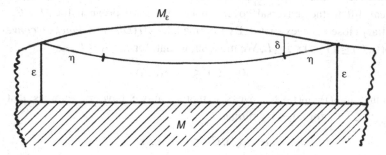

Figure I.2.4.8

I.2.4.9.

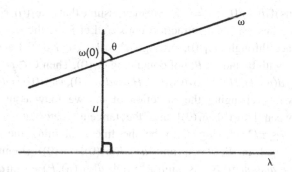

Figure I.2.4.9

Proof. Suppose, for a contradiction, that

$$d(\omega(\alpha), C) > \alpha d(\omega(0), C) + (1 - \alpha)d(\omega(1), C)$$

for some $0 < \alpha < 1$. Let $\gamma: I \to C$ be the geodesic with $\gamma(0) = r\omega(0)$ and $\gamma(1) = r\omega(1)$. We have

$$d(\omega(0), C) = d(\omega(0), \gamma(0)) \quad \text{and} \quad d(\omega(1), C) = d(\omega(1), \gamma(1)).$$

Then

$$d(\omega(\alpha), \gamma(\alpha)) \geq d(\omega(\alpha), C) > \alpha d(\omega(0), \gamma(0)) + (1 - \alpha)d(\omega(1), \gamma(1))$$

which is a contradiction. $\square$

We can now proceed with the proof of the theorem.

Proof. Geodesics dip.   It is sufficient to prove the result for $M \subset \mathbb{H}^n$, since we can lift to the universal cover. In fact we shall prove it for $M = C$, an arbitrary closed convex subset. By Lemma I.2.4.10 (*Distance function convex*), $d(\omega(t), C) \leq \varepsilon$ for $0 \leq t \leq L$. We must show that, for $\eta \leq t_1 \leq L - \eta$,

$$d(\omega(t_1), C) \leq \varepsilon - \delta(\varepsilon, \eta).$$

Defining $\omega_1(t) = \omega(t + t_1)$, we see that what we have to show is that if $d(\omega_1(t), C) \leq \varepsilon$ for $-\eta \leq t \leq \eta$ *then*

$$d(\omega_1(0), C) \leq \varepsilon - \delta(\varepsilon, \eta).$$

By the definition of $\delta(\varepsilon, \eta)$ this is equivalent to showing that

$$\cosh \eta \sinh d(\omega_1(0), C) \leq \sinh \varepsilon.$$

This is obvious if $d(\omega_1(0), C) = 0$. So we may assume that $d(\omega_1(0), r\omega_1(0)) > 0$, where r is the nearest point retraction onto C. Let P be the $(n-1)$ dimensional subspace through $r\omega_1(0)$, orthogonal to $[\omega_1(0), r\omega_1(0)]$ and let H be the half space with boundary P, not containing $\omega_1(0)$. Then $C \subset H$. Hence $\varepsilon \geq d(\omega_1(t), C) \geq d(\omega_1(t), H)$. Also $d(\omega_1(0), H) = d(\omega_1(0), r\omega_1(0)) = d(\omega_1(0), C)$.

After possibly changing the direction of ω, we may assume that the angle θ between $[r\omega_1(0), \omega_1(0)]$ and the tangent vector at $t = 0$, $\omega_1'(0)$, satisfies $0 \leq \theta \leq \pi/2$. Taking λ to be the line containing the orthogonal projection of $|\omega|$ in P, $t = \eta$ and $u = d(\omega_1(0), r\omega_1(0))$ (Lemma I.2.4.7 (*Estimate for distance*)) gives $\sinh \varepsilon \geq \sinh d(\omega_1(\eta), P) \geq \sinh d(\omega_1(0), C)$ $\cosh \eta$ as required.

> **Geodesics dip**

I.2.4.11. Corollary: Convex thickens to strictly convex.   *If M is a convex hyperbolic manifold, M_ε (see Definition I.2.4.3 (Epsilon neighbourhood of manifold)) is strictly convex for all $\varepsilon > 0$.*

I.2.5. Convex thickenings

See Proposition 8.3.3 of Thurston (1979).

I.2.5.1. Theorem. Nearby structures have convex thickenings.   *Let N be a compact manifold with boundary. Let M be a convex hyperbolic structure on N and let M_{Th} be a thickening of M. Then there is a neighbourhood U of M in $\Omega(N)$ such that, if $M' \in U$, then M' can be thickened to a compact convex manifold M''.*

Proof. By Theorem I.1.7.1 (*Neighbourhood is a product*), we may assume that all the manifolds in $\mathcal{U}$ are embedded in N_{Th} by an embedding near to the identity $i\colon N \to N_{\text{Th}}$ and that the hyperbolic structure on N_{Th} inducing the given structure M' is near to that of M_{Th}. It follows that we can restrict our attention to the varying hyperbolic structure on N_{Th}, and not worry about the embedding of N in N_{Th}. We shall use metrics which come from the path metric on even larger thickenings. (We have to be careful here because of examples such as that shown in Figure I.1.5.7.)

We choose $\varepsilon > 0$ so that M_{Th} is a 10ε-thickening of M. We choose $\mathcal{U}$ sufficiently small so that, if $M' \in \mathcal{U}$, then the corresponding hyperbolic structure on N_{Th} makes N_{Th} a 9ε-thickening of N. This is possible by Corollary I.1.7.5 (*Epsilon thickenings exist*). We define $\delta = \delta(\varepsilon, \varepsilon/2)$, where δ is the function defined in Theorem I.2.4.6 (*Geodesics dip*). Recall that $\delta < \varepsilon/2$. Let $K_0 \subset \text{int}$ $K_1 \subset \tilde{N}_{\text{Th}}$ where K_0 and K_1 are compact connected subspaces, and the translates of K_0 by the covering translations cover $\tilde{N}_{\text{Th}}$. We assume that K_1 contains an r-neighbourhood of K_0 where $(r-1)$ is larger than the diameter of M_{Th}. We also assume that $\mathcal{U}$ is small enough so that $d(Dx, D'x) < \delta/4$, for all $x \in K_1$, where $D, D'\colon \tilde{N}_{\text{Th}} \to \mathbb{H}^n$ are developing maps for our fixed convex hyperbolic manifold M and for M' respectively. We shall also assume that $D'|K_1$ is an embedding (see Section I.1.5.4 (*Space of developing maps*)). Then K_1 contains an $(r-\delta)$-neighbourhood of K_0 in the metric d' related to the structure M'.

To define the convex thickening claimed in the statement, we fix $M' \in \mathcal{U}$ and let $\mathcal{A}$ be the collection of $(n+1)$ tuples of points $(x_0, \ldots, x_n)$ with $x_i \in M'_\varepsilon$ and such that $d'(x_i, x_j) \leq \varepsilon$. For each $(n+1)$-tuple $(x_0, \ldots, x_n) \in \mathcal{A}$, we take the convex hull $\mathcal{C}(x_0, \ldots, x_n)$ and set $U = \bigcup_{\mathcal{A}} \mathcal{C}(x_0, \ldots, x_n)$. (The convex hull is defined inside an ε-ball centred on x_0.) We claim that U is the compact convex thickening we seek. The fact that U is compact is an immediate consequence of the compactness of $\mathcal{A}$. The fact that U is convex will be deduced from its local convexity (see Corollary I.1.3.7 (*Local convexity implies convexity*)). Given $u, v \in U$, such that $d'(u, v) \leq \varepsilon$, we show that the geodesic interval $[u, v]$ is in U. Let $u \in \mathcal{C}(u_0, \ldots, u_n)$ and $v \in \mathcal{C}(v_0, \ldots, v_n)$, where $u_i, v_i \in M'_\varepsilon$ for $0 \leq i \leq n$, and $d'(u_i, u_j) \leq \varepsilon$ and $d'(v_i, v_j) \leq \varepsilon$ for $0 \leq i < j \leq n$. Then

$$d(u_i, v_j) \leq d(u_i, u) + d(u, v) + d(v, v_j) \leq 3\varepsilon.$$

It follows that we need only establish the following claim. Let $\{w_0, \ldots, w_n\}$ be contained in an ε-neighbourhood of $D'(K_1 \cap \tilde{N}) \subset \mathbb{H}^n$, and in a 4ε-neighbourhood of some point $x \in D'(K_0 \cap \tilde{N})$, and let $d(w_i, w_j) \leq 3\varepsilon$. Then $\mathcal{C}(w_0, \ldots, w_n) \subset D'(K_1 \cap \tilde{U})$, where $\tilde{U}$ is the inverse image of U in the universal cover.

There is no loss of generality in supposing that $d(w_0, w_1)$ maximizes $d(w_i, w_j)$ $(0 \leq i < j \leq n)$. If $d(w_0, w_1) \leq \varepsilon$ there is nothing to do. So we suppose that $d(w_0, w_1) > \varepsilon$. Let z be the midpoint of $[w_0, w_1]$. We divide $\mathcal{C}(w_0, \ldots, w_n)$ into two pieces, $\mathcal{C}(z, w_1, w_2, \ldots, w_n)$ and $\mathcal{C}(z, w_0, w_2, \ldots, w_n)$. We want to show that each of these pieces satisfies the hypotheses for the claim. By the definition of $\mathcal{U}$, $d(w_i, D(K_1 \cap \tilde{N})) \leq \varepsilon + \delta/4$ for $0 \leq i \leq n$. Since $D\tilde{N}$ is convex, Theorem I.2.4.6 (*Geodesics dip*) implies that

$$d(z, D(\tilde{N})) \leq \varepsilon + \delta/4 - \delta(\varepsilon + \delta/4, \varepsilon/2)$$

$$\leq \varepsilon + \delta/4 - \delta(\varepsilon, \delta/2)$$

$$\leq \varepsilon - 3\delta/4.$$

The second inequality is a result of the monotonicity of the function $\delta(\varepsilon, \eta)$ in the variable ε, pointed out just after Theorem I.2.4.6 (*Geodesics dip*). It follows that

$$d(z, D'(K_1 \cap \tilde{N})) \leq \varepsilon - \delta/2 < \varepsilon.$$

The final point to check is that this chopping in half process gives a figure with all sides having length less than ε after a finite number of steps. This is not obvious, because, although the longest side is divided in two, the other sides will, in general, become longer than they were. Let $m = d(w_0, w_1)$, and fix m for the next few steps. At each step the number of edges of length greater than $0.9m$ decreases. Therefore, after a finite number of steps, all sides have length less than $0.9m$. (This can be seen by doing a computation in euclidean geometry. As we are dealing with a small region, euclidean estimates give rise to hyperbolic estimates, though the hyperbolic estimates are slightly worse.) Continuing in this manner, the length of each side will eventually become less than ε.

> **Nearby structures have convex thickenings**

I.2.5.2. Corollary: Nearby structures strongly complete. *With the same hypotheses as in the theorem above, if $M' \in U$, $D_{M'}$ is a homeomorphism, and M' can be embedded in a unique complete hyperbolic manifold with the same fundamental group.*

I.2.5.3. Remarks: We can see that convexity of the original manifold with boundary is a necessary condition for the above corollary to hold, by considering $M = S^3 \setminus N(K)$ (where K is a knot whose complement admits a complete hyperbolic structure.) By considering Dehn surgery space, we see that nearby hyperbolic structures on M, do not necessarily admit extensions to complete hyperbolic manifolds. See Thurston (1979) for more details.

Chapter I.3
Spaces of Hyperbolic Manifolds

I.3.1. The geometric topology

We shall define a topology on the set of closed subsets of a topological space. We shall thereby derive topologies for both the space of complete hyperbolic manifolds and the space of geodesic laminations (see Section I.4.1 (*Geodesic Laminations*)). This topology was first considered by Chabauty (1950) as a topology on the space of closed subgroups of a locally compact topological group, and later by Harvey (1977) with specific reference to Fuchsian groups. See also Michael, (1951).

I.3.1.1. Definition. Given a topological space X, the *Chabauty topology* on $C(X)$ (the set of all closed subsets of X) has a sub-basis given by sets of the following form:

(1) $O_1(K) = \{A \mid A \cap K = \emptyset\}$ where K is compact;
(2) $O_2(U) = \{A \mid A \cap U \neq \emptyset\}$ where U is open.

If X is compact and metrizable, the Chabauty topology agrees with the topology induced by the Hausdorff metric. The Chabauty topology has the following nice topological properties.

I.3.1.2. Proposition: Properties of Chabauty topology. *Let X be an arbitrary topological space (no particular assumptions), then:*

(1) *$C(X)$ the set of closed subsets of X with the Chabauty topology is compact.*
(2) *If X is Hausdorff, locally compact and second countable, $C(X)$ is separable and metrizable.*

59

Proof. (1) By Alexander's Sub-base theorem (Rudin, 1973, p. 368), we need only show that every covering by sub-basis elements has a finite sub-covering. Let the covering consist of

$$\{O_1(K_i)\}_{i \in I} \quad \text{and} \quad \{O_2(U_j)\}_{j \in J}.$$

Let $C = X \setminus \bigcup_{j \in J} U_j$. C is closed and thus $C \in C(X)$. C is not in $O_2(U_j)$ for any j, therefore $C \in O_1(K_i)$ for some i. $\{U_j\}_{j \in J}$ is a covering for K_i, so there exists a finite sub-covering $\{U_{j(1)}, \ldots, U_{j(n)}\}$. Given a closed subset L, either $L \cap U_{j(k)} \neq \emptyset$ for some $k = 1, \ldots, n$ (i.e. $L \in O_2(U_{j(k)})$), or $L \cap K_i = \emptyset$ (i.e. $L \in O_1(K_i)$). Thus, $C(X) = O_1(K_i) \cup \bigcup_{k=1}^{k=n} O_2(U_{j(k)})$; which is the desired finite sub-covering.

(2) Since X is Hausdorff and locally compact, it is also regular. Suppose K, $L \in C(X)$ and $x \in K \setminus L$. Because X is regular and locally compact, there exists an open set U such that $x \in U \subset \bar{U} \subset X - L$ where $\bar{U}$ is compact. So $K \in O_2(U)$, $L \in O_1(\bar{U})$, and $O_2(U) \cap O_1(\bar{U}) = \emptyset$. It follows that $C(X)$ is Hausdorff.

Let $\{B_1, \ldots, B_n, \ldots\}$ be a countable basis for X, such that $\bar{B}_i$ is compact for each i. We claim that $\{O_1(\bar{B}_i)\} \cup \{O_2(B_i)\}$ is a sub-basis for the Chabauty topology on $C(X)$. Let $O_1(K)$ be a sub-basis element (as in our Definition I.3.1.1 (*Chabauty topology*)), $C \in O_1(K)$, and $\{\bar{B}_{n(1)}, \ldots, \bar{B}_{n(k)}\}$ a covering for K such that $C \cap \bar{B}_{n(i)} = \emptyset$ for $1 \le i \le k$. Then

$$C \in O_1(\bar{B}_{n(1)}) \cap \cdots \cap O_1(\bar{B}_{n(k)}) \subset O_1(K).$$

Now suppose $C \in O_2(U)$ and $x \in C \cap U$. Choose B, a neighbourhood of x such that $B \subset U$. Then $C \in O_2(B) \subset O_2(U)$. It follows that we have a countable sub-basis consisting of sets of the form $O_1(\bar{B}_i)$ and $O_2(B_i)$. Then, by the Urysohn Metrization Theorem $C(X)$ is separable and metrizable.

$\Box$

The following easily proved lemma exposes the essential geometric nature of our topology.

I.3.1.3. Lemma: Geometric convergence. *Suppose X is a locally compact metric space. A sequence $\{A_n\}$ of closed subsets of X converges in $C(X)$ to the closed subset A if and only if:*

(1) *If $\{x_{n(k)}\} \in \{A_{n(k)}\}$ converges to $x \in X$ then $x \in A$.*
(2) *If $x \in A$, then there exists a sequence $\{x_n\}$, where each x_n is an element of A_n, converging to x.*

The proof is left to the reader.

We now restrict ourselves to the case of closed subgroups of Lie groups, and prove Chabauty's original theorem (Chabauty, 1950; see also Harvey, 1977). The set of closed subgroups of a Lie group L is closed in the Chabauty topology and is thus compact and metrizable. There is a right invariant Haar measure on any Lie group, which induces a measure on $\Gamma \backslash L$ when Γ is discrete. Denote the total volume (which may be infinite), by μ_Γ. When L is $\mathrm{Isom}(\mathbb{H}^n)$, we shall consider instead $\mu_\Gamma = \mathrm{vol}(\mathbb{H}^n/\Gamma)$. The two versions of μ_Γ are equal if we normalize correctly.

I.3.1.4. Theorem: A(U) compact. *Let $\mathcal{G}(L)$ be the space of closed subgroups of a Lie group L with the Chabauty topology (so that $\mathcal{G}(L)$ is compact and metrizable). Let U be an open neighbourhood of $\{e\}$ in L, then:*

(1) *$A(U) = \{G \in \mathcal{G}(L) \mid G \cap U = \{e\}\}$ is compact.*

(2) *$B(U) = \{G \in A(U) \mid G$ is torsion free$\}$ is compact.*

(3) *The set of discrete subgroups of L is open in the space of all closed subgroups. It is the union of the interiors in $\mathcal{G}(L)$ of the compact spaces $A(U)$, as U varies over open neighbourhoods of $\{e\}$ in L.*

(4) *If $\{\Gamma(n)\}$ is a sequence of discrete subgroups converging to the discrete subgroup Γ, then*

$$\mu_\Gamma \leq \lim_{n \to \infty} \inf \mu_{\Gamma(n)}.$$

In particular, $A(U, M) = \{G \in A(U) \mid \mu(G) \leq M\}$ and $B(U, M) = \{G \in B(U) \mid \mu(G) \leq M\}$ are compact.

Proof. (1) We observe that $A(U) = \mathcal{G}(L) \backslash O_1(U \backslash \{e\})$ is a closed subset of the compact space $\mathcal{G}(L)$.

(2) Suppose $\Gamma_i \in B(U)$ converges to Γ (which is in $A(U)$ by (1)). If Γ is not torsion free, there is some $\gamma \in \Gamma$ and $n > 0$ such that $\gamma^n = e$. By Lemma I.3.1.3 (*Geometric convergence*) there exist $\gamma_i \in \Gamma_i$ such that $\{\gamma_i\}$ converges to γ. Thus, $\{\gamma_i^n\}$ converges to e, but this would imply that $\gamma_i^n = e$ for large values of i, which is a contradiction.

(3) Let V be a compact neighbourhood of e in L small enough not to contain a non-trivial subgroup and let U be a smaller open neighbourhood of e in L such that $U^2 \subset V$. Let $K = V \backslash U$. Then any closed subgroup Γ such that $\Gamma \cap K = \emptyset$ also satisfies $\Gamma \cap V = \{e\}$. To see this, let $g \in \Gamma \cap V$. Then $g \in \Gamma \cap U$. Let $n > 0$ be the smallest integer, if any, such that $g^n \notin U$. Then $g^n = g^{n-1} g \in U^2 \subset V$. This is impossible. Hence $g^n \in U \subset V$ for all $n > 0$ and so $g^n \in V$ for all $n \in \mathbb{Z}$. But V contains no subgroup, so $g = e$. What we have shown is that $O_1(K)$

consists of discrete subgroups. This proves that any discrete subgroup has an open neighbourhood consisting of discrete subgroups.

We caution the reader that a current important conjecture is that the set of discrete faithful hyperbolic representations of an abstract finitely generated abstract group in $PSL(2, \mathbb{C})$ is closed. This is with respect to a different topology from the Chabauty topology, namely the compact-open topology.

(4) Choose a non-empty open set $W \subset \mathbb{H}^n$ (or, in general, $W \subset L$) such that $W \cap T(W) = \emptyset$ for all $T \in (\Gamma \setminus \{e\})$, and a non-empty compact subset K' of W. Let $C = \{T \in L : T(K') \cap K' \neq \emptyset\}$. C is a compact subset of L. Also let V be an open neighbourhood of e in L such that $\Gamma \cap V = \{e\}$. Then $\Gamma \in O_1(C \setminus V)$. Therefore $\Gamma(n) \in O_1(C \setminus V)$ for $n \geq N$. We can also assume that $\Gamma(n) \cap V = \{e\}$ for $n \geq N$ by the proof of (3). Hence, if $n \geq N$ and $T_n \in \Gamma(n)$, then $T_n = e$ or $T_n(K') \cap K' = \emptyset$.

Now W may be chosen to have the same measure as μ_Γ and K may be chosen to have measure arbitrarily close to μ_Γ. Therefore

$$\mu_\Gamma \leq \lim_{n \to \infty} \inf \mu_{\Gamma(n)}.$$

$\square$

I.3.1.5. Remarks. For those familiar with the theory of Kleinian groups, the Chabauty topology on the space of discrete subgroups of $PSL(2, \mathbb{C})$ is equivalent to the topology induced by convergence of Poincaré (Dirichlet) fundamental polyhedra (with a fixed origin) and their associated face-pairings.

We now obtain a few useful results of Thurston (see Section 8.8 of Thurston (1979) and Thurston (1986)) as corollaries of Chabauty's theorem.

I.3.1.6. We can think of a complete hyperbolic manifold provided with a frame (M, e) as a discrete torsion-free subgroup Γ of Isom $(\mathbb{H}^n)$. To do this, we fix a point $s_0 \in \mathbb{H}^n$ and an orthonormal frame $(s_1, \ldots, s_n)$ for the tangent space to $\mathbb{H}^n$ at s_0. We refer to this fixed choice as the *standard frame* in $\mathbb{H}^n$. Then we choose the developing map for M which takes some lift of e to s, our standard frame in $\mathbb{H}^n$. The holonomy of this developing map gives an unique Möbius group Γ. Then $(\mathbb{H}^n \setminus \Gamma, s)$ is a complete hyperbolic manifold with base-frame which is isometric (in a frame-preserving way) to (M, e). In this way we can topologize the space $\mathcal{MF}^n$ of complete hyperbolic $\mathcal{M}$anifolds of dimension n with base $\mathcal{F}$rame, using the Chabauty topology on the set of subgroups Γ of Isom$(\mathbb{H}^n)$. We call this the *geometric* topology on $\mathcal{MF}^n$. We topologize $\mathcal{MB}^n$, the space of complete hyperbolic $\mathcal{M}$anifolds of dimension n with

Basepoint as the quotient of $\mathcal{MF}^n$, and $\mathcal{MW}^n$, the space of complete hyperbolic n-Manifolds Without basepoint as a further quotient. All of these topologies are known as the geometric topology.

By part (3) of Theorem I.3.1.4 ($A(U)$ *compact*), $\mathcal{MF}^n$ is a locally compact Hausdorff space. $\mathcal{MB}^n$ is the quotient of $\mathcal{MF}^n$ by the compact group $O(n)$, and is therefore Hausdorff. $\mathcal{MW}^n$ is not Hausdorff.

I.3.1.7. Corollary: Set of hyperbolic manifolds compact.

(1) $\mathcal{MF}^n(\varepsilon)$, *the space of complete hyperbolic manifolds with frame, having injectivity radius bounded below by ε at the basepoint is compact for any $\varepsilon > 0$.*

(2) $\mathcal{MB}^n(\varepsilon)$, *the space of complete hyperbolic n-manifolds with basepoint having injectivity radius not less than ε at the basepoint, is compact.*

(3) *The space of complete hyperbolic n-manifolds, $\mathcal{MW}^n$, is compact.*

Proof. Let x_0 be a fixed basepoint for $\mathbb{H}^n$ and let Γ be a torsion-free discrete group of isometries for $\mathbb{H}^n$. To prove the first statement, we simply observe that the injectivity radius at the basepoint is equal to $\inf\{d(x_0, \gamma x_0) \mid \gamma \in \Gamma\}/2$. By the previous theorem the set of such Γ is compact. The second statement follows since $\mathcal{MB}^n(\varepsilon)$ is the image of $\mathcal{MF}^n(\varepsilon)$ under the obvious projection map. To prove (3) we recall that every hyperbolic n-manifold has a point with injectivity radius not less than ε (the Margulis constant for $\mathbb{H}^n$), so $\mathcal{MW}^n = Y(\mathcal{MB}^n(\varepsilon))$, where $Y: \mathcal{MB}^n \to \mathcal{MW}^n$ is the map which forgets the basepoint.

$\square$

I.3.1.8. Corollary: Compact with bounded volume.

(1) *The space of complete hyperbolic n-manifolds with basepoint having injectivity radius at the basepoint not less than ε and volume not more than V, $\mathcal{MB}^n(\varepsilon, V)$, is compact for any $\varepsilon > 0$ and any $V > 0$.*

(2) *The set of complete hyperbolic n-manifolds with volume not more than V, $\mathcal{MW}^n(V)$ is compact for any $V > 0$.*

I.3.1.9. Definition.
A *marked hyperbolic surface* of finite type is a topological surface T of finite type together with an isotopy class of homeomorphisms $h: T \to S$, where S is a complete hyperbolic surface. Two marked surfaces $h_1: T \to S_1$ and $h_2: T \to S_2$ are said to be *equivalent* (or, some times, *equal*) if there is an isometry $\phi: S_1 \to S_2$ such that ϕh_1 is isotopic to h_2.

I.3.1.10. The space of all equivalence classes of marked hyperbolic surfaces of a given homeomorphism type (where we specify also whether a puncture is to be a funnel or a cusp) is called Teichmüller space and we denote it $T(T)$. An equivalence class of marked hyperbolic surfaces clearly determines and is determined by its holonomy map $H_{[h]}: \pi_1(T) \to \text{Isom}(\mathbb{H}^2)$ which is well-defined up to conjugacy. We may thus topologize Teichmüller space as a subspace of $R(\pi_1(T), \text{Isom}(\mathbb{H}^2))$ (the set of conjugacy classes of representations of $\pi_1(T)$ into $\text{Isom}(\mathbb{H}^2)$) with the compact open topology (i.e. $\{\rho_i\}$ converges if $\{\rho_i(x)\}$ converges for every $x \in \pi_1(T)$). Teichmüller space is an open subspace of $R(\pi_1(T), \text{Isom}(\mathbb{H}^2))$ and is in fact homeomorphic to $\mathbb{R}^{6g-6+2p+3f}$ where g is the genus, p is the number of punctures, and f is the number of funnels. For further information on Teichmüller space see Abikoff (1980) or Bers (1960).

To discuss the connection with the geometric topology, we generalize to any number of dimensions. We first remove the annoyance of dealing with conjugacy classes, by taking manifolds with baseframe. We can adapt the geometric topology to this situation, to include the marking. We topologize the space of injective homomorphisms $\pi_1(T) \to \text{Isom}(\mathbb{H}^n)$ onto a discrete subgroup of $\text{Isom}(\mathbb{H}^n)$, by using the Chabauty topology on the image, with a further refinement. We are allowed to specify a finite number (depending on the neighbourhood we are trying to describe) of elements of $\pi_1(T)$ and demand that the images of each of these lies in a certain open subset (depending on the element) of $\text{Isom}(\mathbb{H}^n)$. We call this the *marked geometric topology*. The compact-open topology on the space of homomorphisms is known as the *algebraic topology*.

There is an obvious continuous map from the marked geometric topology to the algebraic topology. This map is not a homeomorphism in general, though we will prove below that it is a homeomorphism for surfaces of finite type and also for manifolds of finite volume.

A counterexample to the map being a homeomorphism can be made for a surface of infinite type. We give a quick sketch. The basic building block is an infinite strip with one handle. The strip has two boundary components. We specify a hyperbolic structure on the building block, by insisting that the boundary components are infinite geodesics, which are asymptotic at each end, and that there is an orientation reversing isometry, interchanging the two ends of the strip, whose fixed point set consists of a geodesic arc joining the two boundary components and a geodesic simple closed curve going around the handle. There are two parameters for the hyperbolic structure, namely the length of the arc and the length of the circle. We glue a countable number of building blocks together in the simplest possible way. In the algebraic limit, the sum of the lengths of the arcs is convergent. The nth surface has all except a finite number of these lengths equal to 1. Details are left to the reader.

The finitely generated case is more important. Here a counterexample can only be given in dimensions greater than two. For example, let $\rho_n: \mathbb{Z} \to PSL(2, \mathbb{C})$ be generated by

$$\rho_n(1) = \begin{bmatrix} e^{w_n} & n \sinh w_n \\ 0 & e^{-w_n} \end{bmatrix} \quad \text{where } w_n = \frac{1}{n^2} + \frac{\pi i}{n}.$$

Then ρ_n converges to $\rho: \mathbb{Z} \to PSL(2, \mathbb{C})$ where

$$\rho(1) = \begin{bmatrix} 1 & \pi i \\ 0 & 1 \end{bmatrix}$$

in the algebraic topology. But $\{\rho_n(n)\}$ converges to

$$\begin{bmatrix} -1 & -1 \\ 0 & -1 \end{bmatrix} = \begin{bmatrix} 1 & 1 \\ 0 & 1 \end{bmatrix}.$$

Since $\{\rho_n(\mathbb{Z})\} \subset B(U)$ for some open neighbourhood U of e, the geometric limit of $\{\rho_n(\mathbb{Z})\}$ must be discrete and torsion free, and thus a parabolic subgroup of rank 2. For a more in-depth discussion of algebraic and geometric convergence see Chapter 9 of Thurston (1979). Here we confine ourselves to the following case where the geometric and algebraic topologies coincide.

I.3.1.11. Proposition: Algebraic equals geometric. *We restrict our attention to complete hyperbolic manifolds (with baseframe) of volume $\leq V < \infty$ fot some V. The map from the space of such manifolds with the geometric topology to the same space with the algebraic topology is a homeomorphism.*

Proof. Suppose $\{\rho_i: \pi_1 M \to \text{Isom}(\mathbb{H}^2)\}$ is a sequence of homomorphisms that converges to $\{\rho: \pi_1 M \to \text{Isom}(\mathbb{H}^2)\}$ in the algebraic topology. We must show it converges also in the geometric topology. Let $\Gamma_i = \rho_i(\pi_1 M)$ and $\Gamma = \rho(\pi_1 M)$. We choose $\gamma \in \Gamma$ to be a hyperbolic element which is not divisible. Let $\alpha \in \pi_1 M$ be defined by $\rho(\alpha) = \gamma$. Then α is indivisible. Let $\gamma_i = \rho_i(\alpha) \in \Gamma_i$. By passing to a subsequence, we may assume that each γ_i is hyperbolic.

We claim that for some neighbourhood U of the identity, $\Gamma_i \cap U = \{e\}$, for i large. For otherwise (taking a subsequence if necessary) there is a sequence of non-trivial indivisible element $\beta_i \in \Gamma_i$, such that β_i converges to the identity. Then $[\gamma_i, \beta_i]$ converges to the identity. But then $[\gamma_i, \beta_i]$ must commute with β_i for i large, by the Margulis lemma.

If β_i is hyperbolic, this means that $[\gamma_i, \beta_i]$ is a power of β_i, so that γ_i normalizes the subgroup generated by β_i. Therefore γ_i is a power of β_i, which is impossible, since γ_i is indivisible and is not small. If β_i is parabolic, then $[\gamma_i, \beta_i]$ is either trivial or must be parabolic with the same unique fixed point p. But then $[\gamma_i, \beta_i]\beta_i^{-1} = \gamma_i^{-1}\beta_i\gamma_i$ fixes p, and so γ_i fixes p. Since γ_i is hyperbolic, this is also impossible. This proves the existence of the claimed neighbourhood U.

By passing to a subsequence, we may assume that Γ_i converges to $G \subset \mathrm{Isom}(\mathbb{H}^n)$ in the Chabauty topology. Then $G \cap U = \{e\}$ and so G is discrete. Clearly $\Gamma \subset G$.

We want to prove $\Gamma = G$. So suppose $g \in G \backslash \Gamma$. Replacing g by $g\gamma^k$ for a large value of k, we see that there is no loss of generality in assuming that g is hyperbolic. Since Γ has finite co-volume, the index of Γ in G must be finite. It follows that $g^k = \rho(h)$ for some $k > 0$ and some $h \in \pi_1 M$.

Let $g_i \in \Gamma_i$ converge to g and let $h_i = \rho_i(h)$. Then $g_i^{-k}h_i$ converges to the identity. But we have already shown that this implies $h_i = g_i^k$ for large i. This means that g_i centralizes h_i. Now the centralizer of a hyperbolic element in a discrete group is infinite cyclic, and so kth roots are unique. Hence h has a unique kth root α' in $\pi_1 M$. Then $g_i = \rho_i(\alpha')$ and so $g = \rho(\alpha')$. Hence $g \in \Gamma$, a contradiction.

$\square$

Remark. In the case of a surface M of finite type, the algebraic topology and the geometric topology coincide, even when we allow surfaces with infinite area. This can be seen by breaking the surface up into a finite number of pairs of pants. The geometry of each pair of pants, and the way they are glued together, is determined by a finite number of elements of $\rho(\pi_1 T) \subset \mathrm{Isom}(\mathbb{H}^2)$.

I.3.2. ε-relations and approximate isometries

We now consider a generalization of the concept of Hausdorff metric which was developed by Gromov. Intuitively, compact metric spaces can be approximated very well by finite subsets of points and locally compact path spaces can be approximated very well by large compact subsets of themselves. This simple idea developed into the more formal notion of (ε, r)-relations, which provide us with a topology on the space of all complete locally compact path spaces. One of the first applications of this was also one of the most amazing; Gromov used ε-relations to show that if a finitely generated group has polynomial growth then it contains a nilpotent subgroup of finite index. Gromov (1981a, b) contains an extensive investigation into the space of Riemannian manifolds with the geometric topology. His definitions are related to, but slightly different from ours. In this section we will establish that the topology induced by (ε, r)-relations when restricted to the space of hyperbolic manifolds with basepoint agrees with the geometric topology.

I.3.2.1. Definition. Let (X, x_0) and (Y, y_0) be two compact metric spaces with basepoint. An ε-relation between (X, x_0) and (Y, y_0) is a relation R with the following properties:

(1) $x_0 R y_0$;
(2) for each $x \in X$, there exists $y \in Y$ such that xRy;
(3) for each $y \in Y$ there exists $x \in X$ such that xRy;
(4) if xRy and $x'Ry'$, then $|d_X(x, x') - d_Y(y, y')| \le \varepsilon$.

I.3.2.2. Lemma: Metric on metric spaces. *If we define $d((X, x_0), (Y, y_0))$ to be the infimum of all values of ε for which there is an ε-relation between (X, x_0) and (Y, y_0) we obtain a metric on the set of isometry classes of compact metric spaces with basepoint.*

Remark: Of course, the class of all compact metric spaces is not a set in standard set theory. However, one of the usual tricks can be used to get around this objection. For example, every compact metric space has a countable dense subset, so we can consider all completions of all metrics on any subset of the natural numbers.

Proof. The only thing that needs proof is that if $d((X, x_0), (Y, y_0)) = 0$, then (X, x_0) is isometric to (Y, y_0). To prove this we fix a countable dense subset $\{x_i\}$ in X. Given a sequence R_n of ε_n-relations between (X, x_0) and (Y, y_0), such that $\varepsilon_n \to 0$, we choose points $y_{i,n} \in Y$ such that $x_i R_n y_{i,n}$. By using the Cantor diagonalization process, we can assume that $\lim_{n \to \infty} y_{i,n}$ exists for each i. We denote the limit by y_i. Then

$$d(x_i, x_j) = \lim_{n \to \infty} d(y_{i,n}, y_{j,n}) = d(y_i, y_j).$$

So the map which sends $\{x_i\}$ to $\{y_i\}$ is an isometry on this countable set. It is easy to see that $\{y_i\}$ is dense in Y. It now follows that this map extends to an isometry between (X, x_0) and (Y, y_0).

I.3.2.3. Definition. Two metric spaces with basepoint (X, x_0) and (Y, y_0) are (ε, r)-*related* if there is an ε-relation between compact subspaces (X_1, x_0)

and (Y_1, y_0) of (X, x_0) and (Y, y_0) respectively, where $B_X(x_0, r) \subset X_1$, and $B_Y(y_0, r) \subset Y_1$. (Recall from Theorem I.1.3.5 (*Hopf Rinow*) that balls of radius r are compact in a complete, locally compact path space.)

We can use this notion to topologize the space of (isometry classes of) complete locally compact path spaces with basepoint, as follows. Let (X, x_0) be a complete locally compact path space with basepoint and let $r > 0$ and $\varepsilon > 0$. We define the neighbourhood $\mathcal{N}(X, x_0, r, \varepsilon)$ to be the set of complete locally compact path spaces with basepoint (Y, y_0) such that (Y, y_0) is (ε, r)-related to (X, x_0).

I.3.2.4. Lemma: Space of path spaces Hausdorff. *The space of complete locally compact path spaces with basepoint is Hausdorff.*

Proof. Let (X, x_0) and (Y, y_0) have no disjoint neighbourhoods. We must show that they are isometric. We fix r. By the method of Lemma I.3.2.2 (*Metric on metric spaces*) we construct an isometry of $(B_X(x_0, r), x_0)$ into (Y, y_0) from a sequence of ε_n-relations. From the method of construction and using the fact that Y is a path space, it is easy to see that $B_X(x_0, r)$ maps onto $B_Y(y_0, r)$. Let $\phi_r: (B_X(x_0, r), x_0) \to (B_Y(y_0, r), y_0)$ be the isometry. Using Ascoli's theorem and the Cantor diagonalization process, we can find $\phi: (X, x_0) \to (Y, y_0)$, such that, for any fixed compact subset K of X, $\phi \mid K$ is the limit of maps of the form $\phi_r \mid K$. Hence ϕ is an isometry.

□

In this way we get another Hausdorff topology on $\mathcal{MB}^n$, the space of (isometry classes of) complete hyperbolic n-Manifolds with Basepoint in addition to the geometric topology. We shall also use $\mathcal{MW}^n$, the space of (isometry classes of) complete hyperbolic n-Manifolds Without basepoint, which is given the quotient topology. Note that $\mathcal{MW}^n$ is not Hausdorff. As an example, take a compact hyperbolic surface of genus three, in which a separating simple closed geodesic becomes shorter and shorter. If the basepoint is chosen on one side of the geodesic, the limit is a punctured torus. If the basepoint is chosen on the other side, the limit is a punctured surface of genus two. If the basepoint is chosen on the geodesic, the limit is the real line. Thus the sequence has two different limit points in $\mathcal{MW}^2$. We also see that $\mathcal{MB}^2$ is not closed in the space of all complete locally compact path spaces.

We shall also consider $\mathcal{MF}^n$, the space of all (isometry classes of) complete hyperbolic n-Manifolds with base $\mathcal{F}$rame. Such a manifold is a pair (M, e), where M is a complete hyperbolic n-manifold and $e = (e_1, \ldots, e_n)$ is an orthonormal frame of the tangent space at a point $e_0 \in M$. To topologize this space using ε-relations we introduce another definition.

I.3.2.5. Definition. Let (M_1, e_1) and (M_2, e_2) be two Riemannian manifolds with baseframe. Then a *framed* (ε, *r*)-*relation* between (M_1, e_1) and (M_2, e_2) is an (ε, *r*)-relation R between (M_1, e_1) and (M_2, e_2) together with the additional requirement that (exp v) R (exp v') if $v = \sum v_i e_i$ is a tangent vector to M at e_0, $v' = \sum v_i e_i'$ is a tangent vector to M' at e_0' and $\sum v_i^2 \le r^2$.

A neighbourhood $\mathcal{N}(M, e, r, \varepsilon)$ of (M, e) in $\mathcal{MF}^n$ consists of all pairs (M', e') such that there exists a framed (ε, *r*)-relation R between (M, e) and (M', e'). Note that $\mathcal{MF}^n$ has a countable basis of neighbourhoods of each point. Until a few lemmas are established the ε-relation topology is annoyingly difficult to handle. The problem is that the ε-relation controls only C^0 behaviour whereas we need to control derivatives as well.

I.3.2.6. Lemma: Injectivity radius continuous. *Let* $g_r(M, e_0)$ *be the infimum of the injectivity radius at x, as x varies over* $B(e_0, r)$. *Then for any fixed value of r, the function* $g_r: \mathcal{MB}^n \to (0, \infty]$ *is continuous.*

Proof. Injectivity radius continuous: Let $x_0 \in B(e_0, r)$ be a point for which the injectivity radius is minimal. Let γ be a geodesic loop in M of length $2g_r(M, e_0)$, starting and ending at x_0. Let $(x_0, x_1, x_2, x_3, x_4)$ be equally spaced points with $x_4 = x_0$ along γ. Then, since the interior of the ball in M with centre x_0 and radius $g_r(M, e_0)$ is isometric to an open round ball in $\mathbb{H}^n$, we have

$$d(x_0, x_1) = d(x_1, x_2) = d(x_2, x_3) = d(x_3, x_0) = \frac{g_r(M, e_0)}{2}.$$

Let (M', e_0') be near (M, e_0), and let x_0', x_1', x_2', x_3', be points such that $x_i R x_0'$ for an ε-relation R, where ε is small. We write $x_4' = x_0'$. We choose geodesic paths $\beta_1, \beta_2, \beta_3, \beta_4$, of minimal length, such that β_i, joins x_{i-1}' to x_i'. We lift in turn $\beta_1, \beta_2, \beta_3, \beta_4$ to geodesics $\gamma_1, \gamma_2, \gamma_3, \gamma_4$ in $\mathbb{H}^n$, such that the end of γ_i in the beginning of γ_{i+1} ($i = 1, 2, 3$). We denote the endpoints of γ_i by y_{i-1} and y_i. Now, for $i = 0, 1, 2$,

$$d(y_i, y_{i+2}) \ge d(x_i', x_{i+2}') \ge g_r(M, e_0) - \varepsilon$$

and $d(y_i, y_{i+1}) \le g_r(M, e_0)/2 + \varepsilon$. It follows from Theorem I.4.2.10 (*Curve near geodesic*) below that, by taking ε very small, we can ensure that the piece-wise geodesic $\gamma_1 \gamma_2 \gamma_3 \gamma_4$ is very close to a geodesic. In particular the endpoints are distinct. This means that $\beta_1\beta_2\beta_3\beta_4$ is an essential loop of length at most $2g_r(M, e_0) + 4\varepsilon$. Therefore the injectivity radius of (M', e'_0) at x'_0 is bounded above by $g_r(M, e_0) + 2\varepsilon$. Since $d(e'_0, x'_0) \leq r + \varepsilon$, we have

$$g_r(M', e'_0) \leq g_{r+\varepsilon}(M', e'_0) + \varepsilon \leq g_r(M, e_0) + 3\varepsilon.$$

I.3.2.7.

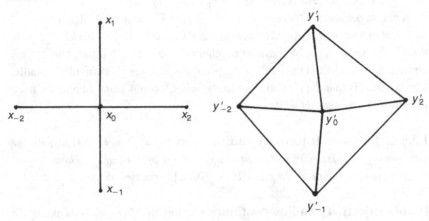

Figure I.3.2.7

We now prove that $\liminf g_r(M', e'_0) \geq g_r(M, e_0)$, where the limit is taken over based manifolds (M', e'_0) converging to (M, e_0), using the topology given by ε-relations. So suppose this is false. We take a sequence of based hyperbolic n-manifolds of the form (M', e'_0) converging to (M, e_0) and a fixed $r > 0$, such that $g_r(M', e'_0) < g_r(M, e_0) - 2\varepsilon_0$, where $\varepsilon_0 > 0$ is small and satisfies $\varepsilon_0 < g_r(M, e_0)/4$.

Let $x'_0 \in B(e'_0, r) \subset M'$ and let $\gamma: [0, t_0] \to M'$ be a non-constant geodesic loop based at x'_0, parametrized according to path length.

We know that there is a correspondence between (M, e_0) and (M', e'_0) which gives an ε-relation between large neighbourhoods of e_0 and e'_0, with ε small. Let $x_0 \in B(e_0, r)$ correspond to x'_0 and let $(u_1, \ldots, u_n)$ be an orthonormal frame at x_0. Let $\delta = g_r(M, e_0)/2$. For $1 \leq i \leq n$, we set $x_{\pm i} = \exp(\pm\delta u_i)$. In M' let the corresponding points be $x'_{\pm i}$. We know approximately their distances apart in M'. By lifting shortest paths from x'_0 to x'_i ($1 \leq |i| \leq n$), we obtain points $y'_0, y'_{\pm i}$ in $\mathbb{H}^n$. For each i, $y'_{-i}y'_0y'_i$ is a once broken geodesic with $d(y'_0, y'_i) = \delta (1 \leq |i| \leq n)$ and

$$d(y'_{-i}, y'_i) \geq d(x'_{-i}, x'_i) \geq 2\delta - \varepsilon.$$

Therefore $y'_{-i}y'_0y'_i$ is very near to a geodesic of length 2δ. Also, if $|i| \neq |j|$, then

$$d(y'_i, y'_j) \geq d(x'_i, x'_j) \geq d(x_i, x_j) - \varepsilon = \text{arccosh}(\cosh^2 \delta) - \varepsilon.$$

It follows that $\angle y'_i y'_0 y'_j$ cannot be much less than $\pi/2$. The same applies to $\angle y'_i y'_0 y'_{-j}$. Since $y'_i y'_0 y'_{-j}$ is almost straight, we can deduce that $\angle y'_i y'_0 y'_j$ is approximately $\pi/2$. Therefore we can find an orthonormal basis $v_1, \ldots, v_n$ for the tangent space to $\mathbb{H}^n$ at y_0 such that $\exp(\pm\delta v_i)$ is approximately equal to $y'_{\pm j}, i \leq j \leq n$. Now let $\tilde{\gamma}$ be a lifting of γ to $\mathbb{H}^n$, with $\tilde{\gamma}(0) = y'_0$. Then $\tilde{\gamma}(t) = \exp(t \sum \alpha_i v_i)$ for some $(\alpha_1, \ldots, \alpha_n) \in \mathbb{R}^n$ with $\sum \alpha_i^2 = 1$. Let $x \in M$ be defined by $x = \exp(t_0/2 + \varepsilon_0) \sum \alpha_i u_i$ and let $x' \in M'$ be a corresponding point. We take a shortest geodesic in M' from x'_0 to x' and lift it to a geodesic in $\mathbb{H}^n$ from y'_0 to a point which we call y'. Then

$$d(y'_0, y') = d(x'_0, x') \leq d(x_0, x) + \varepsilon = \frac{t_0}{2} + \varepsilon_0 + \varepsilon$$

and

$$d(y'_i, y') \geq d(x'_i, x') \geq d(x_i, x) - \varepsilon \quad \text{for } 1 \leq |i| \leq n.$$

These inequalities specify the position of y' reasonably precisely in relation to the points $y'_0, y'_{\pm i}$. Its position is almost the same as that of x relative to $x_0, x_{\pm i}$. This means that $\tilde{\gamma}(t_0/2 + \varepsilon_0)$ is very near to y'. In fact their distance apart converges to zero as ε converges to zero. Now

$$d(x, x_0) \leq d(x', x'_0) + \varepsilon \leq d(y', \tilde{\gamma}(t_0/2 + \varepsilon_0)) + d(\gamma(t_0/2 + \varepsilon_0), \gamma(t_0)) + \varepsilon.$$

The first and third terms on the right tend to zero with ε, and the second term is bounded by $t_0/2 - \varepsilon_0$, so we see that $d(x, x_0) \leq t_0/2 - \varepsilon_0$. But this contradicts $d(x, x_0) = t_0/2 + \varepsilon_0$. This contradiction completes the proof.

> **Injectivity radius continuous**

I.3.2.8. Lemma: Same as quotient. *The topology given by ε-relations on hyperbolic manifolds with basepoint is the quotient of the topology given by framed ε-relations on manifolds with baseframe.*

Proof. Clearly there is a continuous map $\mathcal{MF}^n \to \mathcal{MB}^n$. The injectivity radius is greater than some $\delta > 0$ throughout a neighbourhood of (M, e_0). Let $(e_1, \ldots, e_n)$ be a fixed orthonormal frame for the tangent space to M at e_0, and let $x_i = \exp(\delta e_i) \in M$. If (M', e'_0) is near (M, e_0), we have points $x'_1, \ldots, x'_n$ corresponding to $x_1, \ldots, x_n$. From $e'_0, x'_1, \ldots, x'_n$ we can form an orthonormal basis $(e'_1, \ldots, e'_n)$ in a canonical way as tangent vectors at e'_0 as follows. We define u_i by the equation $x'_i = \exp(u_i)$, where the length of u_i is almost equal to δ, and then we obtain $(e'_1, \ldots, e'_n)$ by applying the Gram–Schmidt process to $(u_1, \ldots, u_n)$. If (M', e'_0) is extremely near to (M, e_0), then it is easy to show that the manifold with baseframe (M', e') is near to (M, e). This shows that the map $\mathcal{MF}^n \to \mathcal{MB}^n$ is open and the lemma is proved.

$\square$

There is an alternative way of defining a topology on $\mathcal{MF}^n$, which we have already encountered. We fix a standard orthonormal frame $(s_1, \ldots, s_n)$ for the tangent space at a fixed point $s_0 \in \mathbb{H}^n$. There is exactly one covering map $\mathbb{H}^n, s \to M, e$ which is a local isometry. The fundamental group of M acts on $\mathbb{H}^n$ as a group of covering translations. Thus we can write $\pi_1(M, e)$, and we get a well-defined subgroup of $\mathrm{Isom}(\mathbb{H}^n)$. Recall that subgroups of $\mathrm{Isom}(\mathbb{H}^n)$ are topologized with the Chabauty topology.

I.3.2.9. Theorem: Map to torsion-free subgroups is a homeomorphism. *The map $\pi_1, : \mathcal{MF}^n \to \mathcal{DJ}^n$, which sends (M, e) to the discrete torsion free subgroup $\pi_1(M, e)$ of $\mathrm{Isom}(\mathbb{H}^n)$, is a homeomorphism.*

Proof. Map to torsion-free subgroups is a homeomorphism: To prove the theorem, first note that the map $\pi_1 : \mathcal{MF}^n \to \mathcal{DJ}^n$ is bijective: given a discrete torsion-free subgroup Γ of $\mathrm{Isom}(\mathbb{H}^n)$, we obtain a complete hyperbolic manifold by taking $M = \mathbb{H}^n/\Gamma$ and e to be the image of s.

We now prove that $\pi_1 : \mathcal{MF}^n \to \mathcal{DJ}^n$ is continuous. We want to show that if (M', e') is near (M, e), then $\pi_1(M', e')$ is near $\pi_1(M, e)$ in the Chabauty topology on the space of subgroups of $\mathrm{Isom}(\mathbb{H}^n)$. By Lemma I.3.2.6 (*Injectivity radius continuous*), we may assume that the injectivity radius is greater than some $\delta > 0$ throughout the regions of interest in the proof we are about to present.

Suppose $\gamma \in \pi_1(M, e)$. Let $B_1, B_2, \ldots, B_k$ be a circular chain of balls in M, where $B_i = B(x_i, \delta)$, $x_1 = e_0, x_i \in \gamma$, and the balls cover γ. We assume that the interior of B_i meets the interior of B_{i+1}. Let $x_i R x'_i$. We obtain a chain $B'(x'_i, \delta)$. The holonomy corresponding to this chain is nearly equal to γ. We see this by taking $(n + 1)$ generic points in $\mathrm{int}\, B_i \cap \mathrm{int}\, B_{i+1}$ and keeping track of the corresponding points in $\mathrm{int}B'_i \cap \mathrm{int}\, B'_{i+1}$. So if $\gamma \in \pi_1(M, e)$, then γ is the limit of elements $\gamma' \in \pi_1(M', e')$.

We also need to show that if $\gamma(i) \in \pi_1(M(i), e(i))$, and $\gamma(i) \to \gamma$, then $\gamma \in \pi_1(M, e)$. Let s_0 be the standard basepoint in $\mathbb{H}^n$. Let $d(s_0, \gamma s_0) = r$. Then the geodesic from s_0 to $\gamma(i)s_0$ has length approximately equal to r, for large i.

This gives an essential loop in $M(i)$, based at $e(i)_0$, of length approximately r. It follows that there is a geodesic loop in M, based at e_0, of length approximately equal to r, and the holonomy of this loop is very near to that of γ. Therefore $\gamma \in \pi_1(M, e)$.

This shows that if (M', e') is near (M, e) in the ε-relation topology, then, by Lemma I.3.1.3 (*Geometric convergence*), $\pi_1(M', e')$ is near $\pi_1(M, e)$ as subgroup of $\mathrm{Isom}(\mathbb{H}^n)$ in the Chabauty topology.

Conversely, suppose Γ_i converges to Γ in the Chabauty topology. We must prove that $(\mathbb{H}^n/\Gamma_i, e(i))$ converges to $(\mathbb{H}^n/\Gamma, e)$ where $e(i)$ and e are the images of the standard point and frame s. Let $A_r = \{g \in \mathrm{Isom}(\mathbb{H}^n): B(s_0, r) \cap gB(s_0, r) \neq \varnothing\}$. Then A_r is compact. We fix r and let id, $\gamma_1, \ldots, \gamma_k$ be the elements of Γ in A_r. Then, for large values of i, there are elements id, $\gamma(i)_1, \ldots, \gamma(i)_k$ in Γ_i with $\gamma(i)_j$ near to γ_j. Moreover, for fixed $\varepsilon > 0$ (where ε is significantly smaller than δ), each element in $\Gamma_i \cap A_{r-\varepsilon}$ appears in this list. We can now deduce that a large compact region of $\mathbb{H}^n/\Gamma$ is almost isometric to a large compact region of $\mathbb{H}^n/\Gamma_i$. In fact the isometry is induced by a map which is C^∞-near the identity for large values of i. The proof is the same as that presented in Section I.1.7.2 (*Holonomy induces structure*).

> **Map to torsion-free subgroups is a homeomorphism**

I.3.2.10. Definition. Let (M_1, e_1) and (M_2, e_2) be two Riemannian manifolds with baseframe. Then a *framed (K, r)-approximate isometry* between (M_1, e_1) and (M_2, e_2) is a diffeomorphism $f: (X_1, e_1) \to (X_2, e_2)$ such that $B_{M_1}(x_1, r) \subseteq (X_1, x_1) \subseteq (M_1, x_1), B_{M_2}(x_2, r) \subseteq (X_2, x_2) \subseteq (M_2, x_2), Df(e_1) = e_2$ and

$$\frac{d(x, y)}{K} \leq d(f(x), f(y)) \leq K d(x, y) \quad \text{for all } x, y \in X_1.$$

We may similarly define (K, r)-approximate isometries, and K-approximate isometries.

There are many possible definitions of an approximate isometry. We have chosen a relatively strong one (as we require differentiability) and (ε, r)-relations may be thought of as the weakest possible notion of an approximate isometry. However, the proof of the above result tells us that these two definitions (and thus many other definitions "between" the two) are equivalent for complete hyperbolic manifolds.

I.3.2.11. Corollary: Approximate isometries. *The identity map on $\mathcal{MF}^n$ is a continuous map from the topology induced by (K, r)-approximate* *isometries to the topology induced by framed (ε, r)-relations (and thus to the Chabuty topology).*

We now return (hopefully with new insight) to two topics which we discussed at the end of the last section. We can extend Corollary I.3.1.8 (*Compact with bounded volume*) by verifying that the volume map is continuous for hyperbolic *n*-manifolds of finite volume ($n \geq 3$). Recall that Theorem I.3.1.4 (*A(U) compact*) only guarantees that this map is lower semicontinuous; recall also that the statement for 2-manifolds, corresponding to the next theorem, is actually false (e.g. a sequence of compact surfaces of genus two may converge to a punctured torus).

I.3.2.12. Theorem (Jorgensen): Volume is continuous. *The map* vol *from the space of hyperbolic n-manifolds ($n \geq 3$) of finite volume with baseframe to $\mathbb{R}$ which takes each manifold to its volume is continuous.*

Proof. Suppose (M_i, e_i) converges to (M, e). Notice that for hyperbolic *n*-manifolds ($n \geq 3$) the thick part is connected, since the boundary of each component of the thin part is connected. Also, we may assume that inj $_{Mi}(e_i) > \bar{\varepsilon}$ for some $\bar{\varepsilon} > 0$ and that vol$(M_i) < V$ for some V and all i. There exists a bound on the diameter of $M_{[\varepsilon,\infty)}$ using only the topology induced by K-approximate isometries (since they are all compact and of uniformly bounded diameter). Thus, $\{\text{vol}(M_{i,[\varepsilon,\infty)})\}$ converges to vol$(M_{[\varepsilon,\infty)})$ for all $\varepsilon > 0$. But since the volume of the thin part converges to 0 as ε converges to 0, we see that $\{\text{vol}(M_i)\}$ converges to vol(M).

$\Box$

When we apply the above analysis to the 2-dimensional case we obtain the following proposition:

I.3.2.13. Proposition: Mumford's Lemma. *The subset $\mathcal{M}(T)_\delta$ of $\mathcal{MW}^2$ consisting of all finite area surfaces homeomorphic to a given surface T with no closed geodesics shorter than some $\delta > 0$ is compact, for any $\delta > 0$.*

Proof. Simply notice that if $S \in \mathcal{M}(T)_\delta$, $S_{[\delta',\infty)}$ is connected for all $\delta' < \delta$. Then the argument above proves that $\mathcal{M}(T)_\delta$ is a closed subset of $\mathcal{MW}^2(A)$ where $A = \text{area}(S)$. But since $\mathcal{MW}^2(A)$ is compact by Corollary I.3.1.8 (*Compact with bounded volume*) so is $\mathcal{M}(T)_\delta$.

I.3.2.14. Remarks. Thurston has further proved that the set of volumes of complete hyperbolic 3-manifolds form a closed, non-discrete set in $\mathbb{R}^+_\infty$. This

set is well-ordered and has ordinal type ω^ω. For a detailed discussion of this result see 6.6 of Thurston (1979) or Gromov (1980). Thurston (see Section 6.6 of Thurston (1979)), has also shown that there are finitely many complete hyperbolic manifolds of any fixed volume, but in Wielenberg (1981) it is shown that there is no bound on the number of complete hyperbolic manifolds of a given volume, by proving that there exist fundamental polyhedra with arbitrarily many associated non-conjugate Kleinian groups. In Wang (1972) it is shown that there are only finitely many hyperbolic manifolds with volume less than any given real number in dimension four and above.

I.3.2.15. We finish this section by returning briefly to the subject of marked hyperbolic surfaces. We may also topologize the Teichmüller space of a surface of finite type using approximate isometries. We may define the K-neighbourhood of a marked surface $[h: T \to S]$ to be the set of all marked surfaces $[h': T \to S']$ such that there exist representatives $\bar{h}$ and $\bar{h}'$ for $[h]$ and $[h']$ and a diffeomorphism $\phi: S \to S'$ which is a K-approximate isometry when restricted to the respective convex cores and such that $\phi \circ \bar{h}$ is isotopic to $\bar{h}'$. Such neighbourhoods form a basis for the topology of Teichmüller space; it is left to the reader to satisfy himself that this topology agrees with the topology defined in the last section. Equivalently, one may also topologize Teichmüller space using ε-relations, but approximate isometries are more frequently used.

Chapter I.4
Laminations

I.4.1. Geodesic laminations

For a more detailed treatment see Casson (1983), or Harer–Penner (1986) or Chapter 8 of Thurston (1979).

I.4.1.1. Definition. Let S be a connected complete hyperbolic surface. Then a *geodesic lamination* on S is a closed subset λ of S which is a disjoint union of simple geodesics of S (which are called *leaves* of the lamination).

Remark: We allow the empty set as a geodesic lamination.

We denote the set of all geodesic laminations on S by $\mathcal{GL}(S)$.

I.4.1.2. Definition. If λ is a lamination on S, then a component of $S - \lambda$ is called a *flat piece* or a *complementary region*.

In general these need not be simply connected and may have a finite or infinite number of sides. The leaves of the geodesic lamination which form the boundary of some complementary region are called *boundary leaves*. For a surface of finite area there is an upper bound on the number of complementary regions, since each has area $n\pi$ for some positive integer n. Moreover each complementary region has finite type. On a surface with finite area, the set of boundary leaves is dense in the geodesic lamination.

I.4.1.3. Definition. A lamination such that each complementary region is isometric to an ideal triangle is said to be *maximal*.

We shall see, from Theorem I.4.2.8 (*Structure of lamination*) that any lamination on a surface of finite area can be extended by adding a finite number of new leaves to obtain a maximal lamination.

76

On a surface of finite area, any geodesic lamination has measure zero in the surface. Equivalently any C^1-curve C transverse to the geodesic lamination intersects it in a set of measure zero in C. Since S is a complete hyperbolic surface, its universal cover is $\mathbb{H}^2$. We may lift a lamination λ to $\mathbb{H}^2$ to obtain a lamination $\tilde{\lambda}$ on $\mathbb{H}^2$, which is invariant under the action of the covering transformations (which are elements of some Fuchsian group Γ).

Consider the closed unit disk $\mathbb{B}^2$ as the compactification of $\mathbb{H}^2$. The space of geodesics is homeomorphic to an open Möbius band M (see Part II).

I.4.1.4. Transferring laminations.
Given a geodesic lamination λ_1 on a complete hyperbolic surface S_1 of finite area, and a homeomorphism ϕ onto a complete hyperbolic surface S_2 of finite area, the lamination λ_1 can be canonically transferred to a geodesic lamination λ_2 of S_2. The reason is that the homeomorphism can be lifted to an equivariant map from $\mathbb{H}^2$ to $\mathbb{H}^2$ which extends to a homeomorphism between boundary circles. Since a geodesic is an unordered pair of elements of $\partial\mathbb{H}^2$, we see how to transfer a geodesic in S_1 to a geodesic in S_2. This transfer induces a canonical homeomorphism $\phi_\#$: $GL(S_1) \to GL(S_2)$, which only depends on the isotopy class of ϕ.

I.4.1.5. Definition.
The Chabauty topology is the topology induced on $\mathcal{GL}(S)$ as a subspace of $C(M)$, the set of closed subsets of the open Möbius band M with the Chabauty topology (see Section I.3.1 (*The geometric topology*) for the definition of the Chabauty topology).

If S has finite area, a lamination on S is determined by its underlying point set (see Proposition I.4.1.6 (*Lamination determined by pointset*)). We can therefore also topologize $\mathcal{GL}(S)$ using the Chabauty topology on subsets of $\mathbb{H}^2$. We shall also see that these two topologies are homeomorphic.

If ε is small, $S_{(0,\varepsilon)}$ consists entirely of cusps. All simple geodesics entering a cusp are asymptotic to each other; they are orthogonal to the horocycles. Therefore a lamination on S is completely determined by its intersection with $S_{[\varepsilon,\infty)}$.

I.4.1.6. Proposition: Lamination determined by pointset.
Let S be a complete hyperbolic surface of finite area. Let $C(S)$ be the space of closed subsets of S with the Chabauty topology and let $L \subset C(S)$ be defined by

$$L = \{X : X = |\lambda| \text{ for some lamination } \lambda \subset S\}.$$

Then L is a closed subset of $C(S)$ and the map $\mathcal{GL}(S) \to L$ defined by $\lambda \longmapsto |\lambda|$ is a homeomorphism. In particular, the topologies induced by regarding $GL(S)$ as a subset of $C(M)$ and $C(S)$ agree.

Proof. We need to only show that $\mathcal{GL}(S) \to L$ is continuous and injective, since a continuous injective map onto a Hausdorff space is a homeomorphism. Geodesic laminations on a surface of finite area are nowhere dense, so no two laminations can share the same underlying pointset. (We note that this is not true for surfaces of infinite area.) Thus, our map is injective.

So suppose that $\{\lambda_i\}$ converges to λ in $\mathcal{GL}(S)$. Then, given a point $x \in |\lambda|$, x lies on some geodesic l, and l is the limit of geodesics $l_i \in \lambda_i$ as we see by looking in the universal cover of S. We choose $x_i \in l_i$, such that x_i converges to x. This is the first condition for the convergence of $|\lambda_i|$ to $|\lambda|$. To prove the second condition, we suppose that $x_i \in |\lambda_i|$ and that x_i converges to some point x. We must show that $x \in |\lambda|$. We have $x_i \in l_i \in \lambda_i$. By lifting to the universal cover, we see that we may assume l_i converges to $l \in \lambda$, and that $x \in l$. Hence $x \in |\lambda|$. This completes the proof of the proposition.

$\Box$

I.4.1.7. Proposition: GL(S) compact. $\mathcal{GL}(S)$ *is compact, metrizable, and separable in the Chabauty topology.*

Note that this is true even if S is not compact.

Proof. By Proposition I.3.1.2 (*Properties of Chabauty topology*) it is sufficient to show that $\mathcal{GL}(S)$ is a closed subset of $C(M)$, where M is the open Möbius band.

Suppose we have a sequence of geodesic laminations $\{\lambda_i\} \subset \mathcal{GL}(S)$ with limit $\lambda \in C(M)$. By Lemma I.3.1.3 (*Geometric convergence*) this means that every geodesic of λ is the limit of a sequence of geodesics, one in each λ_i, and that every convergent sequence of geodesics, one in each λ_i, converges to a geodesic of λ.

Let $S = \mathbb{H}^2/\Gamma$. We need to check two things: that λ is a *disjoint* union of geodesics, and that λ is invariant under Γ. Suppose l and k are two leaves in λ and that l_i converges to l and k_i converges to k where $l_i, k_i \in \lambda_i$. Since l_i and k_i are disjoint or equal for each i, the same is true of l and k. To check that λ is Γ-invariant, recall that if a group acts on a space, then, for any subset X which is pointwise fixed under Γ, its closure $\overline{X}$ is also fixed under Γ. Here the space is $C(M)$ and the subset X is the countable set $\{\tilde{\lambda}_i\} \subset C(M)$. So λ is a lamination and thus $\mathcal{GL}(S)$ is closed.

$\Box$

The following lemma results immediately by applying Lemma I.3.1.3 (*Geometric convergence*).

I.4.1.8. Lemma: Geometric convergence for laminations. *If* (l_i) *and* (k_i) *are two sequences of geodesic laminations converging to* l *and* k, *respectively, with* $l_i \subset k_i$ *for all* i, *then* $l \subset k$.

We set $\mathcal{GLM}(S)$ to be the set of maximal laminations on S. Then the following lemma results.

I.4.1.9. Lemma: GLM closed in GL. *Let* S *have finite area. Then* $\mathcal{GLM}(S)$ *is a closed subset of* $\mathcal{GL}(S)$.

Proof. We shall work in $C(S)$, the set of closed subsets of S.

Suppose λ is not maximal, and $\lambda_i \to \lambda$, with each λ_i maximal. Let P be a complementary region for λ which is not an ideal triangle. First we show that P can have no simple closed geodesics in its interior. For if K is the underlying pointset of such a geodesic, then $\lambda \in O_1(K)$ and so $\lambda_i \in O_1(K)$ for i large. But this means that there is a simple closed geodesic in a complementary region of λ_i which is forbidden. Hence P is a finite sided polygon.

It is easy to construct a connected compact subset K in the interior of P such that each infinite geodesic in P meets the interior of K and such that $\lambda \in O_1(K)$. If i is large, then λ_i contains geodesics near the boundary geodesics of P, and $\lambda_i \in O_1(K)$. Moreover none of the geodesics of λ_i meet K. Let P_i be the complementary region of λ_i in which K lies. We can see that among the boundary geodesics of P_i, there are geodesics very near to each of the boundary geodesics of P. Therefore some leaf of λ_i must pass through K since λ_i is maximal. This contradiction proves the result. □

We now discuss another topology on $\mathcal{GL}(S)$, which we call the *Thurston topology*. The reference for this is Section 8.10 of Thurston (1979), where it is referred to as the geometric topology.

I.4.1.10. Definition. The *Thurston topology* on $\mathcal{GL}(S)$ is the topology induced by once again treating $\mathcal{GL}(S)$ as a subset of $C(S)$, but with the topology generated by sub-basis elements of the form $O_2(V) = \{A \in C(S) | A \cap V \neq \emptyset\}$, where V is open in S.

So we see that a neighbourhood of a geodesic lamination λ contains all geodesic laminations λ' such that $\lambda \supset \lambda'$.

Note that this topology is strictly weaker than the Chabauty topology and is non-Hausdorff, but it is more closely related to the consideration of pleated surfaces. For example, a sequence of surfaces, bent along a single geodesic converges to a geodesic surface if the bending angle converges to zero. The sequence of laminations corresponding to the surface converge to the lamination corresponding to the limit surface in the Thurston topology, but not in the Chabauty topology.

The analogue of Lemma I.3.1.3 (*Geometric convergence*) is as follows.

I.4.1.11. Lemma: Geometric convergence in Thurston topology. *If (λ_i) converges to λ in the geometric topology then, given any geodesic $l \in \lambda$, there is a sequence of geodesics (l_i), where $l_i \in \lambda_i$, which converges to l.*

I.4.2. Minimal laminations

See Section 8.10 of Thurston (1979).

I.4.2.1. Definition. A non-empty geodesic lamination λ is said to be *minimal* if no proper subset of λ is a geodesic lamination.

A single geodesic is a minimal lamination if and only if it is closed subspace of the surface (either a simple closed geodesic, or an infinite geodesic, with each end converging to a cusp).

I.4.2.2. Lemma: One or uncountable. *Let λ be a minimal lamination on S. Then either:*

(1) λ *consists of a single geodesic, or*
(2) λ *is uncountable.*

Proof. Let l be a leaf of λ. If l is isolated, then $\lambda \backslash l$ is again a lamination. Since λ is minimal, $\lambda \backslash l = \emptyset$, so that condition (1) holds. So we assume that l is not isolated. Take some point x on l and a small transverse line L through x. Since l is not isolated, x must be an accumulation point of $L \cap \lambda$ in L. But since $L \cap \lambda$ is closed and each point is an accumulation point, it is a perfect set and is thus uncountable. Since each leaf of λ intersects L at most a countable number of times, λ must have uncountably many leaves. $\quad\square$

We now discuss hyperbolic surfaces of finite area in order to discover precisely the structure of any lamination on such a surface.

Let S be a complete hyperbolic surface without boundary, and let λ be a lamination on S. We cut S along λ. The formal definition of this process is to take $S \backslash \lambda$ with its Riemannian metric, and complete it. We obtain a complete (possibly disconnected) hyperbolic surface with geodesic boundary.

I.4.2.3. Lemma: Building a surface. *Let $\Delta_1, \ldots, \Delta_k$ be a finite set of ideal triangles. Let $(A_1, B_1), \ldots, (A_r, B_r)$ be r pairs made up from $2r$ distinct edges, chosen from the $3k$ edges of $\Delta_1, \ldots, \Delta_k$. Let $h_i: A_i \to B_i$ $(1 \leq i \leq r)$ be an isometry. Identifying using the h_i, we obtain a hyperbolic surface, possibly with boundary. The completion of this surface consists of adding a finite, possibly empty, collection of simple closed boundary geodesics.*

I.4.2.4.

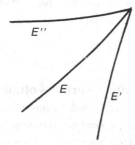

Figure I.4.2.4

Proof. Each end E of an identified edge gives rise to two other ends E' and E'' of identified edges, namely those that occur on either side of E, see Figure I.4.2.4. Possibly $E' = E''$, and possibly $E = E' = E''$. (The last case occurs when, say, two edges of Δ_1 are identified with each other with a certain choice of orientation.) Here E and E' have a common ideal vertex in one of the ideal triangles $\Delta_1, \ldots, \Delta_k$, and similarly for E and E''. Thus, the ends of identified edges can be arranged in cycles $(E_1, \ldots, E_s)$, where E_i and E_{i+1} have a common ideal vertex (interpreting E_{s+1} as E_1).

We can explicitly work out the local geometry of the piece of the surface arising from such a cycle. In the upper half-plane model, it is obtained from the strip

$$\{(x, y) \mid 0 \leq x \leq 1, \quad y \geq k\}$$

modulo the gluing map $z \longmapsto z + 1$, or from the strip

$$\{(x, y) \mid 1 \leq x \leq a, \quad y \geq kx\}$$

modulo the gluing map $z \longmapsto az$. The first case gives us a cusp. The second case is isometric to the sector $\{(x, y) \mid y \geq kx, \ x > 0\}$ modulo $z \longmapsto az$. The completion gives us the geodesic $\{x = 0, \ 1 \leq y \leq a\}/\{z \longmapsto az\}$, and our piece of surface is a nice neighbourhood of this geodesic.

$\square$

I.4.2.5. Corollary: Bound on boundary components.   *Using the same notation as in the preceding lemma, the number of boundary components is at most $3k - r$.*

Proof.   The $2r$ edges $\{A_1, B_1, \ldots, A_r, B_r\}$ give rise to r edges in the surface. So there are at most r cycles of edges of the type described in Lemma I.4.2.3 (*Building a surface*). Hence there are at most r new boundary components in the completed surface. It follows that the glued up and completed surface has at most $3k - 2r + r = 3k - r$ boundary components.

$\square$

I.4.2.6. Lemma: Non-compact surfaces obtainable.   *Every finite area complete hyperbolic surface S with geodesic boundary can be constructed (non-uniquely) by the above process, except for a compact surface without boundary. In the case of a compact surface without boundary we start by cutting the surface along a simple closed geodesic, and then cut what remains into triangles. If S is non-compact and ∂S has only non-compact components, then the completion step, after gluing the triangles together, can be avoided by cutting up into triangles correctly.*

Proof.   By doubling S we obtain a complete finite area surface without boundary. It follows that S has only a finite number of boundary components. Also it is easy to see what S must look like topologically.

If we have a cusp or an ideal vertex x in S, we may cut a finite number of times along geodesics from x to x. We eventually get a disjoint union of surfaces, each of which is an ideal polygon or an annulus. An ideal polygon may be cut into triangles. In the annulus case, we may assume that one boundary is a geodesic from x to x and the other a geodesic circle or a cycle of non-compact oriented geodesics $B_1, \ldots, B_k$, with the positive end of one geodesic asymptotic to the negative end of the next. In the annulus case we can make the surface simply connected by cutting along a geodesic from x to an endpoint of one of the B_i, or by cutting along a geodesic from x which spirals around the simple closed geodesic at the other end of the annulus. This deals with the non-compact case. If S is compact with boundary, we reduce to the previous case by cutting along a geodesic which spirals to the boundary at each end. If S has no boundary, we cut first along a simple closed geodesic.

$\square$

I.4.2.7. Corollary: Another bound on boundary components. *Let S be a complete hyperbolic surface of finite area A with geodesic boundary. Let b be the number of boundary components. Then $\pi b \leq 3A$.*

Proof. Let S be obtained by gluing together k ideal triangles and completing. Then

$$\pi b = \pi(3k - r) = 3A - r\pi$$

using Corollary I.4.2.5 (*Bound on boundary components*).

$\square$

I.4.2.8. Theorem: Structure of lamination. *Let λ be a lamination on a complete hyperbolic surface of finite area with geodesic boundary. Then λ consists of the disjoint union of a finite set of minimal sublaminations of λ together with a finite set of additional geodesics, each end of which either "spirals" onto a minimal lamination or goes up a cusp. Each of the additional geodesics is isolated – it is contained in an open subset which is disjoint from the rest of the lamination. Each cusp contains only a finite number of geodesics of λ.*

Proof. Structure of lamination. First note that minimal sublaminations exist. On a compact surface, this is a consequence of compactness in the usual way. On a non-compact surface, we take the intersection with the thick part of the surface and argue there. Recall that if we cut along any lamination we obtain a finite number of components, since each component has area $n\pi$ for some integer n.

If we cut S along a sublamination λ_1 of λ we obtain a new surface S' with a new lamination λ' obtained from λ in the obvious way. Let λ_1 be a minimal sublamination of λ. We claim that the minimal sublaminations of λ correspond one-to-one with the minimal sublaminations of λ', except that λ_1 itself disappears and is replaced by one or more new boundary leaves of S', each of which is minimal in λ'. (These boundary leaves form the set of points added in the process of completion.) Each boundary leaf is of course minimal. If $\lambda_2 \neq \lambda_1$ and λ_2 is minimal in λ, it is clearly minimal in λ'. A minimal sublamination λ_3 of λ' is either a boundary component of S' or is disjoint from $\partial S'$. In the second case λ_3 is a minimal sublamination of λ. In the first case, λ_3 is either a component of ∂S, in which case it is minimal in λ, or a new component of $\partial S'$. So the claim is established.

To prove the theorem, we cut successively along minimal sublaminations which are not boundary components. By our claim, each such minimal lamination is minimal in λ. Each cut increases the number of boundary components. Since this number is bounded by Corollary I.4.2.5 (*Bound on boundary components*), the process ends. This shows there are only a finite number of minimal sublaminations.

To complete the proof of the theorem, we need a lemma.

I.4.2.9. Lemma: L finite. *If all minimal sublaminations are contained in the boundary, λ is finite.*

Proof. L finite. We define a *corner* of S to be the cusp-like region lying between two asymptotic boundary components of S. A *corner* of a complementary region is defined similarly, by completing the complementary region. If we double S, then the double of a corner is a cusp. First note that each corner of S and each cusp of S can contain only a finite number of geodesics of λ, since the area bounds both the number of complementary regions and the number of corners in each complementary region. Let K be the compact subspace of S obtained by cutting off the corners and cusps. (K is the intersection of S with the thick part of the double of S.) Every simple geodesic in S meets K.

Suppose that λ is infinite. Then there is a sequence of disjoint geodesics in λ converging to a geodesic $l \in \lambda$. By what has already been shown, neither end of l can enter a cusp or corner. Hence l lies in K. So the closure of l is compact.

We claim that $\bar{l} \cap \partial S = \emptyset$. To see this note that if a component B of ∂S is contained in $\bar{l}$, then B is a circle since $\emptyset$ is compact. Each leaf of λ which is near B must spiral onto B. Since there is a bounded number of corners in each complementary region of λ, and each spiral gives rise to a corner, there are only a finite number of leaves spiralling onto B. Since l is a limit of an infinite sequence of distinct geodesics of λ, l cannot spiral onto B. Therefore $\bar{l}$ contains a minimal sublamination disjoint from ∂S. This contradicts our hypothesis.

> **L finite**

This completes the proof of the theorem.

> **Structure of lamination**

I.4.2.10. Theorem: Curve near geodesic. *Let $\varepsilon > 0$ be fixed. Then there is a $\delta > 0$ with the following property. Let α be a piecewise geodesic curve parametrized by arc length, $0 \leq t \leq |\alpha|$, in $\mathbb{H}^n$, whose pieces have length at least ε and such that pieces with a common endpoint meet at an angle greater than $\pi - \delta$. Let $\beta(t)$ be the geodesic parametrized by arc length, $0 \leq t \leq |\beta|$, joining the endpoints of α. Set $\beta^*(t) = \beta(t|\beta|/|\alpha|)$, $0 \leq t \leq |\alpha|$. Then $d(\alpha(t), \beta^*(t)) < \varepsilon$ for all t.*

I.4.2.11.

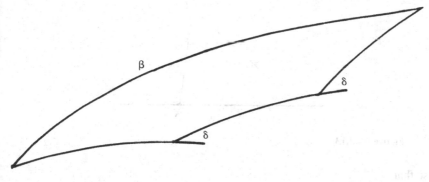

Figure I.4.2.11

Proof. Curve near geodesic. First we need a lemma which determines the situation in a triangle.

I.4.2.12. Lemma: Angle derivative. *Let T be a fixed point and let Y move along a fixed geodesic containing a fixed point X. Set $t = d(X, Y)$ and $x = d(Y, T)$. Set $\theta = \angle XYT$. We regard x and θ as functions of t.*
Then

$$\theta' = \frac{-\sin \theta}{\tanh x}.$$

Proof. Angle derivative. From the hyperbolic law of sines we see that $\sinh x \sin \theta$ is constant. Also $x' = \cos \theta$. The result follows.

| Angle derivative |

Continuation, proof of Curve near geodesic. Let $\theta(t)$ be the angle between the geodesic from $\alpha(0)$ to $\alpha(t)$ and the geodesic subarc on which $\alpha(t)$ lies. At the bends in α, θ is discontinuous, jumping by less than δ. Over the first segment of α, $\theta = \theta' = 0$. Let $0 = t_0 < t_1 < \cdots < t_k$ be the points at which α is not geodesic. Then $t_{i+1} - t_i > \varepsilon$ for each i. We claim that $\theta(t) \le \varepsilon$ for all t. If not, let (t_i, t_{i+1}) be the first interval on which we do not have $\theta(t) \le \varepsilon$. On (t_0, t_1), $\theta = 0$ so that $i \ge 1$. Since θ is monotonic decreasing on each subinterval, the limit of $\theta(t)$ as t decreases to t_i is greater than ε, and $\theta(t) > \varepsilon - \delta$ on (t_{i-1}, t_i). From Lemma I.4.2.12 (*Angle derivative*) we see that, on (t_{i-1}, t_i),

$$\theta' < -\sin(\varepsilon - \delta)$$

I.4.2.13.

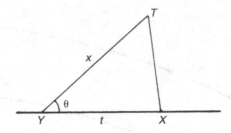

Figure I.4.2.13

so that

$$\theta(t_{i-1}) > (\varepsilon - \delta) + \varepsilon \sin(\varepsilon - \delta).$$

If we choose δ very small, then the right-hand side is bigger than ε, which is a contradiction, proving the claim.

It now follows that, by taking δ small, we can make $\theta(t)$ uniformly small. This ensures that dx/dt is near 1, where $x = d(\alpha(0), \alpha(t))$, so that $d(\alpha(0), \alpha(t))$ is uniformly near t.

> **Curve near geodesic**

The following is a result which is frequently used.

I.4.2.14. Theorem: Finite laminations dense.   *On any fixed hyperbolic surface of finite area, the finite laminations are dense in the space of all geodesic laminations, with the Chabauty topology. Hence the same is true with the Thurston topology.*

Proof. Finite laminations dense: We recall from Theorem I.4.2.8 (*Structure of lamination*), that any lamination is the union of a finite set of disjoint minimal laminations $\mu_1, \ldots, \mu_k$, and a finite set of isolated leaves. We first approximate the minimal laminations.

I.4.2.15. Lemma: Approximating minimal laminations.   *Let μ be a minimal lamination on a complete hyperbolic surface. Then μ can be approximated in the Chabauty topology by a simple closed geodesic.*

Proof. Approximating minimal laminations.   We fix a point $p \in |\mu|$ and consider a geodesic arc in μ starting at p. Given $\varepsilon > 0$, we can get within ε of each point of $|\mu|$, by taking the length of the arc long enough. We then expand this sideways to a strip P, of parallel geodesic arcs in μ of the same length. We assume that this strip has width less than ε at any point along its length.

Let γ be the complete geodesic of μ through p. We orient γ and this orients each arc of γ which lies in the strip P. There are two possible situations. The first is that there are two arcs α_1 and α_2 of the strip, lying in γ, oriented in the same direction along the strip, which are not separated in the strip by an arc α_3 of γ, lying between α_1 and α_2 as we travel along γ.

I.4.2.16.

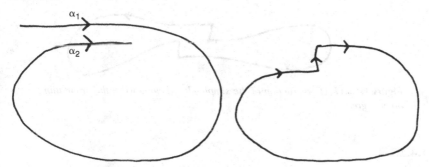

Figure I.4.2.16 *Construction of the simple closed curve when the orientations agree.*

In that case we get a simple closed curve, as shown in Figure I.4.2.16. The second possibility is that the first possibility does not occur. In that case the first three arcs of γ in which γ meets the strip must be arranged as in Figure I.4.2.17. In each case the simple closed curve is essential and the homotopic geodesic approximates μ. To see that the homotopic geodesic approximates μ, we change the transverse arc of the construction to an arc which is almost parallel to the arcs of the strip (see Figure I.4.2.18). We then apply Theorem I.4.2.10 (*Curve near geodesic*).

> **Approximating minimal laminations**

Continuation, proof of finite laminations dense. To complete the proof that any lamination can be approximated by a finite lamination, we approximate each minimal sublamination μ_j by a simple closed geodesic C_j ($1 \leq j \leq k$). Then $S \backslash C_1 \cup \cdots \cup C_k$ is nearly the same as $S \backslash \mu_1 \cup \cdots \cup \mu_k$. The difference is that complementary regions of $S \backslash \mu_1 \cup \cdots \cup \mu_k$ become joined up in $S \backslash C_1 \cup \cdots \cup C_k$ through thin gaps between geodesics which were closed off in $S \backslash \mu_1 \cup \cdots \cup \mu_k$. It is easy to make sure that the geodesics C_i do not intersect each other since each minimal lamination intersects the thick part in a compact set, and these compact sets are then separated by fixed distances.

I.4.2.17.

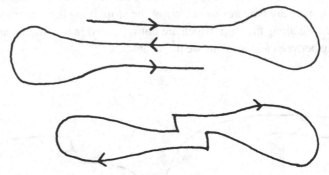

Figure I.4.2.17 *Construction of the simple closed curve when the orientations do not agree.*

I.4.2.18.

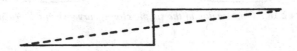

Figure I.4.2.18

Geodesics in $S \backslash C_1 \cup \cdots \cup C_k$ corresponding to the isolated geodesics of $S \backslash \mu_1 \cup \cdots \cup \mu_k$ can now be drawn in. In the complement of a finite lamination they will extend through the new gaps and spiral around one of the geodesics C_j.

> **Finite laminations dense**

We have the following corollary.

I.4.2.19. Corollary: Finite laminations dense in GLM. *The set of finite maximal laminations in $\mathcal{GLM}(S)$ is dense.*

Proof. Suppose that λ is maximal, then by the theorem above there is a sequence (λ_i) converging to λ, where each of the λ_i finite. Each λ_i can be extended to a finite maximal lamination $\hat{\lambda}_i$. Then $(\hat{\lambda}_i)$ converges to a lamination $\hat{\lambda}$ which contains λ. But since λ is maximal, $\hat{\lambda} = \lambda$ and so we have a sequence $(\hat{\lambda}_i)$ converging to λ as required.

$\square$

Chapter I.5
Pleated Surfaces

I.5.1. Introduction

We now discuss pleated surfaces, which are a basic tool in Thurston's analysis of hyperbolic structures on 3-manifolds. See Section 8.8 of Thurston (1979); there, pleated surfaces are called uncrumpled surfaces.

Recall from definition under Section I.1.3.3 (*Isometric map*) that an isometric map takes rectifiable paths to rectifiable paths of the same length.

I.5.1.1. Definition. A map $f: M \to N$ from a manifold M to a second manifold N is said to be *homotopically incompressible* if the induced map $f_*: \pi_1(S) \to \pi_1(M)$ is injective.

I.5.1.2. Definition. *A pleated surface* in a hyperbolic 3-manifold M is a complete hyperbolic surface S together with an isometric map $f: S \to M$ such that every point $s \in S$ is in the interior of some geodesic arc which is mapped by f to a geodesic arc in M. We shall also require that f be homotopically incompressible.

Note that this definition implies that a pleated surface f maps cusps to cusps since horocyclic loops on S are arbitrarily short and f is isometric and homotopically incompressible.

I.5.1.3. Definition. If (S, f) is a pleated surface, then we define its *pleating locus* to be those points of S contained in the interior of one and only one geodesic arc which is mapped by f to a geodesic arc.

An example of a pleated surface is the boundary of the convex core (see Part II).

I.5.1.4. Lemma: Pleating locus is a lamination.   *Let (S,f) be a pleated surface. Then the pleating locus of (S,f) is a geodesic lamination and the map f is totally geodesic in the complement of the pleating locus.*

Proof.   We need only consider pleated maps from $\mathbb{H}^2$ to $\mathbb{H}^3$ since we can always work in the universal covers. Let γ be the pleating locus of a pleated surface $f : \mathbb{H}^2 \to \mathbb{H}^3$. If $x \notin \gamma$, then there are two transverse geodesic arcs through x. Let $[a \ x \ b]$ and $[c \ x \ d]$ be geodesic arcs in $\mathbb{H}^2$, mapped by f to geodesic arcs $[a'x'b']$ and $[c'x'd']$ in $\mathbb{H}^3$. Let $\angle axc = \theta$. Then $\angle bxc = \pi - \theta$. Since f is isometric, $d(a', c') \leq d(a, c)$. Hence $\theta' = \angle a'x'c' \leq \theta$. Similarly $\angle b'x'c' \leq \angle bxc$, so that $\pi - \theta' \leq \pi - \theta$. Hence $\theta' = \theta$. It now follows that $d(a', c') = d(a, c)$ and that f maps the geodesic $[a, c]$ to the geodesic $[a', c']$. This implies that f is totally geodesic in a neighbourhood of x. In particular we see that the pleating locus is a closed set.

I.5.1.5.

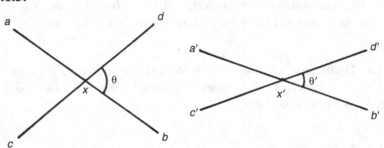

Figure I.5.1.5   *Diagram showing two transverse arcs which are mapped by f to geodesics.*

Now let $x \in \gamma$, the pleating locus of f, and let x lie in the interior of the open arc α, such that $f \,|\, \alpha$ is geodesic. We take α maximal with this property and we claim that α is a complete geodesic. For if not, let y be a finite endpoint of α and let z be a point on α, on the other side of x from y. Let $[a \ y \ b]$ be a geodesic arc $(a, y, b$ distinct$)$, on which f is geodesic. Then $[a \ y \ b]$ meets the geodesic containing α only in y, by the maximality of α.

Since f is isometric, $d(fa, fz) \leq d(a, z)$ and so $\angle fa\,fy\,fz \leq \angle ayz$. Similarly $\angle fb\,fy\,fz \leq \angle byz$. Since these angles add up to π, we must have equality in each

I.5.1.6.

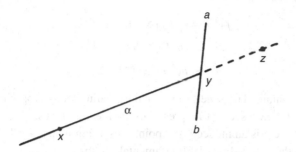

Figure I.5.1.6

case. Hence f maps the geodesic $[a, z]$ to the geodesic $[fa, fz]$. We deduce that f is geodesic on the triangle zab. But this contradicts the fact that $x \in \gamma$. The contradiction shows that α is a complete geodesic.

The same argument, with y defined as any point of α which is not on the pleating locus, shows that $\alpha \subset \gamma$.

$\square$

I.5.2. Compactness properties of pleated surfaces

See Section 8.8 of Thurston (1979).

We now wish to consider the space of all possible pleated surfaces (in all possible 3-manifolds), and derive the various compactness results which will be used later in our paper and are also used elsewhere in Thurston's work.

Let $\mathcal{PSF}$ be the set of $\mathcal{P}$leated $\mathcal{S}$urfaces with base $\mathcal{F}$rame. More precisely, $\mathcal{PSF}$ is the set of triples (Γ_1, Γ_2, f), where Γ_1 is a torsion-free discrete subgroup of Isom $^+\mathbb{H}^2$, Γ_2 is a torsion-free discrete subgroup of Isom $^+\mathbb{H}^3$, and f is a pleated map such that there exists a homomorphism $h \colon \Gamma_1 \to \Gamma_2$ with:

(1) $f(s_0) = s_0$ (recall that s_0 is a fixed basepoint in $\mathbb{H}^2$ and we are thinking of $\mathbb{H}^2$ as embedded in $\mathbb{H}^3$);
(2) $f \circ T = h(T) \circ f$ for $T \in \Gamma_1$.

One often assumes also that h is injective, but if we assume this, we will state it explicitly.

We topologize $\mathcal{PSF}$ by using the Chabauty topology for the groups Γ_1 and Γ_2 and the compact open topology for f.

Let S be a stratum for the pleating locus. Then $f|S = A(S)|S$ for some $A(S) \in \text{Isom}(\mathbb{H}^3)$. (Since $\mathbb{H}^2$ is embedded in $\mathbb{H}^3$, the isometries of $\mathbb{H}^3$ map

$\mathbb{H}^2$ into $\mathbb{H}^3$.) Let $T \in \Gamma_1$. Then

$$f|TS = (fT)|S \circ T^{-1}|TS$$
$$= (h(T)f)|S \circ T^{-1}|TS$$
$$= h(T)A(S)T^{-1}|TS.$$

Hence TS is contained in some stratum of the pleating locus. Applying the same argument to T^{-1}, we see that Γ_1 preserves the pleating locus.

Note that there is an induced basepoint preserving map $\bar{f}\colon \mathbb{H}^2/\Gamma_1 \to \mathbb{H}^2/\Gamma_2$ and that h is the induced map of fundamental groups.

I.5.2.1. Definition. We define $\mathcal{PSF}(A, \varepsilon)$ to be the set of triples $(\Gamma_1, \Gamma_2, f) \in \mathcal{PSF}$, such that $\text{Area}(\mathbb{H}^2/\Gamma_1) \leq A$, the injectivity radius of $\mathbb{H}^3/\Gamma_2$ at s_0 is greater than or equal to ε and the injectivity radius of $\mathbb{H}^2/\Gamma_1$ is greater than or equal to ε at s_0.

Let $|S|$ be the topological type of a surface S. Note that $|S|$ determines Area (S) via the Euler characteristic. Note also that if we assume that h is injective, the assumption on the injectivity radius in the domain in the preceding definition follows from the assumption in the range, because f is isometric.

I.5.2.2. Theorem: Compactness of pleated surfaces. $\mathcal{PSF}(A, \varepsilon)$ *is compact.*

In our experience facts about pleated surfaces are accepted far too readily by readers of Thurston's notes. Here are some examples which show how naive intuition can go wrong and which explain why the hypotheses in the theorem are necessary.

I.5.2.3. Example 1. Let ABC be a small equilateral hyperbolic triangle in $\mathbb{H}^3$. Let α, β, and γ be the geodesics orthogonal to the plane of ABC, through A, B, and C, respectively. On $\mathbb{H}^2$ we fix a geodesic l and mark out lengths along l equal to the sidelength of ABC. We denote these marks by

$$\ldots A_{-1}, B_{-1}, C_{-1}, A_0, B_0, C_0, A_1, B_1, C_1 \ldots$$

Let α_n, β_n, and γ_n $(n \in \mathbb{Z})$ be the geodesics orthogonal to l in $\mathbb{H}^2$ through A_n, B_n, C_n. Let f be the obvious pleated map which sends A_n to A, B_n to B, C_n to C, α_n to α, β_n to β, and γ_n to γ. We are taking $\Gamma_1 = \Gamma_2 = \text{id}$ here, so f is homotopically incompressible.

If we take the limit as the triangle shrinks to a point, we get a limit map which is orthogonal projection of $\mathbb{H}^2$ onto a line. In particular f is not isometric.

This example shows that it is essential for the area of $\mathbb{H}^2/\Gamma_1$ to be bounded.

I.5.2.4. Example 2.   Here is another example, due to Thurston. Let P be a fixed pair of pants. We fix a hyperbolic structure P_0 on P, with geodesic boundary components. We define a map $f: P \to P$ which sends the generators of $\pi_1 P$, which we call x and y, to xyx and yx, respectively. The map on homology is the familiar matrix

$$\begin{bmatrix} 2 & 1 \\ 1 & 1 \end{bmatrix}.$$

The boundary components of P correspond to the elements x, y and xy of $\pi_1 P$. They are sent to xyx, yx, and $xyxyx$, respectively. The map $f^n: P \to P$ sends x and y to loops whose geodesic representatives have exponentially increasing length. Using Theorem I.5.3.6 (*Finite laminations realizable*) we can find a hyperbolic structure P_n on P, with geodesic boundary components and a pleated map f_n representing f^n. Let α_n be a geodesic arc of minimal length joining two distinct boundary curves of P_n. Then the length of α_n tends to zero. We choose as a basepoint for P_n the midpoint of α_n and we choose the image under f_n of this basepoint as a basepoint for P_0. The pleating locus consists of the boundary components plus three spirals which cut P_n up into two ideal triangles.

Now $f_n: P_n \to P_1$ does not converge to a pleated surface. In fact, in the appropriate sense, P_n converges to a straight line. We get an example between closed surfaces of genus 2, by doubling P_n and P_1. However, when extending f_n to the double of P_n, we are obliged to send both copies of P_n to the same copy of P_1. The hypothesis of the theorem which then fails is the incompressibility.

Proof. Compactness of pleated surfaces.   Consider $\{P_i\} = \{(\Gamma_{1,i}, \Gamma_{2,i}, f_i)\}$. By Corollary I.3.1.7 (*Set of hyperbolic manifolds compact*), and Corollary I.3.1.8 (*Compact with bounded volume*), we may choose a subsequence $\{P_j\}$ such that $\Gamma_{1,j}$ converges to Γ_1, a torsion free discrete subgroup of $\mathrm{Isom}(\mathbb{H}^2)$, $\Gamma_{2,j}$ converges to Γ_2, a torsion free subgroup of $\mathrm{Isom}(\mathbb{H}^3)$, and, by Ascoli's theorem, f_j converges to some continuous map $f: \mathbb{H}^2 \to \mathbb{H}^3$. Let the pleating locus λ_j of f_j converge to λ. By Theorem I.3.1.4 (*A(U) compact*), $\mathbb{H}^2/\Gamma_1$ has area bounded by A. It is easy to see that λ is Γ_1 equivariant, and it therefore induces a lamination on a finite area surface. It follows that λ has measure zero. It is easy to see that f is geodesic on each stratum of λ.

I.5.2.5. Lemma: Injective homomorphism.   *The injective homomorphisms $h_j: \Gamma_{1,j} \to \Gamma_{2,j}$ converge to an injective homomorphism $h: \Gamma_1 \to \Gamma_2$*

(possibly after taking a subsequence if Γ_2 does not consist entirely of orientation preserving isometries of $\mathbb{H}^3$).

Proof. Injective homomorphism.   Let S be a complementary region of the lamination λ. Given $T \in \Gamma_1$, we let $\{T_j \in \Gamma_1(j)\}$ be a sequence such that $\{T_j\}$ converges to T. Since f_j converges to f and $f_j T_j(x)$ converges to $fT(x)$ for each $x \in \mathbb{H}^2$, $h_j(T_j)$ converges on $f(S)$, which is an open subset of a hyperbolic plane. But since $\{h_j(T_j)\}$ is a sequence of isometries, it converges to some $U \in \text{Isom}(\mathbb{H}^3)$. By Lemma I.3.1.3 (*Geometric convergence*), $U \in \Gamma_2$. U will be called $h(T)$. Actually U is only determined if we know its orientation. So we have to pick a subsequence such that $h_j T_j$ converges whenever we have a sequence $\{T_j\}$ converging to T. It is easy to check that h is a homomorphism. Furthermore, h is injective, since if $h(T) = h(T')$ there would exist two sequences T_j and T'_j converging to T and T', respectively such that $h_j(T_j)$ and $h_j(T'_j)$ both converge to $h(T)$. But this implies that $h_j(T_j^{-1} T'_j)$ is equal to the identity for large values of j. Then we have $T_j = T'_j$ so $T = T'$.

> **Injective homomorphism**

Continuation, proof of Compactness of pleated surfaces.   To complete the proof of the theorem, we need to check that f is an isometric map. Before doing this we prove the following technical lemma.

I.5.2.6. Lemma: Technical lemma.   *Suppose $r > 0$ is given. Then there is a number K_r with the following property. Let X, Y, Z, U, V, W be points in $\mathbb{H}^2$ such that the distance between any two of them is at most r. Suppose that we have disjoint geodesics α, β, γ with $X, U \in \alpha$, $Y, V \in \beta$ and $W, Z \in \gamma$. Suppose further that X, Y, Z lie on a geodesic which is orthogonal to α, and that $U, V,$ and W are also collinear (see Figure I.5.2.7). Then $d(Y, Z) \le K_r d(V, W)$.*

Proof. Technical lemma.   We may regard Y and β as fixed. We assume the result is false and take a sequence of situations such that $d(Y, Z_i)/d(V_i, W_i)$ is unbounded. By taking subsequences, we assume that all sequences of points, geodesics or real numbers which occur in the proof converge (possibly to infinity).

Since $d(Y, Z_i) \le r$, $d(V_i, W_i)$ tends to zero. Hence γ_i converges to β. Let V_i converge to $V \in \beta$. Then $d(Y, V) \le r$. Using hyperbolic trigonometry, it is not difficult to show, using the fact that β and γ_i are disjoint, that

$$e^{-r} \le \lim_{i \to \infty} \frac{d(Y, \gamma_i)}{d(V_i, \gamma_i)} \le e^r.$$

I.5.2.7.

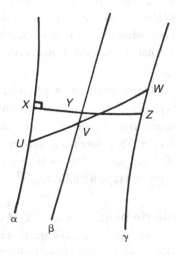

Figure I.5.2.7

Since γ_i and α_i are disjoint, we have

$$|\cos(\angle X_i Z_i W_i)| \leq \tanh d(X_i, Z_i)$$

which tends to zero. By the sine rule for hyperbolic triangles,

$$\lim \frac{d(Y, \gamma_i)}{d(Y, Z_i)} = 1.$$

Also

$$1 \geq \lim_{i \to \infty} \frac{d(V_i, \gamma_i)}{d(V_i, W_i)}.$$

It follows that

$$\lim_{i \to \infty} \frac{d(Y, Z_i)}{d(V_i, W_i)} = \lim_{i \to \infty} \frac{d(Y, Z_i)/d(Y, \gamma_i)}{d(V_i, W_i)/d(V_i, \gamma_i)} \frac{d(Y_i, \gamma_i)}{d(V_i, \gamma_i)} \leq e^r.$$

But this contradicts our assumption that $d(Y, Z_i)/d(V_i, W_i)$ is unbounded. This proves the lemma.

> **Technical lemma**

Remark.   Recently Thurston has circulated a preprint (Thurston, 1986) in which the definition of a pleated surface is changed in order to avoid the necessity for the above technical lemma. In the new definition, a pleated map sends every geodesic to a rectifiable path of the same length, whereas in the old definition, every rectifiable path is sent to a rectifiable path of the same length. Our lemma can be regarded as showing that the new weaker hypothesis implies the old stronger hypothesis, or that the new definition is equivalent to the old one.

Continuation, proof of Compactness of pleated surfaces.   Let l be a geodesic in λ. Then there exists $l_j \in \lambda_j$ such that l_j converges to l. We may assume that l and l_j are parametrized geodesics. Then l_j converges uniformly to l on compact subintervals of $\mathbb{R}$. Therefore $f_j \circ l_j$ converges uniformly to $f \circ l$ on compact subintervals of R. Therefore $f \circ l \colon \mathbb{R} \to \mathbb{H}^3$ is an isometry. Notice that f is lipschitz. In fact $d(fx, fy) \leq d(x, y)$ for all $x, y \in \mathbb{H}^2$.

I.5.2.8. Lemma: Isometric map.   *Let $\lambda \in \mathcal{GL}(\mathbb{H}^2)$ have measure zero, and suppose that $f \colon \mathbb{H}^2 \to \mathbb{H}^3$ is lipschitz with lipschitz constant less than 1, and is an isometry on each stratum of λ. Then f is an isometric map.*

Proof. Isometric map.   Let $p \colon [0, L] \to \mathbb{H}^2$ be a rectifiable path parametrized proportional to arc length and choose $\varepsilon > 0$. Since $p^{-1}(\mathbb{H}^2 \backslash \lambda) \subset [0, 1]$, it is composed of a countable union of open intervals which we call the *complementary intervals*. Note that the sum of the lengths of the complementary intervals may be less than L. Pick a finite number of these intervals, $J_1, \ldots, J_r$, such that the total length of the remaining intervals is less than ε. Given any partition of $[0, 1]$ we enlarge it so that:

(1) every endpoint of $J_1, \ldots, J_r$ is in the partition,
(2) if a point of the partition lies in a complementary interval, then the endpoints of the interval are in the partition.

Now take an open interval (t_j, t_{j+1}) of the partition which meets λ and let the total length of the complementary intervals contained within (t_j, t_{j+1}) be ε_j. Then $\sum \varepsilon_j \leq \varepsilon$. Let $C = p(t_j)$ and $B = p(t_{j+1})$. By our choice of partition, $B, C \in \lambda$. Let C lie on the geodesic $g \in \lambda$. Let A be the intersection of g with the geodesic orthogonal to g through B. See Figure I.5.2.9.

Let $\{(a_i, b_i)\}$ be the set of complementary intervals of $[A, B] \backslash \lambda$. Since λ has measure zero, $\sum d(a_i, b_i)$ is equal to $d(A, B)$. Let g_i' be the geodesic of λ containing a_i and g_i'' the geodesic of λ containing b_i. Let δ_i be the greatest length of any subpath of $\rho|[t_j, t_{j+1}]$ connecting g_i' to g_i''. By Lemma I.5.2.6 (*Technical lemma*), there is a constant $K = K_L$, such that

$$d(A, B) = \sum d(a_i, b_i) \leq K \sum_i \delta_i \leq K\varepsilon_j.$$

I.5.2.9.

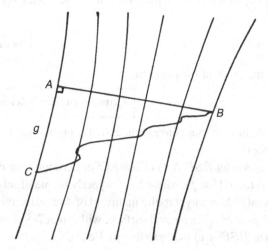

Figure I.5.2.9

Since f is lipschitz with constant not greater than one, $d(f(A), f(B)) \le d(A, B)$. By the triangle inequality

$$|d(f(A), f(C)) - d(f(B), f(C))| \le d(f(A), f(B)) \le K\varepsilon_j.$$

Also by the triangle inequality we obtain

$$|d(A, C) - d(B, C)| \le d(A, B) \le K\varepsilon_j.$$

Since f is an isometry on each stratum and both A and C lie on g, $d(f(A), f(C)) = d(A, C)$. So we may combine the two inequalities above to obtain

$$|d(f(B), f(C)) - d(B, C)| \le 2K\varepsilon_j.$$

Also note that on each of the intervals (t_j, t_{j+1}) of the partition, whose interiors do not meet λ, $d(fp(t_j), fp(t_{j+1})) = d(p(t_j), p(t_{j+1}))$ since f is an isometry on each stratum.

We have thus proved that, given any partition $0 = t_0 < \cdots < t_n = L$ of $[0, L]$ and any $\varepsilon > 0$, then (after possibly refining the partition)

$$\left| \sum_{j=0}^{n-1} d(f(p(t_j)), f(p(t_{j+1}))) - \sum_{j=0}^{n-1} d(p(t_j), p(t_{j+1})) \right| \le 2K \sum \varepsilon_j \le 2K\varepsilon.$$

This shows that $f \circ p$ is a rectifiable path with the same length as p, completing the proof.

Isometric map

This completes the proof of the theorem.

Compactness of pleated surfaces

We can now deduce various consequences of Theorem I.5.2.2 (*Compactness of pleated surfaces*).

In the first version let $\mathcal{PSB}(A, \varepsilon)$ (*P*leated *S*urfaces with *B*asepoint) be the set of quintuplets (S, p, M, q, f), where S is a hyperbolic manifold of dimension two with basepoint p, M is a hyperbolic manifold of dimension three with basepoint q, and $f: S, p \to M, q$ is a pleated surface, with $\text{Area}(S) \leq A$ and $\text{inj}_q M \geq \varepsilon$.

We topologize $\mathcal{PSB}(A, \varepsilon)$ as a quotient of $\mathcal{PSJ}(A, \varepsilon)$.

I.5.2.10. Corollary: PSB compact. $\mathcal{PSB}(A, \varepsilon)$ *is compact.*

This follows since $\mathcal{PSB}(A, \varepsilon)$ is a quotient of $\mathcal{PSF}(A, \varepsilon)$.

Let K be a compact subset of a fixed complete hyperbolic 3-manifold N. Let $\mathcal{PS}(A, K, N)$ be the space of pleated surfaces $f: S \to N$, without basepoint, such that $K \cap fS \neq \varnothing$ and with $\text{Area}(S) \leq A$. We give this space the quotient topology from the space of pleated surfaces with basepoint $f: S, p \to N$, q, where $q \in K$. The topology can be defined directly. Given $k > 1$, $\varepsilon > 0$ and a compact subset K_1 of S, a neighbourhood of $f: S \to N$ consists of all pleated surfaces $f': S' \to N$ such that there is a k-approximate isometry ϕ between K_1 and a compact subset $K_1' \subset S'$, such that $d(fx, f' \circ \phi(x)) < \varepsilon$ for all $x \in K_1$. The details of checking that this is a correct description of the quotient topology is left to the reader.

I.5.2.11. Corollary: Unmarked pleated surfaces compact. *Let K be a compact subset of a fixed hyperbolic 3-manifold N. Let $A > 0$. Then $\mathcal{PS}(A, K, N)$, the set of pleated surfaces without basepoint which meet K, is compact.*

Proof. Let $\delta < \text{inj}_K(N)$. The space of all orthonormal frames over K forms a compact space K_1. There is a continuous map from K_1 into the space of discrete subgroups of $\text{Isom}(\mathbb{H}^3)$, defined by lifting a frame to the standard frame in $\mathbb{H}^3$ and taking the corresponding group of covering translations. This map is clearly continuous. It follows that the relevant set of triples (Γ_1, Γ_2, f) is a closed subset of $\mathcal{PS}(A, \delta)$ and is therefore compact.

□

If we wish to consider only a single homeomorphism type of surface, we have to introduce a condition which will prevent a simple closed geodesic pinching down into two cusps. Since a pleated surface is isometric, it will automatically map cusps of S to cusps of N. This condition means that parabolic elements of $\pi_1 S$ map to parabolic elements of $\pi_1 N$, but another condition is also needed. This condition is that hyperbolic elements in $\pi_1 S$ map to hyperbolic elements in $\pi_1 N$.

I.5.2.12. Definition. We say a homotopically incompressible map $f: S \to N$ is *non-parabolic*, (abbreviated to *np*) if the induced map $f_*: \pi_1(S) \to \pi_1(N)$ is injective and takes hyperbolic elements to hyperbolic elements and parabolic elements to parabolic elements.

I.5.2.13. Corollary: Compactness of pleated surfaces of fixed topological type. *Let T be a fixed topological surface of finite type and N a fixed complete hyperbolic 3-manifold. Let $K \subset N$ be a fixed compact subset. Then $\mathcal{PS}^{(np)}(T, K, N)$ the space of non-parabolic pleated surfaces of finite area without basepoint $f: S \to N$ which meet K, and such that S homeomorphic to T, is compact.*

Proof. Since $\mathcal{PS}^{(np)}(T, K, N) \subset \mathcal{PS}(A, K, N)$ (where $A = \text{area}(T)$) which is compact, we need only prove that $\mathcal{PS}^{(np)}(T, K, N)$ is closed in $\mathcal{PS}(A, K, N)$. Suppose $\{[f_i: S_i \to N]\} \subset \mathcal{PS}^{(np)}(T, K, N)$ converges to $[f: S \to N] \in \mathcal{PS}(A, K, N)$. We need only prove that S has the same (np)-topological type as S_i. We do this by proving that, for some $\alpha > 0$, no closed geodesic in S_i has length less than α and then applying Proposition I.3.2.13 (*Mumford's Lemma*) to obtain our result.

We define the δ-length $l_\delta(\gamma)$ of a path γ in S_i to be $l(\gamma \cap S_{[\delta, \infty)})$. Correspondingly, we define δ-distance and the δ-diameter of a compact subset of the surface. Clearly, the δ-diameter of S_i is less than

$$\frac{4 \, \text{area}(S_i)}{\delta} = \frac{4A}{\delta} = B_\delta.$$

Choose $\delta < \min(\varepsilon_0/2, \min_{x \in K}(\text{inj}(x)))$ where ε_0 is the Margulis constant. Then since loxodromic components of $N_{[\delta/2,\infty)}$ are separated by a δ-distance of at least $\cosh^{-1}[\frac{\cosh \delta + 1}{\cosh \delta}]$, the set K_δ of all points of δ-distance less than or equal to B_δ from K, contains only finitely many such components. Thus K_δ contains no closed geodesics of length less than α for some $\alpha > 0$. But $f(S_i) \subset K_\delta$, since f decreases δ-distance, completing our proof.

$\square$

We now prove the compactness of the space of marked pleated surfaces, under appropriate conditions.

I.5.2.14. Definition.   A *marked pleated surface* is a pair $([h], f)$, where $h: T \to S$ is a homeomorphism from a topological surface T to a complete hyperbolic 2-manifold S of finite area; $[h]$ is the isotopy class of h; and $f: S \to N$ is a pleated map.

More directly, the pair (h_1, f_1) is equivalent to the pair (h_2, f_2) (where $h_i: T \to S_i$ are homeomorphisms and $f_i: S_i \to N$ are pleated surfaces) if there is an isometry $\phi: S_1 \to S_2$ such that the left-hand side of the Figure I.5.2.15 commutes up to isotopy and the right-hand side commutes precisely.

I.5.2.15.

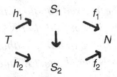

Figure I.5.2.15

Our discussion of pleated surfaces will relate to a fixed choice of T and N. We shall denote the space of $\mathcal{M}$arked $\mathcal{P}$leated $\mathcal{S}$urfaces by $\mathcal{MPS}(T, N)$.

The space of marked pleated surfaces is topologized by saying that $([h_2]: T \to S_2, f_2: S_2 \to N)$ is near to $([h_1]: T \to S_1, f_1: S_2 \to N)$ if $[h_2]$ is near to $[h_1]$ in $\mathbf{T}(T)$ and there exist representatives h'_1 and h'_2 such that $f_1 \circ h'_1$ and $f_2 \circ h'_2$ are nearby in $C(T, N)$ in the compact-open topology. Thus we may consider $\mathcal{MPS}(T, N) \subset \mathbf{T}(T) \times \hat{C}(T, N)$ where $\hat{C}(T, N)$ is the quotient space of $C(T, N)$ induced by the equivalence relation of pre-composition by homeomorphisms of T isotopic to the identity. Note that this is really only a condition on the thick part of S_1 and S_2, because the values of a pleated surface map on a cusp are determined by its values on the boundary of the thick part.

I.5.2.16. Lemma: Marked pleated surfaces Hausdorff.   *The space of marked pleated surfaces is Hausdorff and has a countable basis.*

Proof.   Suppose

$$T \xrightarrow{h_1} S_1 \xrightarrow{f_1} M \quad \text{and} \quad T \xrightarrow{h_2} S_2 \xrightarrow{f_2} M$$

do not have disjoint neighbourhoods in $\mathcal{MPS}$. Then h_1 and h_2 must represent the same element of Teichmüller space. This means that we can take the hyperbolic surfaces $S_1 = S_2 = S$. We may also take h_1 to be the identity, and then $h = h_2$ is isotopic to the identity.

Let L_1 and L_2 be the pleating loci of f_1 and f_2, respectively. We claim that $L_1 = L_2$. For suppose l is a parametrized geodesic in L_1. By our hypothesis, we may assume that $f_2 h$ is arbitrarily close to f_1 in the compact-open topology, if we change h by an isotopy. Let $\tilde{l}$ be a lift of l to $\mathbb{H}^2$. Then $\tilde{f}_1 \tilde{l}$ is a geodesic in $\mathbb{H}^3$. Now $\tilde{f}_2 \tilde{h}(\tilde{l})$ is close to $\tilde{f}_1 \tilde{l}$, and, by changing h by an isotopy, we can make it arbitrarily close to $\tilde{f}_1 \tilde{l}$ as a parametrized path. Since $\tilde{f}_2$ is an isometric map, $\tilde{h}\tilde{l}$ is close to some parametrized geodesic. Since $\tilde{h}\tilde{l}$ has the same endpoints as $\tilde{l}$, $\tilde{h}\tilde{l}$ is close to the parametrized geodesic $\tilde{l}$. It follows that $\tilde{f}_2 \tilde{l}$ is equal to the parametrized geodesic $\tilde{f}_1 \tilde{l}$, except for a change of origin. Hence l is in the pleating locus of f_2. This means we have proved that $L_1 = L_2$.

But now f_1 and f_2 agree on the boundary of each complementary piece of the lamination (or, more precisely, their images agree). It follows that $f_1 = f_2$. This completes the proof that the sequence of marked pleated surfaces is Hausdorff.

$\mathcal{MPS}(T, N)$ has a countable basis since both $\mathbf{T}(T)$ and $C(T, N)$ do.

□

Suppose $M \to S^1$ is a fibre bundle and M has a complete hyperbolic structure of finite volume. We know that such structures exist, firstly by the work of Jørgensen (1977) in special cases, and later, in great generality, by Thurston (1986a). The fibre is a surface, denoted by S. The monodromy $\phi: S \to S$ has infinite order in the mapping class group. We shall see later in Theorem I.5.3.6 (*Finite laminations realizable*) that the fibre can be represented by a pleated surface.

Since the monodromy has infinite order, we get an infinite sequence of homotopic marked pleated surfaces with no convergent subsequence. In fact there is not even a convergent subsequence in Teichmüller space. However, up to finite coverings, this is the only way in which the compactness of the space of homotopic marked pleated surfaces fails, as we now proceed to prove.

I.5.2.17. Definition. Let $f: S \to N$ be a map between manifolds. We say f is a *virtual fibre* of a fibre bundle over N, if N has a finite cover $\tilde{N}$ which is a fibre bundle over the circle, and f lifts to a map into $\tilde{N}$ which is homotopic to the inclusion of the fibre.

I.5.2.18. Theorem: Compactness of marked pleated surfaces. *Let N be a complete hyperbolic 3-manifold, and let K be a compact subspace of*

N. Let T be a topological surface of finite type. Then $\mathcal{MPS}^{np}(N, T, K)_{(\bar{f},\bar{h})}$, *the space of marked np-pleated surfaces* $(h\colon T \to S, f\colon S \to N)$ *meeting K such that* $f \circ h$ *is homotopic to* $\bar{f} \circ \bar{h}$ *(by a cusp-preserving homotopy) is compact, unless* $f \circ h$ *is the virtual fibre of a fibre bundle over N.*

Proof. Compactness of marked pleated surfaces. Suppose (h_i, f_i) is a sequence of homotopic marked pleated surfaces meeting K, with no convergent subsequence. We may assume that $f_i\colon S_i \to N$ converges in the space of unmarked pleated surfaces, to a pleated surface $f\colon S \to N$ where S is homeomorphic to T. Let $\phi_i\colon S \to S_i$ be an approximate isometry, which is an isometry on cusps, such that $f_i\phi_i$ approximates f. Let $L \subset S$ be the complement of the cusps. Since L is compact, the injectivity radius is bounded below on a neighbourhood of L. Therefore $f_i\phi_i$ is homotopic to f by a linear homotopy along shortest geodesics on L and this homotopy extends in a unique standard way to a cusp-preserving homotopy on all of S. Suppose there are only finitely many distinct homotopy classes among the maps $\{\phi_i^{-1}h_i\colon T \to S\}$. In this case we may suppose that this homotopy class is constant and is represented by $h\colon T \to S$. Now $f_i\phi_i$ is approximately equal to f. Therefore $fh \simeq f\phi_i^{-1}h_i \simeq f_ih_i$. It follows that (f, h) is the limit of the sequence $\{(f_i, h_i)\}$. This is a contradiction.

It follows that there are infinitely many distinct homotopy classes $\{\phi_i^{-1}h_i\colon T \to S\}$.

I.5.2.19. Lemma: Torsion-free kernel. *Let A be a free abelian group of finite rank, and let n > 2 be an integer. Then the kernel of* $\operatorname{Aut} A \to \operatorname{Aut}(A \otimes \mathbb{Z}_n)$ *contains no elements of finite order.*

Proof. Torsion-free kernel. Suppose $g\colon A \to A$ is an automorphism of finite order which induces the identity on $A \otimes \mathbb{Z}_n$. By averaging, we construct a g-invariant bilinear positive definite inner product on A. Let N be the set of nearest elements to 0. This is a finite set (as we see by comparing with the associated positive definite g-invariant inner product on the associated vector space $A \otimes \mathbb{R}$), and it is g-invariant. If x_1 and x_2 are distinct elements of N, then x_1 and x_2 are distinct modulo n (for otherwise one of the $n - 1$ intermediate points of A would be non-zero and nearer to 0).

It follows that g is fixed on N. Let B be the subgroup of A generated by N, and let

$$C = \{x \in A\colon kx \in B \quad \text{for some } k \neq 0\}.$$

Then g is fixed on B and hence fixed on C. Now A/C is a free abelian group of smaller rank than A, and g induces an automorphism of finite order on A/C, which becomes the identity on $(A/C) \otimes \mathbb{Z}_n$. It follows by induction that this automorphism is the identity on (A/C). Taking a basis for C and extending to a basis of A, we see that g has matrix

$$\begin{bmatrix} I & X \\ 0 & I \end{bmatrix}$$

and, since g has finite order, $X = 0$.

> **Torsion-free kernel**

I.5.2.20. Lemma: Non-trivial on homology. *Let T be a surface of finite type and let $\chi(T) < 0$. Let $g: T \to T$ have finite order in the mapping class group, and suppose $g \neq \mathrm{id}$. Then the induced map on $H_1(T; \mathbb{R})$ is non-trivial.*

The following proof was explained to us by Peter Scott.

Proof. Non-trivial on homology. Let $n > 0$ be the order of g. We may suppose that n is prime. According to Kerckhoff (1983) (in fact this was known earlier), we may impose on T a hyperbolic structure preserved by an isometry g of finite order. Let $Q = T/g$. Then Q is an orbifold. A component of the singular set of Q is either a point or a circle. (The latter case happens if g fixes a circle and interchanges the two sides of the circle.) Let E be the group which is the finite extension of $\pi_1 T$ by g. Then E is the fundamental group of the orbifold Q. It can be computed from van Kampen's theorem, where the pieces to be glued together are nice neighbourhoods of the components of the singular set and Q_0, the complement of these neighbourhoods. It is easy to see that

$$H_1(|Q|; \mathbb{R}) \cong \frac{E}{[E,E]} \otimes \mathbb{R} \cong \frac{H_1(T; \mathbb{R})}{<g>}$$

by applying the Mayer–Vietoris theorem to the same pieces of $|Q|$. Therefore g induces a non-trivial map on $H_1(T; \mathbb{R})$ if and only if $H_1(|Q|; \mathbb{R})$ has smaller dimension than $H_1(T; \mathbb{R})$.

Suppose, for a contradiction, that these two dimensions are equal and denote the common value by r. Since $\chi(T) < 0$, we must have $r \geq 2$. Let Q have q singular points, and let the inverse image of Q_0 in T be T_0. Amongst the boundary components of T_0, suppose q_1 of these bound disks in T. Then $q_1 \leq nq$. We have

$$n\chi|Q| = n\chi Q_0 + nq = \chi T_0 + nq \geq \chi T_0 + q_1 = \chi T.$$

If $|Q|$ is not a closed orientable surface, then $\chi|Q| = -r + 1$, so that $n(r-1) = -n\chi|Q| \leq -\chi T = r - 1$. Therefore $n \leq 1$, a contradiction. If $|Q|$ is a closed orientable surface, then $r \geq 4$ and $n(r - 2) = -n\chi |Q| \leq -\chi T \leq r - 1$, so that $n \leq (r - 1)/(r - 2) \leq 3/2$. Therefore $n = 1$, again a contradiction.

> **Non-trivial on homology**

The following is a well-known result, which we do not in fact use in the sequel, but which is included for its general interest.

I.5.2.21. Corollary: Finite order implies G finite. *If G is a subgroup of the mapping class group such that every element has finite order, then G is finite.*

Proof. Finite order implies G finite. Let A be the quotient of $H_1(T; \mathbb{Z})$ by its torsion subgroup. The map $G \to \text{Aut } A$ is injective by lemma under Section I.5.2.20 (*Non-trivial on homology*). The composite $G \to \text{Aut } A \to \text{Aut}(A \otimes \mathbb{Z}_3)$ is injective by Lemma I.5.2.19 (*Torsion free kernel*). Therefore G is finite.

> **Finite order implies G finite**

Continuation, proof of compactness of marked pleated surfaces.
There must exist $r \neq s$ such that $h_1^{-1}\phi_1\phi_r^{-1}h_r$ and $h_1^{-1}\phi_1\phi_s^{-1}h_s$ have the same image in $\text{Aut}(H_1(T; \mathbb{Z}_3))$. Then $h_r^{-1}\phi_r\phi_s^{-1}h_s$ is trivial in $\text{Aut}(H_1(T; \mathbb{Z}_3))$ and therefore is of infinite order in the mapping class group. Now h_r, ϕ_r, ϕ_s, h_s can all be assumed to be standard on the cusps. So we can remove the cusps from T to obtain a manifold T_b with boundary and $\psi = h_r^{-1}\phi_r\phi_s^{-1}h_s$ is a homeomorphism of T_b.

Let M be the mapping torus of ψ. We shall define $F: M \to N$. We have already seen that $f_s\phi_s \simeq f_r\phi_r$ by a cusp-preserving homotopy. Therefore $f_r h_r \psi = f_r\phi_r\phi_s^{-1}h_s \simeq f_s h_s$ by a cusp-preserving homotopy. We can use this to construct a map of M into N. Note that the boundary components of M (all of which are tori or Klein bottles) are sent to cusps of N.

Now $F_*: \pi_1 M \to \pi_1 N$ is injective. To see this note that $\pi_1 M$ is the extension of $\pi_1 T$ by $\mathbb{Z}$. Let α be the generator of $\mathbb{Z}$, and suppose that $F_*(\gamma\alpha^n)$ is trivial in $\pi_1 N$ for some $\gamma \in \pi_1 T$. Then $(F_*\alpha)^n$ lies in the image of $\pi_1 T$. This means $n = 0$ or that conjugation by $F_*\alpha$ has finite order in the outer automorphism group of $\pi_1 T$. But the effect of this conjugation is equal to ψ_* which has infinite order. Therefore $n = 0$. Since the pleated surface is incompressible, γ is trivial.

Let $\tilde{N}$ be the covering space of N such that the lift $\tilde{F}: M \to \tilde{N}$ induces an isomorphism of fundamental groups.

First we dispose of the case where M is a closed manifold. Since the higher homotopy groups of M and $\tilde{N}$ are zero, we know that M and $\tilde{N}$ are homotopy equivalent. Hence $\tilde{N}$ is also a closed manifold. From a result of Stallings (Hempel, 1976; Theorem 11.6) we deduce that M is homeomorphic to $\tilde{N}$.

From now on we assume that M has a non-empty boundary. We remove from $\tilde{N}$ uniform horoballs to make a manifold with boundary, which, by abuse of notation, we continue to denote by $\tilde{N}$, and we have $\tilde{F}: M, \partial M \to \tilde{N}, \partial \tilde{N}$. A boundary component of M has fundamental group which is maximal solvable in $\pi_1 M$. Therefore its image is maximal solvable in $\pi_1 \tilde{N}$. Therefore each component of $\partial \tilde{N}$ in the image of ∂M is a torus or a Klein bottle and the map has degree one on each component of ∂M. Since fundamental groups of different components of ∂M are not conjugate in $\pi_1 M$, these components are mapped to distinct components of $\partial \tilde{N}$. Now consider the following diagram:

$$H_3(M, \partial M; \mathbb{Z}_2) \to H_3(\tilde{N}, \partial \tilde{N}; \mathbb{Z}_2)$$
$$\downarrow \qquad\qquad \downarrow$$
$$H_2(\partial M; \mathbb{Z}_2) \to H_2(\partial \tilde{N}; \mathbb{Z}_2).$$

The map on H_2 is injective, and so is $\partial: H_3(M, \partial M; \mathbb{Z}_2) \to H_2(\partial M; \mathbb{Z}_2)$ (assuming for the moment that $\partial M \neq \emptyset$). Hence F_* is injective on H_3. It follows that $H_3(\tilde{N}, \partial \tilde{N}; \mathbb{Z}_2) = \mathbb{Z}_2$. Hence $\tilde{N}$ is a compact manifold and the covering $\tilde{N} \to N$ is finite sheeted. It also follows that each boundary component of $\partial \tilde{N}$ is in the image of ∂M.

M and $\tilde{N}$ are clearly sufficiently large, since they have non-empty boundary. Therefore a result of Waldhausen (Hempel, 1976; Theorem 13.7) can be applied. The twisted I-bundle situation described in the statement of that theorem is ruled out, since there is a bijection between the set of boundary components in our case, but not in the I-bundle situation. Once again, M is homeomorphic to $\tilde{N}$ (with the cusps removed). The theorem follows.

Compactness of marked pleated surfaces

I.5.3. Realizations

See Section 8.10 of Thurston (1979).

Given a pleated surface, we have its pleating locus. Rather surprisingly, it is often possible to reverse the direction of this construction. We shall show how, given a topological surface T of finite type, a homotopy class of maps $f: T \to N$

into a complete hyperbolic 3-manifold, and a maximal lamination λ of T, there often exists a complete hyperbolic structure on T and a pleated map $f: T \to N$ in the homotopy class, whose pleating locus is contained in λ.

First we make a definition.

I.5.3.1. Definition.   Let $f: S \to N$ be a pleated surface and let α be a finite geodesic arc in S. We define a non-negative real number which we call the *pleating* of α as follows. We lift α to $\tilde{\alpha}$ in $\mathbb{H}^3$. The pleating is the difference of the distance between the endpoints of $\tilde{\alpha}$ in $\mathbb{H}^3$ and the length of α.

If α has zero pleating, then it must lie in a single stratum of the pleating locus (except possibly for its endpoints).

We now prove that the pleating locus is a continuous function of the pleated surface.

Let N be a fixed complete hyperbolic 3-manifold. For each marked pleated surface $(h: T \to S, f: S \to N)$ the pleating locus of f is a geodesic lamination on S. By fixing a complete hyperbolic structure on T we obtain a geodesic lamination on T (see Section I.4.1.4 (*Transferring laminations*)).

I.5.3.2. Lemma: Pleating locus continuous.   *The map* $\mathcal{MPS}(T,N) \to \mathcal{GL}(T)$ *is continuous if we use the Thurston topology on* $\mathcal{GL}(T)$.

I.5.3.3. Remark.   If we use the Chabauty topology on $\mathcal{GL}(T)$, the map is not continuous. For example, we can take a Fuchsian group with empty pleating locus and approximate it by a quasifuchsian group with any preassigned lamination of compact support as its pleating locus.

Proof.   Let $(h_i: T \to S_i, f_i: S_i \to N)$ converge to (h,f). Let $\phi_i: S \to S_i$ be an approximate isometry such that $f_i \phi_i$ is approximately equal to f and h_i is homotopic to $h\phi$. We may assume that h and h_i are standard on the cusps and that ϕ_i is an isometry on the cusps. There is no loss of generality in taking $T = S$, $h = \text{id}$, and $h_i = \phi_i$. Let λ and λ_i be the pleating loci of f and f_i. Let α be a geodesic in $\mathbb{H}^2$ in the lamination $\tilde{\lambda}$. We must show that there is a geodesic α_i in $\tilde{\lambda}_i$ such that $\tilde{\phi}_i^{-1} \alpha_i$ has endpoints near to those of α.

By choosing appropriate coverings and lifts, we may assume that $\{\tilde{f}_i: \mathbb{H}^2 \to \mathbb{H}^3\}$ converges uniformly to $\tilde{f}: \mathbb{H}^2 \to \mathbb{H}^3$ on compact sets. Let β be a short arc crossing α. Then the pleating of $(\beta, \tilde{f})$ is non-zero. It follows that the pleating of $(\beta, \tilde{f}_i)$ is non-zero for sufficiently large i. Therefore there exists a geodesic $\alpha_i \in \tilde{\lambda}_i$ which crosses β. We claim that $\{\alpha_i\}$ converges to α. For suppose not. Then by choosing a subsequence, we may assume $\{\alpha_i\}$ converges to $\alpha_\infty \neq \alpha$. Therefore α_∞ intersects α at a point x. But, by uniform convergence,

$f|\alpha_\infty$ must be an isometry, and this would imply that x is not in the pleating locus of f, which is a contradiction.

□

I.5.3.4. Definition. Suppose λ is a lamination on a complete hyperbolic surface T, and that $f: T \to N$ is a given *np* incompressible surface. We say (λ, f) is *realizable* and write $\lambda \in R_f$ if there is a map f' homotopic to f, by a cusp-preserving homotopy, which maps each geodesic of λ homeomorphically to a geodesic of N.

We do not require f' to preserve arclength (not even up to change of scale) on the geodesics of λ. Nor do we require the image of a geodesic of λ under f' to be a simple geodesic in N. It follows that the question of whether a lamination is realizable or not does not depend on the underlying hyperbolic structure of the surface. By an easily constructed homotopy, we may assume that every realization f is isometric on each geodesic ray of S, which lies in a cusp of S and travels straight towards the cusp point. We may also assume that f sends every horocyclic curve in a cusp of S to a piecewise linear curve in a horosphere of N (using the euclidean metric in a horosphere of N to give the meaning of "piecewise linear"). The map f is linear on each horocyclic interval disjoint from λ.

I.5.3.5. Lemma: Image well-defined. *If (λ, f) is realizable, then the image of λ in N under a realization is well defined.*

Proof. Image well-defined. More precisely, we can lift any two realizations f_1, f_2 to two maps $\tilde{f}_1, \tilde{f}_2: \mathbb{H}^2 \to \mathbb{H}^3$ such that $\tilde{f}_1(\alpha)$ and $\tilde{f}_2(\alpha)$ are equal as directed geodesics in $\mathbb{H}^3$ (though the parametrizations induced by $\tilde{f}_1$ and $\tilde{f}_2$ may be different) for every $\alpha \in \tilde{\lambda}$. To see this, we fix a complete hyperbolic structure on T. Let $h: T \times I \to N$ be a cusp-preserving homotopy between f_1 and f_2. We adjust h, f_1, and f_2 so that on each cusp they are "linear", that is isometric on each geodesic straight up the cusp and horocyclic intervals disjoint from λ are sent linearly to straight lines in a horosphere of N. The thick part of T is compact. It follows that by lifting h to $\tilde{h}: I \times \mathbb{H}^2 \to \mathbb{H}^3$ we obtain a homotopy such that the distance in $\mathbb{H}^3$ from $\tilde{h}_0(x)$ to $\tilde{h}_1(x)$ is bounded as $x \in \mathbb{H}^2$ varies. But then $\tilde{h}_0$ and $\tilde{h}_1$ have the same effect on endpoints of geodesics of $\tilde{\lambda}$.

| Image well-defined |

We spend the rest of this section proving that if f preserves parabolicity, then R_f is open and dense. The proof is by a series of lemmas.

I.5.3.6. Theorem: Finite laminations realizable. *Any finite lamination is realizable. If μ is a geodesic lamination which is realizable and λ is any finite extension of μ, then λ is also realizable.*

Proof. Finite laminations realizable. We first do the case when λ is finite (i.e. $\mu = \varnothing$) first. Start by arranging for f to be a geodesic immersion on each cusp of T. Each simple closed geodesic in λ is sent to a well-defined conjugacy class in $\pi_1 T$ and hence to a well-defined closed geodesic. Fix a map on such simple geodesics which is parametrized proportional to arclength. By the Homotopy Extension theorem we change f so that it is such a map on the closed geodesics of λ. We lift f to $\tilde{f} : \mathbb{H}^2 \to \mathbb{H}^3$. Now f realizes the sublamination μ of λ consisting of closed geodesics. Therefore $\tilde{f}$ defines a map on the endpoints of geodesics of $\tilde{\mu}$. $\tilde{f}$ also gives a map on the endpoints of geodesics of $\tilde{\lambda}$ which run up cusps of T. If α is any geodesic of $\tilde{\lambda}$, then each end either spirals onto a geodesic in μ or ends in a cusp. Therefore we know where the endpoints of α should be mapped to. This determines the image of α under the realization.

Now $|\lambda|$ is not a nice subspace of T and the Homotopy Extension theorem cannot be applied to it, so we need to define the realization on a subspace which contains $|\lambda|$ and to which the Homotopy Extension theorem can be applied.

Let N be the set of points x in T such that there are distinct homotopy classes of paths of length less than ε, with one endpoint at x and the other on λ. There is a canonical foliation of N (see Part II). We map each leaf of this foliation to the appropriate horocyclic curve in N, by a map which is proportional to arclength on each horocyclic interval complementary to λ. It is easy to see that this gives a definite map on N. We now extend to the rest of $|\lambda|$ and then use the Homotopy Extension theorem to extend to the rest of T.

The same method is used to extend a realization from any geodesic lamination μ to a finite extension λ of μ. The only point we need to check on is that, when trying to define the image of $\alpha \in \tilde{\lambda} \setminus \tilde{\mu}$, we do not find that the two putative endpoints of $\tilde{f}\alpha$ are equal in $\mathbb{H}^3$. This is clear if μ is a finite lamination, but in general a little work is needed. Let $f : T \to N$ be a realization on μ and let $\tilde{f}$ be a lifting of f. Let λ be a geodesic lamination and let α be a geodesic in $\tilde{\lambda}$ with endpoints at the endpoints of $\tilde{\mu}$. Let $\lambda \setminus \mu$ consist of a finite number of geodesics. We now prove the following lemma.

I.5.3.7. Lemma: Image points not too close. *Let $f: S \to N$ be a cusp-preserving incompressible map which is "linear" on the complement of λ in any cusp and let $\tilde{f}: \mathbb{H}^2 \to \mathbb{H}^3$ be a fixed lifting. Given $r > 0$, there exists $s > 0$ such that if x, $y \in \mathbb{H}^2$ and $d(x, y) > s$, then $d(\tilde{f}x, \tilde{f}y) > r$. In particular, $\tilde{f}$ is proper.*

Proof. Image points not too close. First we show $\tilde{f}$ is proper. Let K be a compact subspace of $\mathbb{H}^3$. Let $\delta > 0$ be chosen small enough so that $\tilde{f}(\pi^{-1}S_{(0,\delta]}) \cap K = \varnothing$. Let $F \subset \mathbb{H}^2$ be a fundamental region for S and let $F_\delta \subset F$ be a fundamental region for $S_{[\delta,\infty)}$. Then F_δ is compact. Since $\pi_1 N$ is a discrete subgroup of $\mathrm{Isom}(\mathbb{H}^3)$, there are only a finite number of covering translations T of $\mathbb{H}^3$ such that $T\tilde{f}F_\delta \cap K \neq \varnothing$. It follows that $\tilde{f}^{-1}K$ is contained in a finite union of translates of F_δ, and is therefore compact. So $\tilde{f}$ is proper.

Now suppose that $x_n, y_n \in \mathbb{H}^2$ and $d(x_n, y_n) \to \infty$. We have to show that $d(\tilde{f}x_n, \tilde{f}y_n) \to \infty$. We may suppose that $x_n \in F$ for each n. We may also suppose that neither x_n nor y_n converges to the end of a cusp, for the result is then obvious by our special assumptions on f. Therefore we may assume that x_n converges to a point $x \in F$. The result then follows since $\tilde{f}$ is proper.

> **Image points not too close**

I.5.3.8. Lemma: Endpoints not equal. *The endpoints of $\tilde{f}\alpha$ are distinct in $\mathbb{H}^3$.*

Proof. Endpoints not equal. Suppose that the two endpoints of $\tilde{f}\alpha$ are equal in $\mathbb{H}^3$. Let the two ends of α be asymptotic to the oriented geodesics α_1 and α_2 at the positive ends of α_1 and α_2 where $\alpha_1, \alpha_2 \in \tilde{\mu}$. These ends are distinct in $\mathbb{H}^2$. Therefore, by Lemma I.5.3.7 (*Image points not too close*), $\tilde{f}\alpha_1$ and $\tilde{f}\alpha_2$ are distinct in $\mathbb{H}^3$. Therefore the endpoints of $\tilde{f}\alpha$ are distinct.

> **Endpoints not equal**

This completes the proof of the theorem.

> **Finite laminations realizable**

Recall from Theorem I.4.2.14 (*Finite laminations dense*) that finite laminations are dense in $\mathcal{GL}(S)$ in both the Chabauty topology and the Thurston topology.

I.5.3.9. Theorem: Existence of realizing structure. *Let* λ *be a maximal lamination in* S *and let* $f: S \to N$ *be a realization of* λ. *Then there is a unique hyperbolic structure* S' *on* S *and a pleated surface* $f': S' \to N$ *homotopic to* f, *with pleating locus contained in* λ.

Proof. Existence of realizing structure. We prove it first for finite laminations. Each ideal triangle in $S \backslash \lambda$ is mapped by a unique isometry to a well-defined ideal triangle in N.

Let $S_1, \ldots, S_k$ be the components of the surface obtained by removing from S all the simple closed geodesics of λ. We get (new) well-defined incomplete hyperbolic structures on each S_i, defined in such a way that we obtain pleated maps on the completion $\overline{S}_i$. S_i is constructed by starting with one ideal triangle, gluing on the next in the unique way so that we get a pleated map into N, and so on. Let α be a simple closed geodesic in λ. Then α corresponds to exactly two boundary curves α_1 and α_2 of the $\overline{S}_i$, and we have to glue α_1 to α_2. Now $\tilde{f}: \tilde{S}_i \to \mathbb{H}^3$ maps horocycles centred at one end of $\tilde{\alpha}_i$ to horospheres centred at the corresponding end of $\tilde{f}\tilde{\alpha}$ in $\mathbb{H}^3$ and similarly for $\tilde{\alpha}_2$. (Which end depends on the direction of spiralling of the lamination around α.) Extending by continuity, we get an isometric map $\tilde{\overline{S}}_i \to \mathbb{H}^3$. This induces a well-defined isometric map $\overline{S}_i \to N$. Now α_1 and α_2 map isometrically to the same closed geodesic of N. We identify α_1 to α_2 in such a way that we obtain a well-defined pleated surface $f': S' \to N$.

For a general maximal lamination, the result follows from the compactness of pleated surfaces. By lifting to a cover, we may assume that $f: S \to N$ induces an isomorphism of fundamental groups. Let μ be a minimal sublamination of S, and let $U \subset N$ be a small closed ball meeting $f\mu$. By Theorem I.4.2.10 (*Curve near geodesic*) and Theorem I.4.2.14 (*Finite laminations dense*), we may assume that we have a sequence λ_i of finite laminations, converging to λ. Since f is uniformly continuous, we may assume that there is a simple closed geodesic C_i in λ_i such that the simple closed geodesic corresponding to fC_i meets U (see Theorem I.4.2.10 (*Curve near geodesic*)). Let f_i be the unique marked pleated surface with pleating locus contained in λ_i. By Theorem I.5.2.18 (*Compactness of market pleated surfaces*), we may assume that f_i converges to a pleated surface $f': S' \to N$ (if f is a virtual fibre we must lift to a cover M such that $\pi_1(M) = f_*(\pi_1(T))$), and by Corollary I.5.2.13 (*Compactness of pleated surfaces of fixed topological type*) S' is of the same type as S. By Lemma I.5.3.2 (*Pleating locus continuous*), the pleating locus λ_0 of f_0 is contained in λ. This shows the existence of $f': S' \to N$. To prove the uniqueness of the pleated surface realizing λ, we give a more abstract version of S'. Let $\mathbb{H}^2 \to S$ be the universal cover of S. We lift λ to a lamination $\tilde{\lambda}$ of $\mathbb{H}^2$. We identify each geodesic of

$\tilde{\lambda}$ to a point and each component of $\mathbb{H}^2/\tilde{\lambda}$ to a point. (Each such component is an open ideal triangle.) The resulting quotient space of $\mathbb{H}^2$ is called P_λ. Geodesics of $\tilde{\lambda}$ give closed points of P_λ and triangular strata give rise to open points. P_λ is not Hausdorff, because a point corresponding to a triangle is in every neighbourhood of each of its three sides. $\pi_1 S$ acts on P_λ via covering translations of $\mathbb{H}^2$.

If S' is a complete hyperbolic surface of finite area and we have a homeomorphism from S to S', we obtain a lifted homeomorphism of $\mathbb{H}^2$ with itself. This homeomorphism extends to the boundary circle, and is unaltered by an isotopy of homeomorphisms between S and S'. (The lift to $\mathbb{H}^2 \cup \partial\mathbb{H}^2 \to \mathbb{H}^2 \cup \partial\mathbb{H}^2$ can be changed by composition with an element of $\pi_1 S$ on the right, or, equivalently, with an element of $\pi_1 S'$ on the left.) A geodesic is represented by an unordered pair of distinct elements of S'. A geodesic of S is, in this way, sent to a geodesic of S' and so λ can be transferred from S to a lamination λ' of S'. Since P_λ can be defined entirely in terms of pairs and triples of points in $\partial\mathbb{H}^2$, we see that P_λ is homeomorphic to $P_{\lambda'}$.

Let $\tilde{f}: \mathbb{H}^2 \to \mathbb{H}^3$ be a lift of f, a realization of λ. If α is a geodesic of $\tilde{\lambda}$, then $\tilde{f}\alpha$ is a geodesic in $\mathbb{H}^3$, and we write $\hat{f}(\alpha)$ for this geodesic. If s is a triangular component of $\mathbb{H}^2 \setminus \tilde{\lambda}$, then its boundary is mapped to an ideal triangle in $\mathbb{H}^3$, which we denote by $\hat{f}(s)$.

Now consider the subset A of $P_\lambda \times \mathbb{H}^3$, consisting of pairs (y, x) such that $x \in \hat{f}y$. It is easy to check that $A/\pi_1 S$ is Hausdorff. Let $f': S' \to N$ be a marked pleated surface homotopic to $f: S \to N$, with pleating locus λ' corresponding to $\lambda \subset S$. Then the lift $\tilde{f'}: \mathbb{H}^2 \to \mathbb{H}^3$ of f' induces a map of $\tilde{S'}$ into A and hence a map $S' \to A/\pi_1 S$. Each cusp of S only contains a finite number of geodesics of λ. From this it is easy to see that the map $S' \to A/\pi_1 S$ is a homeomorphism on the cusps. So we can apply the theorem that a bijective continuous map of a compact space to a Hausdorff space is a homeomorphism, to deduce that $S' \to A/\pi_1 S$ is a homeomorphism. The map $f': S' \to N$ factors as

$$S' \to \frac{A}{\pi_1}S \to N$$

where the second map is induced from the projection $A \to \mathbb{H}^3$. It follows that $f': S' \to N$ is unique, up to composition of an isometry of another complete hyperbolic surface S'' with S'.

Existence of realizing structure

We have shown that R_f is dense (see Theorem I.5.3.6 (*Finite laminations realizable*)); we now wish to show it is open. We need to make use of the theory of train tracks. We shall provide a few of the definitions, but the interested reader is encouraged to seek further detail in (Casson, 1983) or (Harer–Penner, 1986).

A *train track* τ on a hyperbolic surface S, is a finite graph embedded in S, such that all edges of the graph are C^1-embedded in S, all edges are tangent at any vertex (these vertices are called *switches*), and if you double any component of $S \backslash \tau$ along its (open) edges the resulting surface has negative Euler characteristic. A *train route* is a C^1-immersion $\rho \colon \mathbb{R} \to \tau$. A neighbourhood U of τ which is foliated by arcs (called *ties*) transverse to τ, such that each transverse arc meets τ in just one point (except near the switches) is called a *standard neighbourhood* of τ. A geodesic lamination λ is said to be *carried* by a train track τ if there is some standard neighbourhood U of τ such that $\lambda \subset U$ and each leaf of λ is transverse to every tie. So every leaf of the lamination is homotopic to an unique train route. (One way of obtaining such a train track is to take an ε-neighbourhood of λ, and "squeeze" it down into a train track.) Now, we simply note that the set $N(\tau)$ of all laminations carried by a given train track τ, is an open subset of $\mathcal{GL}(S)$ in the Thurston topology, and that every geodesic lamination is carried by some train track.

Suppose λ can be realized by a pleated map $g \colon \overline{S} \to N$ and suppose that λ is carried by a train track τ where $\tau \subset N_\varepsilon(\lambda)$. We choose τ so that an ε-neighbourhood of τ contains λ. By taking ε small enough we can ensure that any train track path can be represented by a sequence of long (i.e. of length greater than some $\alpha > 0$) geodesic arcs, joining each other at angles almost equal to π (we refer to points where these angles occur as bends). Notice that all switches are bends, but that we may have to insert some bends at points which are not switches. Similarly, we may construct a "train track" τ' in N which is "near" to the $g(\tau)$, in particular the bends of τ' are the images of τ's bends, which also has the above properties. The image of every train route of τ is associated to a well-defined geodesic of N, since if we examine a lift of this train route to $\mathbb{H}^3$ it has well defined and distinct endpoints by Lemma I.4.2.10 (*Curve near geodesic*).

Now suppose λ' is carried by τ, by the above λ' has a well-defined image in N which is a collection of geodesics (although it is not necessarily a geodesic lamination as the geodesics may intersect). Now as in Theorem I.5.3.9 (*Existence of realizing structure*) we may choose a sequence of finite laminations λ_i converging to λ' and prove the existence of a realizing map $f' \colon S' \to N$. Hence any lamination carried by τ is realizable, thus proving R_f is open. We summarize the above results in a theorem.

I.5.3.10. Theorem: Realizable laminations open and dense. *If f preserves parabolicity, R_f is open and dense.*

I.5.3.11. Theorem: Lamination realizable. *Given a surface T, a geometrically finite complete hyperbolic 3-manifold N, and $f: T \to N$, an incompressible map taking cusps to cusps, satisfying np and which is not a virtual fibre, then every $\lambda \in \mathcal{GL}(S)$ may be realized by a pleated surface in the homotopy class of f (i.e. $R_f = \mathcal{GL}(S)$).*

Proof. Lamination realizable. First note that if f realizes λ, then $f\lambda \subset C_N$, the convex core of N. The reason for this is that it is true for finite laminations by construction, and these are dense. Let K be the thick part of the convex core of N. Then K is compact, see Morgan–Bass (1984) for example, and every pleated surface meets K. By Theorem I.5.2.18 (*Compactness of marked pleated surfaces*), the space of marked pleated surfaces is compact. The set of finite-leaved maximal laminations is dense in the set of maximal laminations, and every maximal geodesic lamination is realizable by a pleated surface in the homotopy class of f. Since every geodesic lamination is contained in a maximal lamination, every geodesic lamination is realizable.

$\Box$

As an example of the power of the techniques in the last few sections, we state a trivial corollary of Corollary I.5.2.13 (*Compactness of pleated surfaces of fixed topological type*), and the realizability of any (*np*)-surface subgroup of a hyperbolic 3-manifold by a pleated surface. (Of course, the main applications of the above material is in Thurston's proof of his uniformization theorem.)

I.5.3.12. Corollary: Finiteness theorem. *Let S be any surface of finite area and N any geometrically finite hyperbolic 3-manifold. There are only finitely many conjugacy classes of subgroups $G \subset \pi_1(N)$ isomorphic to $\pi_1(S)$ by an isomorphism which preserves parabolicity.*

References

Abikoff, W. (1980), "The real analytic theory of Teichmüller space," *Springer Lecture Notes*, **820**.

Apanasov, B. (1985), "Cusp ends of hyperbolic manifolds," *Ann. Glob. Anal. Geom.*, **3**, 1–11.

Beardon, A.F. (1983), *The Geometry of Discrete Groups*, Springer, New York.

Bers Lipman (1960), "Quasiconformal mappings and Teichmüller's theorem," In *Analytic Functions* (eds. R. Nevanlinna *et al.*), Princeton University Press, Princeton N.J., pp. 89–119.

Casson, A. (1983), *Automorphisms of Surfaces after Nielsen and Thurston*, University of Texas at Austin, Lecture Notes by S. Bleiler.

Chabauty, C. (1950), "Limites d'ensembles et géométrie des nombres," *Bull. Soc. Math. de France*, **78**, 143–151.

Connelly, R. (1971), "A new proof of Brown's Collaring Theorem," *Proc. A.M.S.*, **27**, 180–182.

Douady, A. (1979), "L'espace de Teichmüller," *Astérisque*, **66–67**, pp. 127–137, Société Mathématique de France.

Edwards, R.D. and Kirby, R.C. (1971), "Deformations of spaces of imbeddings," *Ann. Math.*, **93**, pp. 63–68.

Epstein, D.B.A. (1983), *Isomorfismi Conformi e Geometria Iperbolica*, University of Cagliari, Lecture notes, written by G. D'Ambra.

Epstein, D.B.A. (1984), "Transversely hyperbolic 1-dimensional foliations," *Astérisque*, **116**, pp. 53–69, Société Mathématique de France.

Epstein, D.B.A. and Marden, A. *Convex Hulls in Hyperbolic Space, a Theorem of Sullivan, and Measured Pleated Surfaces*, Cambridge University Press (these proceedings).

Godement, R. (1958), *Topologie algébrique et théorie des faisceaux*, Hermann, Paris.

Gromov, M. (1980), "Hyperbolic manifolds according to Thurston & Jørgensen," In *Séminaire Bourbaki 1979/80*, Springer-Verlag.

Gromov, M. (1981a), "Groups of polynomial growth and expanding maps," *Publ. Math. I.H.E.S.*, **53**, 53–78.

Gromov, M. (1981b), *Structures Metriques pour les Varietes Riemanniennes*, CEDIC/ Fernand Nathan, Paris.

Haefliger, A. (1958), "Structure Feuilletées et Cohomologie à Valeur dans un Faisceau de Groupoides," *Comm. Math. Helv.*, **32**, 248–329.

Harer, J. and Penner, R. (1986), "Combinatorics of train tracks," *Ann. Mathematics Studies*, Princeton University Press, Princeton, N.J. (with an Appendix by N. Kuhn), 1992.

Harvey, W.J. (1977), "Spaces of discrete groups," In *Discrete Groups and Automorphic Functions* (ed. W.J. Harvey), Academic Press, London.

Hempel, J. (1976), *3-Manifolds*, Annals of Mathematics Studies, Princeton University Press, Princeton, N.J.

Jørgensen, T. (1977), "Compact 3-manifolds of constant negative curvature fibering over the circle," *Ann. Math.*, **106**, 61–72.

Kazdan, D.A. and Margulis, G.A. (1968), "A proof of Selberg's conjecture," *Math. USSR Sbornik*, **4**, 147–152.

Kerckhoff, S.P. (1983), "The Nielsen realization problem," *Ann. Math.*, **117**, 235–265.

Kobayashi, S. and Nomizu, K. (1963), *Foundations of Differential Geometry I*, Interscience.

Lok, W.L. (1984), *Deformations of locally homogeneous spaces and Kleinian groups*, Ph.D. Thesis, Columbia University.

Michael, E. (1951), "Topologies on spaces of subsets," *Trans. A.M.S.*, **71**, 152–182.

Morgan, J. and Bass, H., (1984), *Proceedings of the Smith Conjecture Symposium: Columbia University 1979*, Academic Press.

Rudin (1973), *Functional Analysis*, McGraw-Hill, New York.

Scott, P. (1983), "The Geometries of 3-manifolds," *Bull. Lond. Math. Soc.*, **15**, 401–487.

Siebenmann, L. (1972), "Deformations of homeomorphisms on stratified sets," *Comm. Math. Helv.*, **47**, 123–136.

Thurston, W.P. (1979), *The Geometry and Topology of Three Manifolds*, Princeton Lecture Notes, 1979. Reprinted *http://msri.org/publications/books/gt3m/*

Thurston, W.P. (1986), "Hyperbolic structures on 3-manifolds I: Deformations of acylindrical structures," *Ann. Math* **124** (1986), 203–246.

Thurston, W.P. (1986a), "Hyperbolic Structures on 3-manifolds II: Surface groups and 3-manifolds which fiber over the circle," mathGT/9801045.

Thurston, W.P. (1997), edited by S. Levy, *Three Dimensional Geometry and Topology*, Princeton Univ. Press, 1997.

Wang, H.C. (1972), "Topics in totally discontinuous groups," In *Symmetric Spaces* (ed. Boothby-Weiss), New York.

Weil, A. (1960), "On discrete subgroups of lie groups," *Ann. Math.*, **72**, 369–384.

Whitney, H. and Bruhat, F. (1959), "Quelques propriétés fondamentales desensembles analytiques-réels," *Comm. Math. Helv.*, **33**, 132–160.

Wielenberg, N.J. (1981), "Hyperbolic 3-manifolds which share a fundamental polyhedron," In *Riemann Surfaces and Related Topics: Proceedings of the 1978 Stony Brook Conference*, Princeton University Press, Princeton, N.J.

PART II

Convex Hulls in Hyperbolic Space, a Theorem of Sullivan, and Measured Pleated Surfaces

D.B.A. Epstein, Mathematics Institute, University of Warwick,
Coventry CV47AL, UK

A. Marden, School of Mathematics, University of Minnesota,
Minneapolis, MN 55455, USA

Introduction

Our purpose here is to give a detailed exposition of three closely related fundamental topics in the Thurston theory of hyperbolic 3-manifolds.

We work with hyperbolic 3-space $\mathbb{H}^3$ and its "visual" boundary, the sphere $S^2 = \partial \mathbb{H}^3$ at infinity. A closed subset $\Lambda \subset S^2$, not contained in a round circle, gives rise to its hyperbolic convex hull $C(\Lambda)$ in $\mathbb{H}^3$. Each component S of $\partial C(\Lambda)$ is, in essence a "ruled surface" of differential geometry, except that in general it is not smooth. Indeed, S consists of perhaps an uncountable number of geodesics called "bending lines", interspersed with at most a countable number of "flat pieces", each a sector of the hyperbolic plane whose edges are bending lines. Instead of tangent planes to S, one must work with support planes, and there is a "bending measure" which describes how these evolve over the surface. There is also a path metric on S, induced from $\mathbb{H}^3$. In terms of this, we prove (following Thurston) that, if S is simply connected, there exists an isometry $h: \mathbb{H}^2 \to S$ from the hyperbolic plane to S.

Given a simply connected component S of $\partial C(\Lambda)$ in $\mathbb{H}^3$, there is a component $\Omega(S)$ of $S^2 \setminus \Lambda$, that faces S. Chapter II.2. contains a proof of the following remarkable theorem of Dennis Sullivan (1981) there is a K-bilipschitz map (i.e. a K-quasi isometry) between S with its path metric and $\Omega(S)$ with its Poincaré metric, extending to the identity map on the common pointset boundary of S and $\Omega(S)$, where K is a universal constant *independent of* Λ. Sullivan's theorem plays an important role in Thurston's Hyperbolization Theorem for fibre bundles over the circle with fibre a surface. It is conjectured that $K = 2$, but there it is only shown that $K \leq 67$.

The proof is technically very difficult, involving some delicate averaging between S and the smooth C^1-surface S_ε. at distance ε from S. To do the averaging, we have found it necessary to develop the theory of lipschitz vector fields on non-smooth surfaces. It has also been necessary to prove an analogue on a

Riemannian manifold of the formula, well known in euclidean space, estimating the error in an approximate solution of a differential equation (see especially the first three sections of the Appendix, which may have independent interest). This represents the work of the first named author.

Under the isometry $h: \mathbb{H}^2 \to S$ constructed in Chapter II.1, a geodesic lamination $\mathcal{L}$ of $\mathbb{H}^2$ is mapped to the bending lines of S. The pull-back to $\mathbb{H}^2$ of the bending measure, is a transverse measure on $\mathcal{L}$. The purpose of Chapter II.3. is to greatly generalize this situation to an arbitrary complex valued measure μ defined with respect to an arbitrary geodesic lamination $\mathcal{L}$ of $\mathbb{H}^2$. From this data we construct a piecewise isometry $h: \mathbb{H}^2 \to S$ on to a "measured pleated surface" S in $\mathbb{H}^3$. Corresponding to h is a "bending cocycle" for S that describes how S is the envelope of its "support" planes. In the special cases that μ is real and positive, h is the corresponding left earthquake map, and when $\mu = i\beta$ is pure imaginary, with $\beta \geq 0$, h is pure bending, as in the convex hull case of Chapter II.1. Thus our work is, in particular, a detailed exposition of these cases.

If both the transverse measure μ and the lamination $\mathcal{L}$ are invariant under a fuchsian group G, then there is a homomorphism $\theta: G \to PSL\,(2, \mathbb{C})$, so that h is θ-equivariant. If now μ is replaced by $t\mu$, $t \in \mathbb{C}$, one gets a one-parameter family of holomorphic deformations of G. The first and second derivatives of these deformations at $t = 0$ can be expressed by closed formulas, which are related to certain important formulas found earlier by Kerckhoff and Wolpert.

We have taken a long time in writing this paper, and we have had the advantage of conversations with many people, in fact with so many that it is not really possible to list them all. Bill Thurston was very helpful in talking about the overall strategy of the proof. Also Thurston (1979) has been an invaluable reference. Adrien Douady discussed a 1-dimensional analogue of Sullivan's theorem with us at some length. Bill Goldman was one of the first people to understand Thurston's bending construction, and has shared his knowledge with us.

Our style is expository and possibly long winded. Anyone preferring a more rapid treatment is referred to the works of Sullivan and Thurston.

Chapter II.1

Convex Hulls

II.1.1. Introduction

In this chapter we discuss the geometric properties of convex subsets of $\mathbb{H}^3$. In particular we discuss the hyperbolic convex hull of a closed subset of S^2, regarded as the boundary of $\mathbb{H}^3$. We show that, with respect to the metric induced by the length of rectifiable paths, the boundary of such a convex hull is a complete hyperbolic 2-manifold. Following (Thurston, 1979) we show that we obtain a measured lamination from the boundary, where the measure tells one how much the surface is bent.

II.1.2. Hyperbolic convex hulls

We will consider the open unit ball in $\mathbb{R}^n$ as the hyperbolic space $\mathbb{H}^n$, giving it the Poincaré metric $2dr/(1-r^2)$, where r is the euclidean distance to the origin and dr is the euclidean distance element. Hyperbolic isometries act on the closed unit ball conformally. Thus we get hyperbolic geometry inside the ball and conformal geometry on the boundary S^{n-1}. We are mainly interested in the case $n = 3$, though we will also need to discuss $n = 2$ from time to time. We denote the closed unit ball, with its conformal structure and with the hyperbolic structure on its interior, by $\mathbb{B}^n$.

II.1.2.1. Definition. A non-empty subset X of $\mathbb{B}^n$ is said to be *convex* or, more precisely, *hyperbolically convex* if, given any two points of X, the geodesic arc joining them also lies in X.

In particular, we will regard the empty set as convex. We allow X to contain points of the boundary, and allow geodesic arcs to have endpoints on the boundary. The closure of a convex subset of $\mathbb{B}^n$ is also convex. Our convex sets will normally be closed.

121

A hyperbolic subspace in $\mathbb{B}^n$ is a euclidean sphere or plane which is orthogonal to the boundary S^{n-1}. If the subspace has codimension 1, the corresponding sphere or plane splits $\mathbb{R}^n$ into two components.

II.1.2.2. Definition. By a *closed half space* we mean the intersection of the closed unit ball $\mathbb{B}^n$ with the closure of one of these two components. Given a closed convex set X, a *supporting half space* is a closed half space which contains X and is such that the boundary of the half space meets X. A *support plane* P at a point $x \in \mathbb{H}^n \cap X$ is a hyperbolic subspace of codimension 1, containing x, which is the boundary of a supporting half space.

II.1.2.3. Orientation of support planes. If X has non-empty interior (i.e. not contained in a hyperbolic subspace) then a support plane has a natural orientation, with the negative side containing X, and the positive side disjoint from X. We will assume such an orientation for all support planes in this chapter (provided the interior of the convex set is non-empty).

II.1.2.4. Projective model. A more familiar picture of convexity is obtained by using the Klein projective model, instead of the Poincaré disk model just described. In the Klein model a closed half space is just the intersection of the unit ball with a euclidean closed half space, and a geodesic is an interval on a straight line. Convexity in this model is just convexity in the usual euclidean sense. A number of results which we are about to prove could be proved by appealing in this way to standard results about convex subsets of euclidean space. However, we prefer to give direct proofs based on the hyperbolic structure.

II.1.3. The nearest point retraction

Given a closed non-empty convex subset X of $\mathbb{B}^n$, we will follow Thurston and define a canonical retraction $r: \mathbb{B}^n \to X$. If X consists of a single point in S^{n-1}, then r is the unique retraction onto that point. So assume that X contains at least one point of $\mathbb{H}^n$. First note that, by convexity, if a hyperbolic ball (i.e. a finite metric ball with respect to the hyperbolic metric) has interior distinct from X, its boundary can meet X at most once. The same thing is true for a horoball.

If $x \in X$, we define $rx = x$. If $x \in \mathbb{H}^n \setminus X$, we find the largest hyperbolic ball centred at x, whose interior is disjoint from X. We define rx to be the unique point of contact of this ball with X. Similarly, if x lies in $S^{n-1} \setminus X$, we find the largest horoball centred at x, whose interior is disjoint from X, and take rx

II.1.3.1.

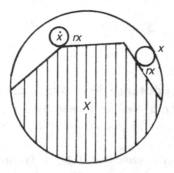

Figure II.1.3.1 *The shaded set is the convex subset X. The construction of the point rx is shown for two points x, one on $\partial\mathbb{H}^n$, and one inside $\mathbb{H}^n$. The largest open ball or horoball, centred at x and disjoint from X, has closure which meets X at the unique point rx.*

to be the unique point of contact. We denote the largest such ball or horoball by $B(x)$.

II.1.3.2. Lemma: Continuous r. *The retraction r is continuous. Given a point $x \in \mathbb{B}^n \backslash X$, there is a support plane P_x, which separates x from $X \backslash P_x$.*

Proof. If $x \in X \cap \mathbb{H}^n$, then continuity of r at x is trivial. If $x \in X \cap S^{n-1}$, let $L \subset X$ be a geodesic arc with one endpoint at x. If y is near x in the euclidean topology on $\mathbb{B}^n$, then either $y \in X$, so that $ry = y$ is near $rx = x$, or $y \notin X$, in which case $B(y)$ is small in the euclidean topology, because it has to miss L. So r is continuous at all points of X.

Now suppose that $x \notin X$. Let P_x be the tangent plane to $B(x)$ at rx. By convexity, we see that P_x separates $B(x)$ from X. Let $\{x_i\}$ be a sequence converging to x in the euclidean topology on $\mathbb{B}^n$. Without loss of generality, let $rx_i \in X \cap \partial B(x_i)$ converge to $y \in X$. We will prove that $y = rx$. We may suppose without loss of generality that the sequence $B(x_i)$ converges, when considered as a sequence of euclidean balls. Let B be the limit ball or point. B has positive radius, because each $B(x_i)$ intersects P_x. Since the interior of each $B(x_i)$ is disjoint from X, the interior of B must be disjoint from X. The centre of B as a hyperbolic ball or horoball is equal to x, and B contains the point $y \in X$ in its boundary. Therefore B is the maximal ball or horoball $B(x)$ centred on xv, and y is equal to rx.

□

II.1.3.3. Note that $r^{-1}(S^{n-1}) = X \cap S^{n-1}$. If we restrict to

$$r: \mathbb{B}^n \backslash (X \cap S^{n-1}) \to X \backslash S^{n-1},$$

we get a proper map.

II.1.3.4. Lemma: r is distance decreasing. *Let $X \subset \mathbb{B}^n$ be a non-empty closed convex set and let $r: \mathbb{B}^n \to X$ be the nearest point retraction defined above. If x, $y \in \mathbb{H}^n$, then $d(r(x), r(y)) \leq d(x, y)$, where d refers to the hyperbolic metric.*

Proof. We may as well assume that $r(x)$ and $r(y)$ are distinct. Let L be the line containing these two points, oriented from rx to ry, and let A be the closed subinterval of L joining them. Then $A \subset X$. Let $\pi: \mathbb{H}^n \to L$ be orthogonal projection. We must have $\pi(x) \leq r(x) < r(y) \leq \pi(y)$, since rx is the nearest point of X to x, and is therefore the nearest point of A to x, and similarly for y. The result follows since π is distance decreasing.

II.1.3.5.

Figure II.1.3.5 *The points πx and πy are the nearest points to x and y on the line containing rx and ry.*

$\Box$

The following lemma seems to be a standard result. However the elegant proof presented here, due to Brian Bowditch, seems not to be well known.

II.1.3.6. Lemma: Smooth epsilon surface. *Let $f: \mathbb{H}^n \backslash X \to (0, \infty)$ be the distance of a point from X. Then f is C^1, and its gradient vector is the unit vector at x, along the geodesic from rx to x, pointing away from rx.*

Proof. (Bowditch). The proof depends on the following observation. Suppose that g_1 and g_2 are two C^1 real valued functions, whose values and derivatives agree at x_0, and suppose g is another function, such that $g_1 \leq g \leq g_2$ near x_0. Then g is continuous and differentiable at x_0. In order to apply this observation, let P be the plane through rx, orthogonal to the geodesic from rx to x. Let $g_1(y)$ be the distance from y to P. Let $g_2(y)$ be the distance from y to rx. Then g_1 and g_2 are analytic functions, whose values at x are both equal to $d(rx, x)$. The derivatives at x are also equal – each is given by inner product with the vector given in the statement. The above observation can be applied, since $g_1 \leq f \leq g_2$. Finally note that the gradient vector field of f described in the statement is continuous, because r is continuous. (In fact the vector field is lipschitz, as will be shown later in Section II.2.11 (*Three orthogonal fields*).)

$\square$

II.1.4. Properties of hyperbolic convex hulls

We are now in a position to prove the main characterization of closed convex subsets of $\mathbb{B}^n$.

II.1.4.1. Proposition: Intersection of half spaces. *A closed non-empty subset X of $\mathbb{B}^n$ is convex if and only if it is the intersection of all its supporting half spaces.*

Proof. Clearly, the intersection of closed half spaces is convex. To prove the converse, suppose X is convex and that $x \notin X$. We need to find a half space which contains X but not x. But such a half space has already been found in Lemma II.1.3.2 (*Continuous r*).

$\square$

II.1.4.2. Corollary: Countable intersection. *A closed non-empty convex subset X of $\mathbb{B}^n$ is the intersection of a countable family of supporting half spaces.*

Proof. Let $\{x_i\}$ be a countable dense subset of $\mathbb{B}^n \backslash X$. For each i, we choose a supporting half space which does not contain x_i. The intersection of these half spaces is equal to X.

$\square$

The next result is standard in euclidean space.

II.1.4.3. Theorem: Standard convexity properties. *Let X be a closed convex subset of $\mathbb{B}^n$, and suppose that X is not contained in any proper hyperbolic subspace. We denote by ∂X the boundary of $X \cap \mathbb{H}^n$ in $\mathbb{H}^n$. Let x_0 be a fixed interior point of X. Given a unit vector v at x_0, let R_v^* be the ray from x_0 with tangent vector v:*

(1) X has non-empty interior.
(2) R_v meets ∂X in at most one point.

(3) *Let $f(v) \in (0, \infty]$ be the (hyperbolic) length of the finite or infinite ray $R_v \cap X$. Then f is a continuous function of v.*

(4) *$X \cap \mathbb{H}^n$ is a manifold with boundary ∂X and with interior homeomorphic to an open disc.*

(5) *Radial projection from x_0 maps ∂X onto $S^{n-1} \setminus X$ homeomorphically.*

II.1.4.4.

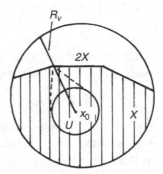

Figure II.1.4.4   *Radical projection from a point x_0 in the interior of X. X is shaded. The dotted lines indicate the convex hull of the ball U and the ray R_v.*

Proof.   Suppose we have found, by induction on k, points $x_0, x_1, \ldots, x_k \in X$, which do not lie in any $(k-1)$-dimensional hyperbolic subspace. They do lie in a k-dimensional hyperbolic subspace V_k. If $k < n$, then there is a point $x_{k+1} \in X \setminus V_k$. Any hyperbolic subspace W, which contains $x_0, \ldots, x_{k+1}$, must contain V_k, otherwise $W \cap V_k$ would contain $x_0, \ldots, x_k$ and have dimension less than k. Therefore the dimension of W must be at least $k+1$, and the induction step is completed. Taking $k = n$, we obtain a hyperbolic n-simplex in X, and this has non-empty interior.

Let $U \subset X$ be an open ball with centre x_0. By considering the convex hull of U and R_v, we see that the interior of X contains R_v, with the possible exception of one of its endpoints. Therefore the only point of R_v which could be in the boundary of X is the endpoint of R_v.

To see that f is continuous, note that examination of the convex hull of R_v and U proves that f is lower semi-continuous. To prove upper semi-continuity, let v_i be a sequence of unit tangent vectors at x_0 converging to v, and suppose that the ray R_i corresponding to v_i has length greater than t for each i. Since X is closed, it follows that $f(v) \geq t$.

Parts (4) and (5) now follow because we can define a homeomorphism which stretches each ray R_v to a ray which reaches the boundary sphere of hyperbolic space.

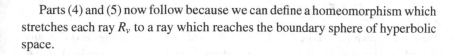

II.1.4.5. Lemma: Support planes exist. *Let X be a closed convex subset of $\mathbb{B}^n$. Let x be a point in the frontier of X in $\mathbb{B}^n$.*
Then x lies in some support plane.

Proof. Let $x_i \notin X$ be a sequence of points in $\mathbb{B}^n$ converging to x. By Proposition II.1.4.1 (*Intersection of half spaces*), there is, for each i, a half space H_i which contains X and does not contain x_i. The sequence of half spaces must converge to some half space H, which contains X. The boundary of H must contain x.

II.1.4.6. Lemma: Distance convexity. *Let α and γ be distinct geodesics in $\mathbb{H}^3$, parametrized proportionally to arclength by a parameter $t \in \mathbb{R}$. Then $d(\alpha(t), \gamma(t))$ is a strictly convex function.*
Note that this result is true, even if the speeds of the geodesics are unequal. This is a standard result in the differential geometry of spaces with non-positive curvature. A proof for the hyperbolic plane is given in (Douady, 1979). Douady's proof works equally well in higher dimensions.

II.1.4.7. Lemma: Neighbourhoods of convex sets. *Let X be a closed convex set in $\mathbb{H}^n$ and let $\varepsilon > 0$. Let X_ε be the set of points $x \in \mathbb{H}_n$ such that $d(x, X) \le \varepsilon$, where d is hyperbolic distance. Then X_ε is strictly convex (i.e. any geodesic arc in X_ε meets ∂X_ε at most at its endpoints).*

Proof. Let $x, y \in X_\varepsilon$ be distinct and let γ be the geodesic joining them. We wish to show that γ lies entirely inside X_ε. Let α be the geodesic joining rx to ry. Then α lies in X. By Lemma II.1.4.6 (*Distance convexity*), $f(t) = d(\alpha(t), \gamma(t))$ is a convex function of t. But $f(0)$ and $f(1)$ are each bounded above by ε. Therefore $f(t) \le \varepsilon$ for $0 \le t \le 1$, and X_ε is convex. If $f(t) = \varepsilon$ for some t with $0 < t < 1$, then $d(\alpha(t), \gamma(t)) = \varepsilon$ for all t. It follows that there is a single geodesic containing both α and γ. Examining the possible configurations in this single geodesic, we obtain a contradiction.

II.1.5. Metric on convex hull boundary

We have shown in Theorem II.1.4.3 (*Standard convexity properties*) that any closed convex set X in $\mathbb{H}^n$ is a manifold with boundary, provided it has non-empty interior. In particular, the components of the boundary are path connected.

II.1.5.1. Lemma: Metric on convex hull boundary.   *Let X be a closed convex subset of $\mathbb{H}^n$. Let x, y be two points in the same component of ∂X. Then there is a rectifiable path in ∂X joining x to y.*

Proof.   Let p_0 be a point in the interior of X. The space $\mathbb{H}^n \backslash X$ is homeomorphic to $\partial X \times (0, \infty)$, where the first coordinate is where a ray from p_0 crosses ∂X, and the second coordinate is the distance along the ray from ∂X. Let C be the component of ∂X containing x and y. The corresponding component of $\mathbb{H}^n \backslash X$ is homeomorphic to $C \times (0, \infty)$. This space is clearly connected by rectifiable paths.

In order to connect x and y by a rectifiable path, we push them out a little by a radial push from p_0 to points x' and y'. Then we join x' to y' by a rectifiable path in $\mathbb{H}^n \backslash X$. Finally we apply the nearest point retraction r. Under r, any rectifiable path is mapped to a rectifiable path, because r is distance decreasing by Lemma II.1.3.4 (*r is distance decreasing*).
                                                                              ⬚

It follows from Part I that any two points in the same component of ∂X can be connected by a shortest rectifiable path in ∂X. However that result will not be used in this chapter.

II.1.5.2. Definition.   The lemma just proved allows us to define the *intrinsic metric d_S* on any component S of ∂X, where X is a closed convex subset with non-empty interior. This metric is just the infimum of lengths of rectifiable paths.

Note that it is equal to the infimum of lengths of rectifiable paths which are disjoint from X except at their endpoints. We will compare d_S to d, the distance in $\mathbb{H}^n$.

We will work with the following open subsets of $\mathbb{H}^n$, which are well adapted to the investigation of radial projection from a point.

II.1.5.3. Definition.   Let v and x be distinct points in $\mathbb{H}^n$ and let ε and δ be small positive numbers. An *open shell $S(x, v, \varepsilon, \delta)$* is the intersection of the

spherical shell

$$\{y \in \mathbb{H}^n : |d(v,y) - d(v,x)| < \varepsilon\}.$$

with the hyperbolic cone with cone point at v, axis the ray xv and vertex angle equal to δ. We further assume that ε and δ are sufficiently small so that any geodesic arc between two points of the shell lies at a distance from v which is greater than $d(x,v)/2$. An open shell has a *radial boundary* where the angle is exactly δ, and a *spherical boundary*, which consists of two components. We call these two components the *inner* and *outer* spherical components. We will say that an open shell $S(x, v, \varepsilon, \delta)$ is *adapted* to a closed convex set X if the interior of X contains v and the inner spherical boundary, and if the outer spherical boundary is disjoint from X. We also say that X is adapted to the open shell. For a fixed x and v, the open shells form a basis for the neighbourhoods of x.

II.1.5.4.

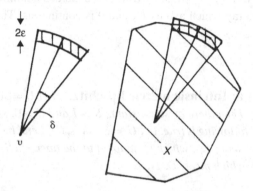

Figure II.1.5.4 *The open shell $S(x, v, \varepsilon, \delta)$ and a shell adapted to a closed convex subset X.*

II.1.5.5. Lemma: Bounded angle.
Let $S(x, v, \varepsilon, \delta)$ be an open shell. Let $\delta' < \delta$. Then there is a positive number η with the following property. Let X be a convex set in $\mathbb{H}^n$ such that the spherical shell is adapted to X. Let $x_1, x_2 \in \partial X \cap S(x, v, \varepsilon, \delta')$ and let z lie on the geodesic arc joining them.
Then the angle between vz and $x_1 x_2$ is at least η.

Proof. Let U be the intersection of the open disc, whose centre is v and radius is $d(x, v)/4$, with the open cone, whose apex is v, whose axis is vx, and which has cone angle δ. The $U \subset \mathrm{int} X$. By part (2) of Theorem II.1.4.3 (*Standard convexity properties*), a geodesic through x_1 and x_2 must be disjoint from U. Recall from Definition II.1.5.3 (*Open shell*) that $d(z, v) \geq d(x, v)/2$. So for each

II.1.5.6.

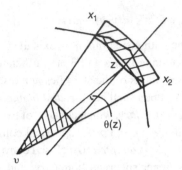

Figure II.1.5.6   *The point z lies on the geodesic arc between x_1 and x_2. The line through z at angle $\theta(z)$ to zv just misses U.*

point z such that $d(x,v)/2 \le d(z,v) \le d(x,v)+\varepsilon$ there is a minimum angle $\theta(z)$ between zv and a ray which avoids U, and θ is continuous. The lower bound for θ gives us η.

$\square$

II.1.5.7. Lemma: Intrinsic metric lipschitz.   *Let $U = S(y, v, \varepsilon, \delta)$ be an open shell in $\mathbb{H}^n$. Then there exist constants $K > 1$ and $\delta' > 0$, such that $\delta > \delta'$, and such that the following is true. Let $U' = S(y, v, \varepsilon, \delta')$. Let X be any non-empty closed convex subset of $\mathbb{H}^n$, which is adapted to the open shell U. Let S be the component of ∂X which meets U. Then*

$$\frac{d_S(p,q)}{K} \le d(p,q) \le d_S(p,q)$$

for all $p, q \in U' \cap S$. Furthermore, the shortest d_S-path between any two points of $U' \cap S$ lies in $U \cap S$.

Proof.   We have $d|S \le d_S$, so we only need to prove $d_S|U' \cap S \le K d|U' \cap S$. Let y_1 be the midpoint of the geodesic from v to y, and let D be an $(n-1)$-dimensional disk, centred at y_1, orthogonal to γ and contained in the interior of X. We suppose that D fits neatly into the cone with apex v and "base" U, touching the surface of the cone. Then the map $\rho: S \cap U \to D$, defined by taking $\rho(z)$ to be the unique point in D on the geodesic from v passing through z, defines a homeomorphism.

Let U_1 be the open shell $S(x, v, \varepsilon, \delta/2)$. From Lemma II.1.5.5 (*Bounded angle*), we deduce that, for some constant $K_1 > 0$, if $y_1, y_2 \in U_1 \cap S$, then ρ

induces a K_1-bilipschitz map from $[y_1, y_2]$ to $\rho[y_1, y_2]$. Here K_1 depends only on the relative positions of U and U_1. The word "bilipschitz" in the preceding sentence refers to the metric d induced from $\mathbb{H}^3$ in both domain and range. This shows that $\rho \colon (S \cap U_1, d) \to (D, d)$ is K_1-bilipschitz.

Let $\varepsilon_1 > 0$. Given a rectifiable path α in $S \cap U_1$ from y_1 to y_2, let

$$\alpha(t_0) = y_1, \alpha(t_1), \ldots, \alpha(t_N) = y_2$$

be points such that

$$\text{length } \alpha - \varepsilon_1 < \sum_{i=1}^{N} d(\alpha(t_{i-1}), \alpha(t_i)) \le \text{length } \alpha.$$

Then

$$\frac{\text{length } \alpha - \varepsilon_1}{K_1} \le \sum_{i=1}^{N} d(\rho\alpha(t_{i-1}), \rho\alpha(t_i)) \le K_1 \text{ length } \alpha.$$

Hence $\rho\alpha$ is rectifiable with

$$\text{length } \alpha / K_1 \le \text{length } \rho\alpha \le K_1 \text{ length } \alpha.$$

This shows that $\rho \colon (S \cap U_1, d_{S \cap U_1}) \to (D, d)$ is K_1-path bilipschitz. Combined with what we already know, this shows that $(S \cap U_1, d_{S \cap U_1})$ and $(S \cap U_1, d)$ are K-path bilipschitz related, with $K = K_1^2$.

It remains to choose $\delta' < \delta/2$, so that $d_{S \cap U_1}$ agrees with d_S on $S \cap U'$. We choose δ' small enough to ensure that the round disc $\rho(U')$ in D has diameter less than $2r/K_1$, where r is the distance in $\mathbb{H}^3$ of the radial boundary of U' to the radial boundary of U_1. Note that δ' is chosen independently of X. Given two points $z_1, z_2 \in S \cap U'$, we take the geodesic path in D which joins ρz_1 to ρz_2. The path has length less than $2r/K_1$. The inverse image in $S \cap U'$ of this path has length less than $2r$. Therefore $d_S(z_1, z_2) < 2r$. The definition of r then ensures that any short path in S from z_1 to z_2 will not exit from U_1. This completes the proof.

$\square$

II.1.5.8. Corollary: Same topology. *Let X be a closed convex subset of $\mathbb{B}^n$ and let S be a component of ∂X. Then d_S induces the subspace topology on S.*

II.1.6. Hyperbolic convex hulls

II.1.6.1. Definition. Any closed subset Λ of $\mathbb{B}^n$ has a *hyperbolic convex hull* which we denote by $\mathcal{C}(\Lambda)$. This is the smallest closed convex subset of $\mathbb{B}^n$ which contains Λ.

It is the intersection of all closed hyperbolic half spaces containing Λ. Note that $\mathcal{C}(\Lambda) \cap S^{n-1} = \Lambda \cap S^{n-1}$.

The notation $\partial\mathcal{C}(\Lambda)$ will denote that part of the boundary of $\mathcal{C}(\Lambda)$ which lies in the open unit n-ball, $\mathbb{H}^n$.

II.1.6.2. Lemma: Shape of faces. *Let X be a closed non-empty subset of $\mathbb{B}^n$, and let $\mathcal{C}(X)$ be its closed hyperbolic convex hull. Let P be a support plane for $\mathcal{C}(X)$. (We will take P to be a closed subset of $\mathbb{B}^n$.) Then $P \cap \mathcal{C}(X) = \mathcal{C}(P \cap X)$.*

Proof. First note that $P \cap X$ cannot be empty. For then compactness of $\mathbb{B}^n$ would show that the closed supporting half space defined by P could be moved into its own interior, and remain a half space which contains X. But, as a support plane, P must meet $\mathcal{C}(X)$.

Clearly $\mathcal{C}(P \cap X) \subset P \cap \mathcal{C}(X)$, since the right-hand side is closed and convex. Suppose $x \in P \backslash \mathcal{C}(P \cap X)$. Then there is a closed half space H of P, which contains $P \cap X$ but not x. It follows immediately from Lemma II.1.4.5 (*Support planes exist*) that there is an $(n-2)$-dimensional hyperbolic subspace Q of P, which separates x from $P \cap X$ in P, and which is disjoint from $P \cap X$. Since P is a support plane for $\mathcal{C}(X)$, at least one of the two components of $\mathbb{H}^n \backslash P$ contains no point of X. It follows that a slight rotation of P about Q will give a plane P', which is disjoint from X, and which separates x from X. Hence $x \notin \mathcal{C}(X)$. This proves that $\mathcal{C}(P \cap X) = P \cap \mathcal{C}(X)$.

$\square$

II.1.6.3. Corollary: Shape of facets

(a) *Let Λ_0 be a closed subset of S^1 containing at least three points. Then $\mathbb{H}^2 \cap \mathcal{C}(\Lambda_0)$ is a 2-dimensional submanifold of $\mathbb{H}^2$, bounded by geodesics, and each boundary geodesic has both its endpoints in Λ_0.*

(b) *Let Λ be a closed subset of S^2, which is not contained in a round circle. Let P be a support plane for $\mathcal{C}(\Lambda)$. Then $P \cap \mathcal{C}(\Lambda)$ is either a geodesic with both endpoints in Λ, or has the form explained in (a).*

Proof. Suppose x is in $\partial\mathcal{C}(\Lambda)$. Let P be a support plane containing x (cf. Lemma II.1.4.5 (*Support planes exist*)). By Lemma II.1.6.2 (*Shape* *of faces*), $C(P \cap \Lambda) = P \cap C(\Lambda)$. Since $P \cap C(\Lambda) \neq \emptyset$, we have $P \cap \Lambda \neq \emptyset$. Now $P \cap \Lambda$ must contain more than one point, because of the existence of $x \in C(P \cap \Lambda)$. In case (a), P is 1-dimensional and both endpoints must lie in Λ. In case (b), $\Lambda_0 = P \cap \Lambda$ either contains two points, in which case $P \cap C(\Lambda)$ is a geodesic, or $P \cap \Lambda$ contains more than two points, in which case its shape can be understood from case (a).

$\square$

Throughout this chapter, the case we will be interested in is $\mathbb{H}^n = \mathbb{H}^3$ and $\Lambda \subset S^2$, and we will henceforth assume this to be the case unless otherwise stated. By Theorem II.1.4.3 (*Standard convexity properties*), the set $C(\Lambda)$ will have empty interior in $\mathbb{B}^3$ if and only if Λ is contained in some round circle in the boundary 2-sphere S^2 of $\mathbb{H}^3$. To exclude uninteresting cases, we will always assume that Λ does not lie in any such circle, unless there is specific indication to the contrary.

We will also assume that $\Lambda \neq S^2$.

II.1.6.4. Definition. If $P \cap C(\Lambda)$ is a line, it is called a *bending line*. Otherwise the interior of $P \cap C(\Lambda)$ in P is called a *flat piece*. The boundary geodesics of a flat piece are called *edges*. An edge is also called a bending line (there may or may not be a support plane whose only intersection with $C(\Lambda)$ is this edge). We use the term *facet* to denote either a bending line or a flat piece. Note that a facet does not contain its boundary edges.

Thus each point of $\partial C(\Lambda)$ lies on a bending line or in a flat piece. These facets are mutually disjoint.

If $p \in \partial C(\Lambda)$ is in the interior of a flat piece, then it has exactly one support plane. If l is a bending line, and P is a support plane at some $p \in l$, then $P \supset l$, and either $P \cap C(\Lambda) = C(P \cap \Lambda) = l$ or l is a boundary edge in the 2-dimensional facet $P \cap C(\Lambda)$. Let $\Sigma(l)$ denote the set of support planes at l. Recall that support planes are oriented – we supply each plane with the unit normal vector pointing away from the side on which $C(\Lambda)$ lies (cf. Section II.1.2.3 (*Orientation of support planes*)). Note that $\Sigma(l)$ cannot contain antipodal planes, since Λ is not contained in a round circle. Therefore any two elements of $\Sigma(l)$ divide $\mathbb{B}^3$ into four quadrants, only one of which contains Λ. Let $S(l)$ be the circle of oriented planes containing l. If P_1 and P_2 are both support planes for $C(\Lambda)$ which contain l, let A be the shortest arc in $S(l)$ with endpoints P_1 and P_2. It is immediate that $A \subset \Sigma(l)$. It follows that $\Sigma(l)$ is either a point or a closed arc in the circle $S(l)$. Let P_1 and P_2 be the two extreme planes (possibly $P_1 = P_2$). If l is oriented, we talk of these as the *left* and *right* extreme support planes for l.

II.1.6.5. Definition.   The *bending angle* at *l* is the angle between P_1 and P_2 (i.e. between their normal vectors).

We have $0 \leq \theta < \pi$. The case $\theta = 0$ corresponds to $P_1 = P_2$. If we regard $\mathcal{C}(\Lambda)$ as "inside", then θ is the exterior dihedral angle between the extreme planes.

II.1.7. Limits of lines and planes

II.1.7.1.

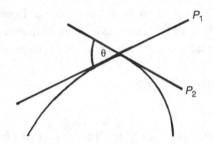

Figure II.1.7.1   *The bending angle θ between support planes P_1 and P_2.*

II.1.7.2. Lemma: Limits of bending lines.   *Let $\Lambda \subset S^2$. Let l_n be a sequence of bending lines in $\partial \mathcal{C}(\Lambda)$, and let $x_n \in l_n$ be a sequence of points converging to $x \in \partial \mathcal{C}(\Lambda)$. Then x lies on a bending line l, and l is the limit of the l_n.*

Proof.   Clearly x cannot lie in (the interior of) a flat piece. We prove by contradiction that the limit of the l_n must be l. For suppose not. By taking a subsequence, we may assume that l_n converges to some geodesic $m \neq l$. Then $x \in m \subset \partial \mathcal{C}(\Lambda)$. But then x lies on two distinct lines in $\partial \mathcal{C}(\Lambda)$, and therefore lies in (the interior of) a flat piece, which is impossible.

$\square$

If we orient a bending line, we can talk about its "left" side and its "right" side on $\partial \mathcal{C}(\Lambda)$. It is not possible to orient the bending lines coherently in general, and so this language is restricted to a small neighbourhood of the bending line.

II.1.7.3. Lemma: Limits of support planes.   *Let x lie on a bending line b, which we orient. Let $x_n \in \partial \mathcal{C}(\Lambda)$ be a sequence of points converging to x from*

*the left. Let the extreme support planes at b be P_l on the left and P_r on the
right. Let P_n be a support plane at x_n. Then P_n converges to P_l. Similarly, if x_n
converges to x from the right, then P_n converges to P_r.*

Proof. We can assume that the oriented planes P_n converge to a support plane
P_∞. Note that P_∞ contains x and therefore must contain the bending line b. We
claim that the oriented plane P_∞ is equal to P_l. Suppose not, we may assume
that each P_n intersects P_l in a line near b. Let H_n be the half of P_n lying on the
same side of P_l as $\mathcal{C}(\Lambda)$. Then H_n converges to H, the half plane of P_∞ lying
below P_l. But this gives rise to the wrong orientation of P_∞, and so the result
is proved by contradiction.

$\square$

II.1.8. Ridge lines

II.1.8.1. Definition. We define a *ridge line* to be the intersection of two
distinct support planes.

II.1.8.2. Lemma: Disjoint geodesics. *Let α be a fixed geodesic in $\mathbb{H}^2$
and let α_n be a sequence of geodesics, each disjoint from α. Let $x_n \in \alpha_n$ be a
sequence of points converging to $x \in \alpha$. Then α_n converges to α.*

Proof. We prove this by contradiction. We may suppose that α_n converges
to a geodesic β which is distinct from α. But then $x \in \beta$, and so α and β meet
transversely. It then follows that α_n meets α for n large. This is a contradiction.

$\square$

II.1.8.3. Lemma: Ridge lines exist. *Suppose $x \in \partial\mathcal{C}(\Lambda)$ lies on a bending
line l. Then x has a neighbourhood $U \subset \partial\mathcal{C}(\Lambda)$, such that if two bending lines
l_1 and l_2 meet U, then any support plane to l_1 meets any support plane to l_2.
Moreover, if W is a neighbourhood of x in $\mathbb{H}^3$, we may choose U small enough
so that the intersection of the support planes meets W. Finally, if a neighbourhood N of l in the space of geodesics (see Definition II.1.14.2* (Convergence of
laminations definition)*) is preassigned, then, by taking U small enough, we may
assume that any ridge line, which is formed by the intersection of two distinct
support planes at points of U, lies in N.*

Proof. We will leave the final statement to the end of the proof. We prove the other statements by contradiction, assuming the support planes do not meet in W. Then there would be sequences x_n and y_n of points in $\partial\mathcal{C}(\Lambda)$ converging to x, and support planes P_n at x_n and Q_n at y_n, such that $P_n \cap Q_n \cap W = \varnothing$. Without loss of generality, we may assume that the sequences P_n and Q_n tend to support planes P and Q, respectively, and the limit planes P and Q contain l. P and Q cannot meet transversely, for otherwise P_n and Q_n would intersect in W for large n. Therefore P and Q are equal as planes, though their orientations could be different.

Let $v_0 \in \operatorname{int} \mathcal{C}(\Lambda)$. Then the ray from v_0 to x_n passes through P_n in the positive direction (see Section II.1.2.3 (*Orientation of support planes*)). Therefore the ray from v_0 to x passes through P in the positive direction. Similarly for Q. Therefore the orientations of P and Q are equal, so $P = Q$. Since $P_n \cap W$ and $Q_n \cap W$ are disjoint, they divide W into three components. Suppose, without loss of generality, that $Q_n \cap W$ is on the positive side of P_n (i.e. on the opposite side from $\mathcal{C}(\Lambda)$). Then $y_n \in Q_n \cap W \cap \mathcal{C}(\Lambda)$, which is impossible. This contradiction proves all of the lemma, except for the last statement.

Now let us prove the result about the neighbourhood N of l, once again by contradiction. We can now assume that P_n meets Q_n near x, and that the limits P and Q exist and contain l. Replacing a flat piece by one of its edges if necessary, we may assume that x_n and y_n lie on bending lines b_n and c_n, respectively. By Lemma II.1.7.2 (*Limits of bending lines*), b_n and c_n are near l.

If $l_n = P_n \cap Q_n$ meets b_n, then Q_n meets b_n. But Q_n is a support plane and $b_n \subset \mathcal{C}(\Lambda)$. Hence $b_n \subset Q_n$. Also $b_n \subset P_n$ by definition. Therefore $l_n = b_n$ and so l_n is near l. It follows from this argument that we may assume that l_n is disjoint from b_n. But then l_n is near b_n by Lemma II.1.8.2 (*Disjoint geodesics*), since both geodesics pass near x. Therefore l_n is near l. This contradiction completes the proof.

$\square$

II.1.9. The roof

Given a point x in a bending line l, we define U to be a spherical shell adapted to $\mathcal{C}(\Lambda)$. Such a neighbourhood has the advantage that its intersection with any bending line is an interval. If l_1 and l_2 are distinct bending lines which meet U, they bound an open "strip" in $U \cap \partial\mathcal{C}(\Lambda)$. Two such lines together with support planes P_1 at l_1 and P_2 at l_2 define a *roof*. This is the union of two pieces. The first piece is the portion of P_1 which lies between l_1 and the ridge line $P_1 \cap P_2$. The second piece is the portion of P_2 which lies between l_2 and $P_1 \cap P_2$. Note that, by Lemma II.1.8.3 (*Ridge lines exist*), P_1 must meet P_2, and the intersection is near l.

If $P_1 = P_2$, we still obtain a roof, in this case the portion of $\partial\mathcal{C}(\Lambda)$ lying in $P_1 = P_2$ between l_1 and l_2, but we will not have a ridge line. It is possible for the ridge line to lie on $\partial\mathcal{C}(\Lambda)$.

In fact, if the ridge line contains a point x of $\mathcal{C}(\Lambda)$, then the whole roof lies in $\partial\mathcal{C}(\Lambda)$. For then P_1 and P_2 are both support planes to x. If $P_1 = P_2$ then the roof consists of the flat piece bounded by l_1 and l_2. If $P_1 \neq P_2$, then x must lie on a bending line, and this bending line must be the ridge line. Once again the convex hull must contain the roof.

We will talk of the open strip of $U \cap \partial\mathcal{C}(\Lambda)$ between l_1 and l_2 as lying "under the roof". We will often omit mention of U, by abuse of notation, and talk of a line which meets U between $l_1 \cap U$ and $l_2 \cap U$, as simply "lying between l_1 and l_2".

II.1.9.1.

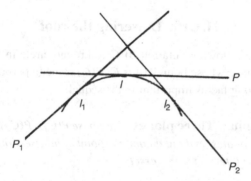

Figure II.1.9.1 *Two support planes P_1 and P_2 forming a roof, and a third plane P, which, together with P_1 and P_2 forms a lower roof.*

II.1.9.2. Lemma: Bending line disjoint from roof. *If any ridge line meets any bending line, then they are equal. Suppose the bending line l lies under the roof formed by support planes P_1 to l_1 and P_2 to l_2. Then l is either equal to or disjoint from the ridge line. If P is a support plane to l, then P is either disjoint from the ridge line, or else contains it.*

Proof. Suppose the bending line l meets the ridge line $P_1 \cap P_2$, where P_i is a support plane for the bending line l_i ($i = 1, 2$). At first we will not assume any relationship between the three bending lines so far introduced. Since $l \subset \mathcal{C}(\Lambda)$

and P_i is a support plane, $l \subset P_i$, $(i = 1, 2)$. But this means that the bending line is equal to the ridge line. This proves the first statement of the lemma.

Now suppose we are working in a small open shell U of $\mathbb{H}^3$ and that l lies under the roof formed by P_1 and P_2. We know that l_1, l and l_2 occur in that order in $U \cap \partial \mathcal{C}(\Lambda)$. Let P be a support plane at l which does not contain the ridge line $P_1 \cap P_2$. Let H, H_1 and H_2 be the supporting half spaces defined by P, P_1 and P_2 and let $K' = H \cap H_1 \cap H_2$. Then $l_1, P_1 \cap P, l, P_2 \cap P$ and l_2 occur in that order on $U \cap \partial K'$, although we may have $l = P_1 \cap P$ or $l = P_2 \cap P$. However, we cannot have $P_1 \cap P = l = P_2 \cap P$, since otherwise $P_1 \cap P_2$ would equal l and be contained in P, which we are assuming is not the case. So either l separates $P_1 \cap P$ from $P_2 \cap P$ in P, or $l = P_1 \cap P$ is disjoint from $P_2 \cap P$, or $l = P_2 \cap P$ is disjoint from $P_1 \cap P$. (Recall that ridge lines and bending lines are equal or disjoint by the first paragraph of this proof.) In each case, $P_1 \cap P_2$ is disjoint from P.

$\square$

II.1.10. Lowering the roof

We introduce the following notation. If C is a round circle in S^2, let $P(C)$ be the hyperbolic plane whose boundary is C. $P(C)$ is the hyperbolic convex hull of C. The following fact is important in the sequel.

II.1.10.1. Lemma: Three planes. *Suppose $P(C_1)$, $P(C_2)$, and $P(C_3)$ are three hyperbolic planes, without a common point of intersection in $\mathbb{H}^3$, and are such that any two intersect transversely:*

(a) *If the circles C_1, C_2, C_3 have no common point of intersection, then there exists a unique circle C^*, which is orthogonal to all the C_i. The plane $P(C^*)$ is orthogonal to all the $P(C_i)$. Consequently, the three lines $P(C^*) \cap P(C_i)$ $(i = 1, 2, 3)$ determine a triangle $\triangle$ in $P(C^*)$, and the vertex angles of $\triangle$ are dihedral angles between the various planes $P(C_i)$, $P(C_j)$.*

(b) *Suppose C_1, C_2, C_3 have a common point of intersection $\zeta \in S^2$. If σ is any horosphere centred at ζ, the three curves $\sigma \cap P(C_i)$, $i = 1, 2, 3$, determine a euclidean triangle $\triangle$ in σ, and the vertex angles of $\triangle$ are dihedral angles between the various planes $P(C_i)$, $P(C_j)$.*

Proof. We use the upper half space model with boundary $\mathbb{C} \cup \{\infty\}$. We may assume that C_1 and C_2 cross at zero and infinity. That is, C_1 and C_2 are straight lines in $\mathbb{C}$ through the origin.

When proving (a), C_3 is a proper circle not enclosing the origin. Let l be one of the two tangent lines to C_3 passing through the origin. Then C^* is the circle centred at the origin, which passes through the point of tangency of l with C_3.

II.1.10.2.

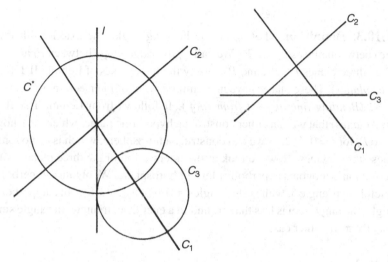

Figure II.1.10.2 *C^* is the common orthogonal to C_1, C_2, and C_3.*

To prove (b), we may assume that C_3 also passes through ∞. That is, C_3 is a straight line. The three straight lines C_1, C_2, C_3 determine a proper euclidean triangle in $\mathbb{C}$, and we get the same picture on any horosphere centred at ∞.

$\Box$

We continue to work in an open shell $U = S(x, v_0, \varepsilon, \delta)$ adapted to $\mathcal{C}(\Lambda)$. Then $U \cap \mathcal{C}(\Lambda)$ corresponds under central projection from $v_0 \in \mathrm{int}\,\mathcal{C}(\Lambda)$ to a small round disc in $S^2 \backslash \Lambda$. Let l_0 and l_1 be bending lines which meet U, with support planes P_0 and P_1. Given any two points $x_0 \in l_0$ and $x_1 \in l_1$, there is a unique shortest path between x_0 and x_1 in the roof formed by P_0 and P_1. This is constructed as follows. Rotate P_0 about the ridge line through the bending angle between P_0 and P_1, until it coincides with P_1. Draw the geodesic between x_1 and the result x_0' of rotating x_0. Then rotate x_0' back to x_0. The geodesic segment between x_0' and x_1 becomes bent in $\mathbb{H}^3$. The unique shortest path from x_0 to x_1 in $P_0 \cup P_1$ lies in the roof. It consists of a geodesic segment from x_0 to a uniquely determined point x on the ridge line, followed by a geodesic segment from x to x_1. Let $d_1(x_0, x_1)$ denote the length of this shortest path. Clearly this definition can be extended to any pair of points in the roof.

Next suppose that the roof between l_0 and l_1 is not contained in $\partial\mathcal{C}(\Lambda)$. Then there is a facet l_2 (i.e. a flat piece or bending line), which is properly under the roof. (We recall that the word "under" refers to the situation in U only.) By Lemma II.1.9.2 (*Bending line disjoint from roof*), l_2 does not meet the roof outside U either. Let P_2 be a support plane for l_2.

II.1.10.3. Definition. Let θ_{ij} be the bending angle (the exterior dihedral angle) between planes P_i and P_j. We call θ_{ij} the *ridge angle* between P_i and P_j.

The three planes P_0, P_1, and P_2 satisfy the hypotheses of Lemma II.1.10.1 (*Three planes*). These planes have no common point of intersection by Lemma II.1.9.2 (*Bending line disjoint from roof*). It follows from Lemma II.1.10.1 (*Three planes*) that we can either construct a hyperbolic plane. which is orthogonal to P_i for $i = 0, 1, 2$, or we can construct a horosphere, which is orthogonal to these three planes. Now we look at the intersection of the three planes with the common orthogonal hyperbolic plane or horosphere. We obtain a hyperbolic or euclidean triangle δ, with vertex angles $\pi - \theta_{01}$, θ_{02}, and θ_{12}. In a hyperbolic triangle, the angle sum is less than π, and in a euclidean triangle, the angle sum is exactly π. In either case:

II.1.10.4.

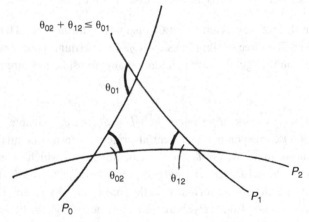

Figure II.1.10.4 *The hyperbolic triangle traced out on the common orthogonal plane.*

We compute also the shortest distance $d_2(x_0, x_1)$ on $P_0 \cup P_1 \cup P_2$ between $x_0 \in l_0$ and $x_1 \in l_1$. This time the shortest path is the union of three geodesic segments, one lying in each of the planes (if the facet l_2 is a flat piece with one of its edges l_0 or l_1, then one or two of these geodesic segments may have length zero). By the triangle inequality,

$$d_2(x_0, x_1) < d_1(x_0, x_1).$$

We have lowered the original roof between l_0 and l_1 to a new polyhedral roof. The new roof has two ridge lines $P_0 \cap P_2$ and $P_2 \cap P_1$. It consists of parts of the support planes P_0, P_1, and P_2 (one or two of these parts may reduce to l_0 and/or l_1).

Clearly the above process can be continued indefinitely, unless it stops with the roof in $\partial \mathcal{C}(\Lambda)$. At the nth stage, we get a distance $d_n(x_0, x_1)$ and

$$d_n(x_0, x_1) < d_{n-1}(x_0, x_1) < \cdots < d_1(x_0, x_1).$$

Also the sum of the bending angles between the planes forming the roof will not increase as we add extra support planes.

II.1.11. The bending measure

A bending line with zero bending angle has exactly one support plane. But if the bending angle, which we now write as 2θ, is positive, there are two extreme support planes, with exterior dihedral angle 2θ. We can uniquely specify a support plane by selecting the *medial support plane*. This intersects each of the two extreme support planes with dihedral angle θ. On a line with zero bending angle, or at an interior point of a flat piece, the medial support plane reduces to the unique support plane at that point.

II.1.11.1. Transverse measures. Before defining the bending measure, let us explain what kind of object we are trying to define. The bending measure will be a regular countably additive measure (in the usual sense of measure) theory, defined on every embedded interval in $\partial \mathcal{C}(\Lambda)$, which is transverse to every bending line it meets. The measure will satisfy two conditions which are obviously necessary for a useful notion. Given any subinterval of a transverse interval, there are two ways of obtaining a measure on it. The first is by restriction, and the second exists because the subinterval is transverse in its own right. These measures will agree. Secondly, any isotopy of $\partial \mathcal{C}(\Lambda)$ which sends each bending line to itself will send a transverse interval to another transverse interval. We require the measure to be invariant under the isotopy.

To define bending measure, it is clearly sufficient to define it locally. To this end we fix a spherical shell U which is adapted to $\mathcal{C}(\Lambda)$. Of course, U is to be small enough so that we can apply Lemma II.1.8.3 (*Ridge lines exist*). And we may assume that the interval I on which we wish to define our measure lies in U and crosses each bending line in U at most once.

II.1.11.2. Definition. Let x be a point of I. Then we define $\beta(x)$, the *bending measure* of x, to be the bending angle at x. This is zero unless x lies on a bending line with positive bending angle. Let A be an open subinterval of I. We define the *bending measure* of A to be

$$\beta(A) = \inf \left(\sum \text{ridge angles} \right)$$

where we vary over all roofs between the endpoints of A. The bending measure of a closed or half closed subinterval of I is defined to be the sum of the measure of its endpoint(s) and the measure of its interior.

Thus $\beta(A) = 0$ if A lies in a flat piece.

The infimum can be computed as follows. Let the endpoints of the open interval A be x_0 and x_1. Let x_i lie in the facet l_i ($i = 0, 1$). We take orientations so that x_0 is to the left of x_1. Let P_0 be the extreme right support plane at x_0 and P_1 the extreme left support plane at x_1. Let S_1 be the roof obtained from P_0 and P_1. We define a sequence $S_2, S_3, \ldots$ of successive lowerings of the roof. That is, S_{n+1} is formed from S_n by inserting a new support plane P_{n+1} to a facet l_{n+1} that lies under S_n. If at some time, we have $S_n \subset \partial \mathcal{C}(\Lambda)$, the process stops. The condition $S_n \subset \partial \mathcal{C}(\Lambda)$ is equivalent to $A \subset S_n$. In general, this does not happen and we can, for example, choose l_{n+1} to contain the midpoint of the largest subinterval of A (according to some parametrization of A), which does not meet S_n. The point is that the sequence of subsets $\{l_n \cap A\}$ should be dense in A, and the particular method of achieving this is not important. The sequence S_n will approximate the strip U_0 of $U \cap \partial \mathcal{C}(\Lambda)$ from l_0 to l_1. If τ is the ray from a point in the interior of $\mathcal{C}(\Lambda)$ through $x \in U_0$, then $\lim(\tau \cap S_n) = x$.

For each index n, compute the sum of the dihedral angles at the ridge lines of S_n. As we have seen, this angle sum decreases (weakly monotonically) as n increases.

The limit as n tends to infinity is independent of the roof lowering sequence. To see this, suppose that $\{S_n\}$, $\{S'_n\}$ are any two lowerings of S_1. Let n and m be fixed. Then there is a polyhedral roof S^* from l_0 to l_1, formed by using all the facets and all the support planes used in either S_n or in S'_m. Let θ^* be the sum of ridge angles for S^* and let $\varepsilon > 0$. We want to see that the sum of ridge angles for the sequence S_i is eventually less than $\theta^* + \varepsilon$. We first possibly decrease θ^* as follows. Each plane of S^* has associated to it a certain facet. If that facet is a bending line with positive bending angle, we replace the plane by two planes, namely the two extreme support planes at that bending line. Suppose there are

N planes in (the new) S^*. If we now take a large value of i, each of these N planes can be approximated by some plane of S_i – see Lemma II.1.7.3 (*Limits of support planes*) – so that the ridge angle is less than the corresponding ridge angle for S^* plus ε/N. Rearranging the order of the planes in S_i, we obtain from the first N planes of S_i a sum of ridge angles which is less than $\theta^* + \varepsilon$. Adding in all the other planes of S_i can only further decrease the sum of ridge angles.

II.1.11.3. Theorem: Bending measure. *The bending measure, as defined above, is a transverse measure on the lamination of all bending lines.*

Proof. The bending measure has been defined on transverse intervals, open or closed at either end. (A closed interval is said to be transverse, if it is a subinterval of a larger transverse open interval.)

The first task is to prove that bending measure does indeed define a measure on transverse intervals. So far we have a set function, defined on the semi-ring of points and intervals, which may be closed or open at either end. The only relevant property we need to prove is countable additivity. So let A_i be a finite or infinite sequence of points or intervals ($i = 1, 2, \ldots$), and suppose that the disjoint union of the A_i is an interval A. If the sequence is finite, we put $A_i = \emptyset$ for large values of i. We want to show that

$$\beta(A) = \sum_i \beta(A_i).$$

We will first show that for each n

$$\sum_{i=1}^n \beta(A_i) \leq \beta(A).$$

Given $\varepsilon > 0$, we find a finite set of support planes, such that the sum of the ridge angles is less than $\beta(A) + \varepsilon$. We add to this collection the two extreme support planes at each endpoint of each A_i ($1 \leq i \leq n$). (The two extreme planes at an endpoint may be equal to each other.) The sum of ridge angles is still less than $\beta(A) + \varepsilon$, and is greater than or equal to $\sum_{i=1}^n \beta(A_i)$. It follows that

$$\sum_i \beta(A_i) \leq \beta(A).$$

To prove the opposite inequality, we first deal with the case where there are only a finite number of A_i, say $A_1, \ldots, A_N$. For each A_i, we find a finite set of support planes, from which we can compute $\beta(A_i)$ to within ε/N. The collection of all these planes gives an upper bound for $\beta(A)$, and so

$$\beta(A) \leq \sum_{i=1}^{N} \beta(A_i) + \varepsilon$$

and we have proved finite additivity.

Now we suppose that there are infinitely many A_i. Note first that we can assume that each A_i is either a point or an open interval. We may also assume that A is a closed interval. There are a finite or countable number of i such that A_i is a point. We enlarge the kth such point $A_{i(k)}$ to an open interval, with bending measure less than $\beta(A_{i(k)}) + 2^{-k}\varepsilon$. This is possible by Lemma II.1.7.3 (*Limits of support planes*). This gives a finite cover of A by open intervals, and the sum of the bending measures of this finite collection is bounded by

$$\sum_i \beta(A_i) + \varepsilon.$$

Using finite additivity, we can reduce this quantity by removing overlaps, and change the cover to a finite disjoint cover by open intervals and points. Using finite additivity again, we find that

$$\beta(A) \leq \sum_i \beta(A_i) + \varepsilon.$$

This completes the proof of countable additivity.

To prove the two properties of transverse measures (see Section II.1.11.1 (*Transverse measures*)) note that the first property is obvious. The second one follows, because any isotopy is the product of small isotopies. A small isotopy does not affect the measure of a transverse interval A, because this is computed in U, referring only to a knowledge of which facets contain the endpoints of A. Since the isotopies are assumed to preserve the set of bending lines, each point will remain in the same facet.

$$\Box$$

II.1.12. The boundary is a complete hyperbolic manifold

Let Λ be a closed subset of S^2, and let $C(\Lambda)$ be the hyperbolic convex hull. Recall that we are assuming that Λ is not contained in a round circle in S^2.

We have already defined the metric on any component of the boundary of a closed convex set with non-empty interior. In particular, we have the intrinsic metric d_S on each component S of $\partial \mathcal{C}(\Lambda)$. We now prove that this metric is approximated by the roof metric, obtained from a sequence of finite outer approximations. Actually this is a consequence of a more general fact. For any closed convex subset of $\mathbb{H}^n$, the intrinsic metric on the boundary is the limit of the metrics obtained from finite outer approximations to the set. The reader may observe that the proof given below goes through in that generality.

Given x, y in the same component of $\partial \mathcal{C}(\Lambda)$, let S_n be a sequence of finite outer approximations, each containing both x and y in its boundary, whose limit is S. That is to say, using Lemma II.1.4.2 (*Countable intersection*), we write $\mathcal{C}(\Lambda)$ as an intersection of supporting half spaces H_i. We suppose that $x \in \partial H_1$ and $y \in \partial H_2$. The convex body $\cap_{i=1}^n H_i$ has a finite number of boundary components, and x and y are in the boundary. We define S_n to be the component containing x.

Let $\varepsilon > 0$, and let α be a path from x to y in $\partial \mathcal{C}(\Lambda)$, with length less than $d_S(x, y) + \varepsilon$. Push α a little away from $\mathcal{C}(\Lambda)$, using a radial push from a point v_0 in the interior of $\mathcal{C}(\Lambda)$, to obtain a path β, of length less than $d(x, y) + 2\varepsilon$. Join up at the endpoints of β to x and y by means of short radial arcs, to obtain a path γ of length at most $d(x, y) + 4\varepsilon$. Note that each of the two radial arcs is disjoint from S_n, except for one endpoint of each arc, by Theorem II.1.4.3 (*Standard convexity properties*) (radial arcs meet once only). Compactness ensures that if n is sufficiently large then S_n will not meet β. Let γ_n be the image of γ under the nearest point retraction to S_n. Then γ_n has length less than $d(x, y) + 4\varepsilon$ by Lemma II.1.3.4 (*r is distance decreasing*).

II.1.12.1. Theorem: Complete hyperbolic manifold. *With respect to the intrinsic metric, each component of $\partial \mathcal{C}(\Lambda)$ is a complete hyperbolic 2-manifold.*

Proof.   For each point $x \in \partial \mathcal{C}(\Lambda)$, we need to find an isometry of some open neighbourhood of x into $\mathbb{H}^2$. If x is contained in a flat piece of $\partial \mathcal{C}(\Lambda)$ then this is obvious. So we assume that x is contained in a bending line l. We will work in a fixed small open shell U in $\mathbb{H}^3$, centred at x. We will map $U' = U \cap \partial \mathcal{C}(\Lambda)$ isometrically into the hyperbolic plane.

We fix an orientation of l, of $\mathbb{H}^2$ and of $\mathbb{H}^3$. Our conventions determine an orientation of $\partial \mathcal{C}(\Lambda)$ (cf. Section II.1.2.3 (*Orientation of support planes*)). We fix once and for all an isometric embedding of l onto a geodesic l' in $\mathbb{H}^2$ and we give l' the corresponding orientation. The map $g \colon U' \to \mathbb{H}^2$, which we are about to define, will map the part of U' to the left of l into the part of the hyperbolic plane to the left of l', and similarly for the right side. On l, g is the given embedding onto l'. We fix distinct points $y_1, y_2 \in l \cap U$. Given $z \in U'$, we define gz to be the unique point in $\mathbb{H}^2$, which is on the appropriate side of l', and such that $d(gz, gy_i) = d_S(z, y_i)$. Here, d on the left of the equation is the usual hyperbolic distance function in $\mathbb{H}^2$, and d_S is the intrinsic metric in S, the component of $\partial C(\Lambda)$ in which U' lies. The reason that it is possible to determine gz is that the distances in S between the three points y_1, y_2, and z satisfy the various triangle inequalities.

If z is fixed, we can find approximations $g_n z$ to gz, by using polyhedral roofs S_n, which contain l and z. The map g_n is defined analogously to g, but using the finite approximation S_n instead of $\partial C(\Lambda)$. Recall from Section 1.10 (*Lowering the roof*) that $d_n(z, y_i)$ decreases monotonically to $d_S(z, y_i)(i = 1, 2)$. Also $g_n z$ converges to gz because the apex of a triangle in $\mathbb{H}^2$, with fixed base, depends continuously on the lengths of the other two sides.

To check that g is an isometry, let z_1 and z_2 be two points in U'. We take a sequence of polyhedral roofs which contain z_1, z_2, and l. It is clear that g_n is an isometry from $U \cap S_n$ into the hyperbolic plane. Then $d_n(z_1, z_2) = d(g_n z_1, g_n z_2)$. Letting n tend to infinity, we obtain the desired result.

II.1.12.2. There is a technical point here, which becomes important when S is not simply connected. We have to use Lemma II.1.5.7 (*Intrinsic metric lipschitz*) to make sure that the path distance on $U \cap S_n$ agrees with d_n, the path metric on S_n. In fact, according to the dictates of that lemma, we do not use the path metric on $S \cap U$, but rather the path metric on $S \cap U_1$, where U_1 is small, but is rather bigger than U. (Names here are not the same as in that lemma.) The above argument, except for the final paragraph, should be carried out with U_1 replacing U throughout.

Thus $\partial C(\Lambda)$ is a hyperbolic 2-manifold with respect to the intrinsic metric. Moreover the manifold is complete. For suppose that $x_n \in \partial C(\Lambda)$ is a Cauchy sequence with respect to the intrinsic metric. Then it is also a Cauchy sequence with respect to the metric of $\mathbb{H}^3$ and therefore has a limit x. Since $\partial C(\Lambda)$ is closed, x is in that set. By Lemma II.1.5.7 (*Intrinsic metric lipschitz*), x_n also converges to x in the intrinsic metric. This completes the proof of the theorem.

$\square$

Note that the mapping g just constructed is unique up to an isometry of $\mathbb{H}^2$. In the preceding proof, we have also proved the following useful fact.

II.1.12.3. Lemma: Convergence of local charts. *Let* g: $U \cap S \to \mathbb{H}^2$ *and* g_n: $U \cap S_n \to \mathbb{H}^2$ *be the charts found in the proof of the preceding Theorem. Then* $g_n(z)$ *converges to* $g(z)$, *if* $z \in U \cap S \cap S_k$ *for some value of* k. *(It then follows that* $z \in S_n$ *for all* $n \geq k$.*)*

II.1.12.4. Lemma: Convergence of inverses. *Let* g: $S \cap U \to \mathbb{H}^2$ *and* g_n: $S_n \cap U \to \mathbb{H}^2$ *be the maps defined in the proof of Theorem* II 1.12.1 *(Complete hyperbolic manifold). Given* $x \in S \cap U$, *there is an open neighbourhood* V *of* gx, *which is contained in the image of each of these maps, provided* n *is large, and* $g_n^{-1}|V: V \to \mathbb{H}^3$ *converges uniformly to* $g^{-1}|V: V \to \mathbb{H}^3$.

Proof. We use again the open shell $U = S(x, v, \varepsilon, \delta)$, which is a neighbourhood of a point $x \in S$. Let r be the distance in $\mathbb{H}^3$ from x to the radial boundary of U. We define V to be the disc in $\mathbb{H}^2$ with centre $x' = g(x)$ and radius r. Note that the g_n have different images, but since, for large n, S_n cuts through U, meeting its boundary only in the radial part, $g_n(S_n \cap U)$ contains V. Consider the map $g_n^{-1} \circ g$. Let $z \in S_n \cap \partial \mathcal{C}(\Lambda) \cap U$. Then $g_n z$ is near gz for n large. Since g_n^{-1} is an isometry onto its image, it is distance decreasing as a map into $\mathbb{H}^3$. Therefore z is near $g_n^{-1} gz$. The set of $z \in \partial \mathcal{C}(\Lambda)$ such that z lies in one of the polyhedral roofs S_n (for some fixed sequence of roofs) is dense. Therefore $g_n^{-1} g(z)$ converges pointwise to z for z in a dense subset of $\partial \mathcal{C}(\Lambda)$. But the maps $g_n^{-1} g$ are distance decreasing, when regarded as going from the intrinsic metric d_S on S to the standard metric on $\mathbb{H}^3$. Since the domain is compact, it follows that the convergence to the identity map is uniform. The stated result follows.

$\square$

II.1.12.5. Mapping the hyperbolic plane to S. We have seen how to construct local isometries g: $U \cap S \to \mathbb{H}^2$. We now show how to piece neighbouring maps together. We choose a countable dense set of facets in S. It will be convenient to assume that any one of these facets is eventually contained in all the surfaces S_n. Let U' be a second open shell which intersects U. For large n, S_n cuts through $U \cup U'$, once again avoiding the spherical boundaries. Choose a facet l of S which intersects $U \cup U'$ and so that l lies in S_n for all large n. Pick distinct points y_1, $y_2 \in U \cap U' \cap l$. Having defined g: $S \cap U \to \mathbb{H}^2$ and g_n: $S_n \cap U \to \mathbb{H}^2$, we can uniquely define g': $S \cap U' \to \mathbb{H}^2$ and g_n': $S_n \cap U \to \mathbb{H}^2$, so that g' and g_n' agree with g and g_n, respectively on y_1 and y_2. Hence g and g' agree on $U \cap U_1 \cap S$ and g_n agrees with g_n' on $U \cap U_1 \cap S_n$. In Lemma II.1.12.3 (*Convergence of local charts*), we have shown that g_n converges to g on y_1 and y_2. Let u_n be the orientation preserving isometry of $\mathbb{H}^2$, such that $u_n g_n(y_i) = g(y_i)$ for $i = 1, 2$. Then u_n converges to the identity. By the same lemma, $u_n g'_n$ converges to g'. Therefore g'_n converges to g'.

Continuing in this manner, the developing map can be constructed. We refer to Thurston (1979) or Epstein (1984) for details. The above remarks enable us to prove the following proposition.

II.1.12.6. Proposition: Covering maps from the hyperbolic plane

(1) *Let S be a component of $\partial C(\Lambda)$, where Λ is a closed subset of S^2. Then there is an orientation preserving covering map h: $\mathbb{H}^2 \to S$ which is a local isometry. Any two such covering maps differ by composition with an orientation preserving isometry of $\mathbb{H}^2$. That is, the group of orientation preserving isometries of $\mathbb{H}^2$ acts transitively on the set of covering maps which are local isometries and the stabilizer of a fixed covering map h is the set of covering translations for h.*

(2) *Let l be a facet common to S and S_n for all n. Let h: $\mathbb{H}^2 \to S$ and h_n: $\mathbb{H}^2 \to S_n$ be covering maps with $hh_n^{-1}|l = \mathrm{id}$. Then h_n converges to h, uniformly on compact sets.*

II.1.12.7. Corollary: Isometry to the hyperbolic plane. *If a component S of $\partial C(\Lambda)$ is simply connected and is given its intrinsic metric, then it is isometric to the hyperbolic plane.*

II.1.13. Finite approximations to the convex hull boundary

Let Λ be a closed subset of the 2-sphere at infinity, S^2, and let Ω be its complement in S^2. Suppose that Ω is a topological disc. This means that Λ is connected, and that the boundary S of its convex hull $C(\Lambda)$ is also a topological disc. One of the main tools in this chapter is to approximate any such surface by a finitely bent surface. We have already used finite outer approximations (see, e.g., Section II.1.10 (*Lowering the roof*)). However, the approximations we have constructed so far will not in general give us simply connected surfaces. Now the simply connected property is in fact crucial in some of our considerations. Sullivan's theorem is false without the assumption of simple connectedness – (see Section II.2.16 (*Counterexample*)). We therefore need a method of finding a sequence of simply connected outer approximations, given a simply connected boundary S. It is the object of this section to provide such a sequence.

There is a (one-to-one) correspondence between Ω and the set of pairs (s, P), where $s \in S$ and P is a support plane at s. Given $x \in \Omega$, we take the geodesic $\gamma(x)$ from x to $s = rx$. We then take the plane P through rx, which is orthogonal to

$\gamma(x)$. Conversely, given a pair (s, P), we take the geodesic orthogonal to P at s. This geodesic ends in Ω, as we see immediately from convexity. This (one-to-one) correspondence is clearly a homeomorphism. Note that any point x in the complement of the convex hull of Λ determines a pair (s, P), in the same way.

II.1.13.1. Definition. We will show how to choose a sequence of pairs (s_i, P_i), which we call *special pairs*. We call the corresponding points *special points*, and the corresponding planes *special planes*. In any actual situation, we will be particularly interested in a finite number of points of S (or points in the complement of the convex hull). In such a situation, we will always assume that such points in S figure amongst the special points. If a certain finite set of points of interest is in the convex hull complement, we assume that the pairs it determines are special pairs. The defining property of a sequence of special pairs is that the intersection of the corresponding half spaces is equal to the convex hull of Λ and there is a sequence of integers $\{m_i\}$, converging to infinity, such that the intersection of the first m_i half spaces is a convex body with a simply connected, finitely bent boundary. We call such a sequence a *standard sequence* of outer approximations.

As a first approximation to our choice of a sequence of special pairs, we choose any countable dense subset of Ω and take the corresponding pairs. Each plane P_i is the boundary of a supporting hyperbolic half space H_i, which contains Λ. The boundary of P_i is a circle C_i lying in the 2-sphere at infinity. Let D_i be the open round disc lying in Ω, with boundary C_i. We change the order of the sequence of special pairs by insisting that the union of the first n D_i's should be connected for each n. We denote this union by V_n. Topologically, V_n is an open disc with a finite (or zero) number of holes, and $V_n \subset \Omega$. The surface in $\mathbb{H}^3$ which is the boundary of the convex hull of the complement of V_n is an approximation to S, but it may not be simply connected.

To rectify this, we make the following construction. Let Ω_n be formed by adding to V_n all its complementary components which do not contain Λ. Then Ω_n is a topological disc, bounded by a finite number of circular arcs, which are also contained in the boundary of V_n. We claim that the boundary of the convex hull of the complement of Ω_n is finitely bent. Maximal circles in Ω_n correspond to support planes for the convex hull of the complement of Ω_n, and it suffices to consider only support planes which are left or right extremal. Such a maximal circle either contains a circular arc on the boundary, in which case it is also a maximal circle for V_n and for the complement of Λ, or it passes through at least three vertices of the boundary of Ω_n. In each case, there are only a finite number of possibilities.

We have thus proved the following result.

II.1.13.2.

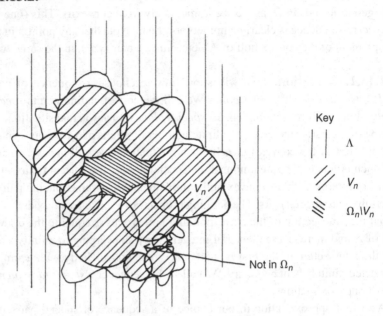

Key

Λ

V_n

$\Omega_n \backslash V_n$

V_n

Not in Ω_n

Figure II.1.13.2 *Maximal disks in the complement of Λ. Their union is V_n, which is not necessarily simply connected. V_n is enlarged to Ω_n, which is simply connected.*

II.1.13.3. Lemma: Finite approximation. *Let Λ be a closed subset of S^2 and let S be a simply connected component of the boundary of its convex hull. Let $F_1, F_2, \ldots$ be a countable collection of facets of S. Then there exists a sequence S_n of finite outer approximations to S which contains $F_1, \ldots, F_n$, and which is simply connected for every n. (F_i is contained in S_n for n large, as a subset, not necessarily as a facet.)*

Note that the convex hull of a general connected compact subset Λ of the 2-sphere, bounded by a finite collection of circular arcs, does not have a finitely bent convex hull in general. The simplest example is to take for Λ the union of two closed disks which meet at a point. In this case the convex hull boundary is a simply connected surface and the bending lamination (i.e. the set of bending lines) covers the whole surface.

II.1.14. Convergence of laminations

Given a standard sequence of outer approximations S_n to a simply connected convex hull boundary component S, we obtain a sequence of isometries

h_n: $\mathbb{H}^2 \to S_n$ converging to an isometry h: $\mathbb{H}^2 \to S$. The bending laminations then induce laminations $\mathcal{L}_n$ and $\mathcal{L}$ on $\mathbb{H}^2$. We will explain in what sense we can regard $\mathcal{L}_n$ as converging to $\mathcal{L}$.

II.1.14.1.

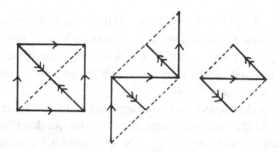

Figure II.1.14.1 *The left-hand picture represents $S^1 \times S^1$ with the diagonal removed. The middle picture shows that this is homeomorphic to a cylinder. The right-hand picture shows the quotient Möbius band under the involution. The first two pictures represent the space of oriented geodesics and the third represents the space of unoriented geodesics.*

The space of all geodesics in the hyperbolic plane can be identified with the open Möbius band. To see this, note that the set of oriented geodesics is equal to the set of ordered pairs of elements in the boundary circle, minus the diagonal. This is an open cylinder. There is an involution, which interchanges the order of a pair, and the space of geodesics is the quotient of the open cylinder under this involution. One easy way to see the cylinder and Möbius band is to think of the product of two circles as a square with opposite sides identified. We cut the square into four triangles, using the diagonals of the square. Then the four triangles together form a cylinder, if we remove the diagonal and then identify opposite edges. Two of the triangles form a Möbius band, if we remove the diagonal, identify opposite edges, and then identify under the involution, which is reflection in the diagonal. Another way of looking at this matter is to use Klein's projective model. The hyperbolic space is open round disc in the real projective plane. The Möbius band is the complement of its closure.

A geodesic lamination is a particular kind of closed subset of the open Möbius band. The restriction arises from the fact that two geodesics in a lamination are not allowed to intersect.

We will use the following definition for the notion of the convergence of a sequence of geodesic laminations $\mathcal{L}_n$ to a geodesic lamination $\mathcal{L}$.

II.1.14.2. Definition. We say that $\mathcal{L}_n$ converges to $\mathcal{L}$ if the following two conditions are satisfied:

(1) Let γ be a geodesic of $\mathcal{L}$. Then, for each n, there is a geodesic $\gamma_n \in \mathcal{L}_n$ such that γ_n converges to γ.
(2) Let $\gamma_n \in \mathcal{L}_n$ be a sequence of geodesics converging to a geodesic γ. They $\gamma \in \mathcal{L}$.

Note that this definition of convergence is the same as that induced from the topology on the space of closed subsets of the open Möbius band. The details of this topology are given in Part I. In that topology the space of geodesics laminations is a closed subset of the space of all closed subsets, and is therefore compact, separable, and metrizable.

Let S_n be a standard sequence of outer approximations for S. Let $h_n \colon \mathbb{H}^2 \to S_n$ and $h \colon \mathbb{H}^2 \to S$ be the associated sequence of isometries described in Section I.1.12 (*The boundary is a complete hyperbolic manifold*). We assume that these maps agree on one particular line. We have the bending lamination $\mathcal{L}$ on $\mathbb{H}^2$ induced by h and the bending laminations $\mathcal{L}_n$ on $\mathbb{H}^2$ induced by h_n.

II.1.14.3. Lemma: Convergence of bending laminations. *The sequence $\mathcal{L}_n$ converges to $\mathcal{L}$.*

Proof. Let l' be a bending line for $h\mathcal{L}$ and let $x \in l'$. Let (s, P) be a special pair with s near x. If s lies in a flat piece, then we define l'' to be an edge which is nearest to l'. If s lies in a bending line, let l'' be this bending line. Thus there is a bending line l'', which lies in S_n for n large, and which is arbitrarily near l'. Then, in the notation of Section II.1.12 (*The boundary is a complete hyperbolic manifold*), gl'' is a geodesic in $\mathcal{L}$, which lies very near to gl'. Also $g_n \, l''$ is a geodesic in $\mathcal{L}_n$. We know from Lemma II.1.12.3 (*Convergence of local charts*) that for n large, $g_n \, l''$ is near gl'', and therefore near to gl'. This proves one of the two conditions for convergence.

To prove the other condition, suppose that l_n is a geodesic of $\mathcal{L}_n$ for each n, and that l_n converges to a geodesic l. We must prove that $h(l)$ is a bending line for $h\mathcal{L}$. Now $h_n(l_n)$ is a geodesic of $\mathbb{H}^3$, which is a ridge line for S. These geodesics have a subsequence which converges to a geodesic l' of $\mathbb{H}^3$. Since h_n converges uniformly to h on compact subsets, we must have $l' = h(l)$.

Let F be a flat piece in S. Then $F \subset S_n$ for large values of n, and so $h_n l_n \cap F = \emptyset$ for n large. (This uses the hypothesis that we are working with a standard sequence of outer approximations.) Hence $l' \cap F = \emptyset$. Since l' is a geodesic in S which does not meet any flat piece of S, l' must be a bending line.

□

Chapter II.2

Foliations and the Epsilon Distant Surface

II.2.1. Introduction

In this chapter we investigate the geometry of the surface at distance ε from the convex hull boundary. We show that the nearest point map induces a bilipschitz homeomorphism between an open simply connected domain on the 2-sphere at infinity and the associated convex hull boundary. After that we show how to extend a lamination on the hyperbolic plane to a foliation in an open neighbourhood of the support of the lamination. By averaging the nearest point map along leaves of the orthogonal foliation, one obtains a bilipschitz homeomorphism ρ between the ε distant surface and the convex hull boundary. The properties of ρ are determined by means of a careful investigation of the foliation, and the properties of the foliation are determined by progressively complicating the situation. The crucial lemmas are given similar names in the progressively more complicated situations, in order to make the parallels clearer.

II.2.2. The epsilon distant surface

The proof of Sullivan's theorem, relating the convex hull boundary with the surface at infinity, is carried out by factoring through S_ε, the surface at distance ε from the convex hull boundary component S. For this reason, we need to know quite a lot about S_ε. In this section we investigate its geometry, in the case of a finitely bent convex hull boundary component. We analyze the structure of geodesics and show that the distance function is C^1.

Let Λ be a closed subset of S^2, whose complement is a topological disk. Let S be the boundary of $\mathcal{C}(\Lambda)$. Let $S_\varepsilon \subset \mathbb{H}^3 \backslash \mathcal{C}(\Lambda)$, be the surface at distance $\varepsilon > 0$ from S. By Lemma I.1.3.6 (*Smooth epsilon surface*), S_ε is a C^1-surface.

It divides $\mathbb{H}^3$ into two components, one of which, containing $C(\Lambda)$, will be referred to as lying under S_ε. $\mathbb{H}^3 \setminus C(\Lambda)$ is the disjoint union of the surfaces S_ε as ε varies $(0 < \varepsilon < \infty)$.

First consider the case of the ε distant surface from a hyperbolic plane in $\mathbb{H}^3$. The nearest point projection map onto the hyperbolic plane changes scale by a factor $1/\cosh \varepsilon$, so that the ε distant surface has constant negative curvature $-1/\cosh^2\varepsilon$.

Now we consider in detail the surface S_ε, where S is a finitely bent convex hull boundary component. S_ε is the union of a finite number of pieces, of which there are two types. The first type of piece is part of the equidistant surface from the plane containing a flat piece of S. We will call this a negatively curved piece. (The curvature with respect to the induced metric is in fact $-1/\cosh^2\varepsilon$.) The second type of piece is part of the equidistant surface from one of the bending lines of S. We will call this a euclidean piece. (The curvature with respect to the induced metric is in fact 0.)

Apart from a change of scale by the factor $\cosh \varepsilon$, a negatively curved piece is an isometric copy of the corresponding flat piece. It is bounded by a finite number of edges, each of which is geodesic in the geometry of the negatively curved piece, though not in the geometry of $\mathbb{H}^3$, where it is a line equidistant from a bending line of S. A euclidean piece is bounded by two edges, which are parallel in the euclidean geometry of the euclidean piece. We call such edges *gluing edges*. If the corresponding bending line has bending angle 2θ, the distance in the intrinsic geometry of the euclidean piece between the two lines is $2\theta \sinh \varepsilon$.

We give S_ε the metric induced from paths which are entirely contained within S_ε. In proving the next lemma, discussions with Wilfred Kendall and David Elworthy were very helpful.

II.2.2.1. Proposition: Geometry of a piecewise smooth manifold.

Any two points in S_ε are joined by a unique geodesic (i.e. a path of shortest length). This path consists of a finite number of geodesic arcs each lying in either a euclidean piece or in a negatively curved piece. As this geodesic crosses gluing edge, there is no refraction (i.e. the angles are the same on the two sides). The distance between two points is a convex C^1-function of the two points (provided they are distinct). (In fact we have strict convexity except for a number of situations which can be explicitly described.) In particular, a metric ball is convex.

Proof. We start by considering the special case where there is only one gluing edge. We make a model of the surface in the upper half plane, which we denote by H. To make the situation more familiar, we change the scale in order to

II.2.2.2.

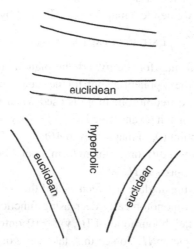

Figure II.2.2.2   *A surface formed by gluing hyperbolic pieces with geodesic boundary to euclidean pieces bounded by parallel straight lines.*

assume that the negatively curved piece has curvature -1. On the right of the y-axis, we take the usual metric of constant negative curvature, multiplied by a factor to give the change of scale

$$\frac{dx^2 + dy^2}{y^2}.$$

On the left of the y-axis, we take the metric

$$\left(\frac{dx}{y} - \frac{x\,dy}{y^2}\right)^2 + \frac{dy^2}{y^2}.$$

This metric is flat, since it is induced from the euclidean plane by the diffeomorphism $\phi\colon H \to \mathbb{R}^2$, defined by

$$\phi(x, y) = \left(\frac{x}{y}, \log y\right).$$

The two metrics agree on the y-axis and give a C^0-metric on the real analytic manifold H.

The geodesics on the left are inverse images under ϕ of straight lines, and the geodesics on the right are, as usual, vertical lines or semicircles orthogonal to the x-axis. If a geodesic crosses the y-axis, then there is no refraction, as we see by arguing locally – refraction would allow a shortcut.

Given a smooth unit vector field U defined on the y-axis, which is nowhere tangent to the y-axis, we define a map Ex: $\mathbb{R} \times \mathbb{R} \to H$ by the formula

$$\text{Ex}(t, y) = \exp(0, y)tU(y)$$

where the exponential map for the hyperbolic plane is used when tU points towards the right, and the exponential map for the euclidean plane (after suitable transformation by the diffeomorphism ϕ) is used when tU points to the left. These two definitions match up and give $\text{Ex}(0, y) = (0, y) \in H$. Furthermore the derivative of Ex at a point $(0, y)$ maps $\partial/\partial y$ to $\partial/\partial y$ and $\partial/\partial y$ to $U(y)$. It follows that Ex is C^1, and gives a diffeomorphism from a neighbourhood of $t = 0$ to a neighbourhood of the y-axis in H.

Now suppose that the angle θ between U and the positive direction of the y-axis $(0 < \theta < \pi)$ is a monotonically decreasing function of y. Then Ex is a diffeomorphism of a neighbourhood of $\{(t, y):t \geq 0\}$ onto a neighbourhood of one of the two quadrants in H. To see this, take any point z in the interior of the appropriate quadrant. As y moves from 0 to ∞ along the y-axis, the angle between the positively oriented y-axis and the geodesic from y to z increases from 0 to π. Therefore, for some value of y, $U(y)$ is tangent to the geodesic through y and z. This shows that Ex maps onto the appropriate quadrant. Our hypothesis also allows us to show that the derivative of Ex is non-singular for all $t \geq 0$. (We can make an estimate comparing with the case where θ is constant.)

We now consider the general case, where there are a finite number of gluing edges. By using the considerations above, we have a real analytic structure on S_ε, with a C^0 metric.

We assume we have a geodesic, and investigate its nature. Look at the open intervals in the geodesic which lie in the interior of pieces of S_ε. Any such interval has its endpoints on distinct gluing edges. An easy induction then shows that the geodesic does not return to the same gluing edge after leaving it. The geodesic is therefore given by a finite sequence of points, each one on a gluing edge, except that the first and the last are not necessarily on a gluing edge. The case of a single gluing edge, already considered, shows that no refraction takes place at a gluing edge. A geodesic is therefore given by a sequence of points lying on successive gluing edges and these points are subject to certain restrictions already explained.

Using the fact that the pieces have negative or zero curvature and that there is no refraction, it is easy to see from the Gauss–Bonnet theorem that two geodesics can meet at most once. Let $x \in S_\varepsilon$ be a fixed point not lying on a gluing edge. Consider the geodesics emanating from x. At a point y lying on such a geodesic, we define the vector $U_x(y)$ to be the unit vector tangent to the geodesic, pointing away from x.

We claim that U_x is defined everywhere in S_ε, except at x itself, and that it is analytic on each piece. To see this we orient the gluing edges so that as one moves along the gluing edge in a positive direction, the point x is on the left. We prove by induction on the number of gluing edges $\gamma_1, \ldots, \gamma_{k-1}$, which one has to cross in going from x to y, that U_x is analytic in a neighbourhood of y (provided one stays in the same piece), and that, along any gluing edge γ_k, the angle between U_x and the positive direction of γ_k is strictly monotonically decreasing from π to 0.

Using the map Ex, defined relative to the vector field U_x on γ_{k-1}, we see that U_x is defined and analytic throughout the piece whose edges include γ_{k-1} and γ_k. The induction step for the monotonicity of the angle uses the Gauss–Bonnet theorem applied to a quadrilateral, two of whose sides lie on geodesics through x and two of whose sides lie on the gluing edges γ_{k-1} and γ_k.

We now want to prove that $d(x,z)$ is a C^1-function of $x, z \in S_\varepsilon$, provided that $x \neq z$.

Suppose first that x and z do not lie in gluing edges. Let $x = x_0, \ldots, x_k = z$ be a sequence of points in S_ε, with $x_i \in \gamma_i$, $1 \leq i < k$, where γ_i is a gluing edge, and x_{i-1} can be joined to x_i by a geodesic path in a single piece. We consider the function $\sum d(x_i, x_{i+1})$, which is an analytic convex function of $x_1, \ldots, x_{k-1}$, if x_0 and x_k are fixed. The geodesic from x_0 to x_k is given by the unique solution of $f = 0$, where f is obtained from the above sum of distances by differentiating with respect to $x_1, \ldots, x_{k-1}$ only. The derivative of f with respect to $x_1, \ldots, x_{k-1}$ is the second derivative of a sum of distance functions, and this is positive definite by an easy argument (see Lemma I.1.4.6 (*Distance convexity*)). From the Implicit Function theorem, we can solve $f = 0$ for $(x_1, \ldots, x_k)$, obtaining analytic functions of the pair (x_0, x_k). Thus, for a geodesic, $x_1, \ldots, x_{k-1}$ are analytic functions of x_0 and x_k, provided the latter do not cross any gluing edge. The convexity result in the statement also follows from the convexity of the distance functions.

Let us now consider the case where one of x and z, say z, does lie on a gluing edge γ. We first assume that x does not lie in γ. We want to show that the distance is a C^1-function; our method is to show that the partial derivatives exist and are continuous. If x does not lie on a gluing edge, and z is held fixed, we get a real analytic function of x from the argument above, since the geodesic from x to z has no way of knowing whether z lies on a gluing edge or not. If x does lie on a gluing edge, then the situation between x and z is symmetrical.

So we will hold $x \notin \gamma$ fixed, and show that the derivative of $d(x, u)$ exists and is continuous for u in a neighbourhood of $z \in \gamma$. Let $\gamma_1, \ldots, \gamma_{k-1}$ be the successive gluing edges met by the geodesic from x to z, and set $\gamma_k = \gamma$. Let G be the set of geodesics through x which intersect γ. Let V be the set of all unit tangent vectors at x, which are tangent to some geodesic of G and point towards γ. Let I_j be the open subinterval of γ_j containing all points where geodesics in G meet γ_j. Let $\rho_1 \colon V \to I_1$ map a vector v to the intersection of the corresponding geodesic with γ_1. Let $\rho_j \colon I_{j-1} \to I_j (1 < j \le k)$ be defined by insisting that y and $\rho_j(y)$ lie on the same geodesic through x. Then $\rho_1, \ldots, \rho_k$ are analytic functions. We define analytic functions $\psi_1 \colon \mathbb{R} \times V \to \mathbb{R} \times I_1$ by

$$\psi_1(t, v) = (t - d(x, \rho_1(v)), \rho_1(v))$$

and $\psi_j \colon \mathbb{R} \times I_{j-1} \to \mathbb{R} \times I_j$ $(1 < j \le k)$ by

$$\psi_j(t, y) = (t - d(y, \rho_j(y)), \rho_j(y)).$$

Now consider the map

$$\psi = \mathrm{Ex} \circ \psi_k \circ \cdots \circ \psi_1$$

where Ex is based on the vector field U_x, defined on γ_x. This composite gives rise to a C^1 diffeomorphism of a small open subset of $\mathbb{R} \times V$ onto a small neighbourhood of z. The maps have been defined so that the first coordinate of $\psi^{-1}(u)$ is equal to $d(x, u)$. Therefore $d(x, u)$ is a C^1-function of u.

We finally need to consider the case where x and z lie on the same gluing edge γ. In that case we may assume that there is only one gluing edge, the case discussed at the beginning of the proof. We can compute the derivative of $d(x, z)$ explicitly in this case. Let v_x be the unit tangent vector to γ, pointing away from z, and let v_z be the unit tangent vector to γ, pointing away from x. Then the derivative of the distance function at (x, z) is the map from the product of the tangent space at x with the tangent space at z to $\mathbb{R}$, given by taking the inner product with (v_x, v_z). This is clearly a correct computation of the partial derivative if one restricts to one side of γ, and is therefore correct without restriction.

$\square$

The derivative of the distance function can be worked out at any pair of distinct points $x, z \in S_\varepsilon$, and the answer is exactly the same as that given in the last part of the proof. Explicitly, let γ be the geodesic containing x and z, so that γ is no longer necessarily a gluing edge. Let v_x be the unit tangent vector to γ at x, pointing away from z, and let v_z be the tangent vector to γ at z, pointing away from z. Then the derivative of d and (x, z) is given by the inner product with (v_x, v_z). To see this note that the derivative is given by the inner product with some pair of tangent vectors w_x and w_z at x and z, respectively. The triangle inequality shows that these vectors have length at most one. The derivative obtained by varying x and z in γ show that the projections of w_x and w_z to γ have unit length, and therefore w_x and w_z already lie in γ.

II.2.3. From infinity to the epsilon surface

To prove Sullivan's theorem, we need to find a homeomorphism from an open subset at infinity (i.e. on S^2) to the convex hull boundary. This homeomorphism is defined by factoring through the surface S_ε, at distance ε from the convex hull boundary. In this section we discuss the map from the surface at infinity to S_ε, and compute its quasiconformal and bilipschitz constants.

Let Λ be a closed subset of S^2, whose complement is a topological disk. Let S be the boundary of $C(\Lambda)$. Let $S_\varepsilon \subset \mathbb{H}^3 \backslash C(\Lambda)$ denote the surface at distance ε from S, where $\varepsilon > 0$. We recall from Lemma I.1.3.6 (*Smooth epsilon surface*) that S_ε is C^1. The surface divides $\mathbb{H}^3$ into two components; we will refer to the one containing $C(\Lambda)$ as lying *under* S_ε. $\mathbb{H}^3 \backslash C(\Lambda)$ is the disjoint union of the surfaces S_ε, as ε varies.

Let $\Omega(S)$ be the component of $S^2 \backslash \Lambda$ facing S. This means that $\Omega(S)$ is the inverse image in S^2 of S under the nearest point mapping $r: \mathbb{B}^3 \to C(\Lambda)$. The closure in $\mathbb{B}^3$ of the ε-neighbourhood of $C(\Lambda)$ is a closed convex subset of $\mathbb{B}^3$ (ε is measured with respect to the hyperbolic metric on the interior of the ball), whose intersection with S^2 is Λ (*cf.* Lemma I.1.4.7 (*Neighbourhoods of convex sets*)). The nearest point mapping onto this convex set induces a homeomorphism σ: $\Omega(S) \to S_\varepsilon$. The inverse of σ is constructed as follows. Given a point $x \in S_\varepsilon$, we take the unit normal vector at x, pointing away from S. This vector defines a geodesic ray, which starts at x and ends in $\Omega(S)$. The endpoint is equal to $\sigma^{-1}(x)$.

Since S is simply connected, so are S_ε and $\Omega(S)$. The Uniformization theorem says that $\Omega(S)$ is conformally equivalent to the unit disk. We use this conformal equivalence to transfer to $\Omega(S)$ the Poincaré metric on the unit disk and we denote by d_Ω this metric on Ω. The transferred metric is also called the Poincaré metric. The metric d_ε used on S_ε is the intrinsic metric induced by the length of rectifiable paths.

II.2.3.1. Theorem: Epsilon surface to infinity. *The map $\sigma: \Omega(S) \to S_\varepsilon$ is $\coth(\varepsilon)$-quasiconformal, and $4\cosh(\varepsilon)$- lipschitz. If $\varepsilon < \log 2$, then the map σ^{-1} is $\operatorname{cosech} \varepsilon$-lipschitz.*

The proof of this result will occupy the remainder of this section.

We fix a standard sequence of outer approximations, as described in Section I.1.13 (*Finite approximations to the Convex Hull Boundary*). The approximation S_n lies above S, so its ε-distant surface $S_{n\varepsilon}$ lies above S_ε.

Let $V \subset \mathbb{H}^3$ be the component of the complement of $C(\Lambda)$ which is mapped onto S by the nearest point retraction r. Analogously, we have open subsets V_n, which map to S_n under the appropriate nearest point retraction r_n. We have

$V \supset V_{n+1} \supset V_n$, and $\bigcup V_n = V$. Any compact subset of V is contained in V_n for large enough values of n. Let R be the unit vector field in V, whose value at x is the vector which points directly away from rx, and let R_n be the vector field defined on V_n pointing directly away from $r_n x$.

II.2.3.2. Lemma: Uniform convergence of R.   *On any compact subset of V, r_n converges uniformly to r and the vector fields R_n converge uniformly to R.*

Proof. If (s, P) is a special pair (see Section I.1.13.1. (*Special pairs*) for the definition), and if x lies on the geodesic through s orthogonal to P, then $r_n x = rx = s$. The set of special pairs is dense, and so this argument works for a dense subset of V. Since r and r_n are distance decreasing, we deduce that r_n converges uniformly to r on compact subsets of V. Using the fact that a compact subset of V has a positive minimal distance to S, we deduce that R_n converges uniformly to R.

☐

II.2.3.3. Lemma: Constants for a line.

(a) *Let l be a geodesic in $\mathbb{B}^3$ and let T be S^2 with the two endpoints of l removed. Let S_ε be the surface at distance $\varepsilon > 0$ from l, with the Riemannian metric induced from $\mathbb{H}^3$. Let $\sigma: T \to S_\varepsilon$ be the diffeomorphism given by the nearest point retraction onto the closed convex subset of $\mathbb{B}^3$ which is the closed ε-neighbourhood of l. Then σ is $\coth \varepsilon$-quasiconformal.*

(b) *Let D be a round disk in S^2, whose boundary contains the endpoints of l. Take the Poincaré metric on the interior of D. Let γ be the geodesic in D connecting the two endpoints of l. Then, along γ, using the Poincaré metric in D along γ and the induced Riemannian metric on S_ε, the map σ has derivative with norm equal to $\cosh \varepsilon$, and its inverse σ^{-1} has derivative with norm equal to $\operatorname{cosech} \varepsilon$.*

Proof. This computation is performed most easily in the upper halfspace model:

$$\{x, y, t: t > 0\} \text{ with metric } ds/t$$

where ds is the euclidean distance element. We represent l as the line $x = y = 0$. Then S_ε is the conical surface with apex the origin, and cone angle θ, where $\sin \theta = \tanh \varepsilon$. We take D to be the plane $\{(x, y, 0): x > 0\}$. Then γ is the ray $(x, 0, 0)$ $(x > 0)$. Since multiplication by a positive scalar is an isometry for the various metrics, there is no loss of generality in working out the constants at

(1, 0, 0). In this model, $\sigma(x,y,0)$ is defined by taking the unique semicircle through $(x,y,0)$, which is orthogonal to the conical surface. Then $\sigma(x,y,0)$ is the point of intersection of the semicircle with the conical surface.

The derivative of σ in the direction of γ has length sec θ, and the derivative in the orthogonal direction has length cot θ. Moreover the images of these particular orthogonal vectors are also orthogonal. Now sec $\theta = \cosh \varepsilon$ and tan $\theta = \sinh \varepsilon$. The quasiconformal constant is given by the ratio of these two numbers. The norm of the derivative of σ is the maximum of these two numbers, and the norm of the derivative of σ^{-1} is the maximum of their inverses.

$\square$

II.2.3.4. Lemma: Constants for a plane.

(a) *Let H be a closed half space in $\mathbb{B}^3$. Let S_ε be one of the two components of the surface at distance $\varepsilon > 0$ from H, with the Riemannian metric induced from $\mathbb{H}^3$. Let D be the component of $S^2 \backslash H$ corresponding to S_ε. Let σ: $D \to S_\varepsilon$ be the diffeomorphism given by the nearest point retraction onto the closed convex subset of $\mathbb{B}^3$ which is the closed ε-neighbourhood of H. Then σ is conformal.*

(b) *We take the Poincaré metric on the interior of D. The map σ has derivative with norm equal to $\cosh \varepsilon$, and its inverse σ^{-1} has derivative with norm equal to sech ε.*

This is proved using the same technique as that used in the preceding lemma.

II.2.3.5.

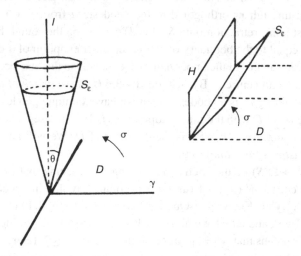

Figure II.2.3.5 *The cone S_ε on the left is the ε-distant surface from the line l. The plane S_ε on the right is the ε-distant surface from the halfspace H.*

II.2.3.6. Inverse image of a facet.   In order to relate the results just proved to compute the constants for an arbitrary simply connected surface S, which is a boundary component of $C(\Lambda)$ for some closed subset Λ of S^2, we need some preliminary discussion.

Suppose S is finitely bent and as above. Let S_ε be the surface at distance ε from S. Let Ω be the component of the complement of Λ corresponding to S, and let $\sigma\colon \Omega \to S_\varepsilon$ be the nearest point retraction onto $C(\Lambda)_\varepsilon$, the ε-neighbourhood of $C(\Lambda)$. Let $r\colon \mathbb{B}^3 \to C(\Lambda)$ be the nearest point retraction. Let f be a facet of S. If f is a flat piece (recall that the boundary edges of f are not part of f), then σ induces a diffeomorphism between $\Omega \cap r^{-1}f$, and $S_\varepsilon \cap r^{-1}f$.

If f is a geodesic, then $\Omega \cap r^{-1}f$ and $S_\varepsilon \cap r^{-1}f$ are each the union of a family of disjoint curves. We get one such curve for each support plane P at f. P determines a disk $D \subset \Omega$ with the same boundary as P. On S_ε, there is a unique curve which is at a distance ε from P. The corresponding curve γ in Ω is the unique circular arc in D which is orthogonal to the boundary of D, and whose endpoints are equal to the endpoints of f. This curve is the geodesic in the Poincaré metric of D.

II.2.3.7. Lemma: Finitely bent case.   *Theorem II.2.3.1* (Epsilon surface to infinity) *is true if S is finitely bent.*

Proof.   The computation of the quasiconformal constant follows from the preceding two lemmas and we say no more about that. Let $x \in \Omega(S)$. Let P be the plane through rx orthogonal to the geodesic α from x to rx. (Here r is the nearest point retraction onto S.) Let $D \subset \Omega(S)$ be the round disk whose boundary is equal to the boundary of P. Let the other endpoint of α on S^2 be y. We choose coordinates in the upper half-space model, so that $y = \infty$, $x = 0$ and D has radius 1 and centre 0. By Section II.2.3.6 (*Inverse image of a facet*), if the facet containing rx is a geodesic, then we have a unique geodesic γ in the Poincaré metric of D with the same endpoints as f. In this case f is a semicircle through the line $x = y = 0$ and γ is the diameter of D. We will apply Lemma II.2.3.3 (*Constants for a line*) to γ.

Let $\phi\colon D \to \Omega(S)$ be the Riemann mapping. We assume that $\phi(0) = 0$. By Schwarz's Lemma $|\phi'(0)| > 1$ (unless S is completely flat, in which case σ is an isometry). By the Koebe One-Quarter theorem, $|\phi'(0)| \le 4$. Let v be a vector tangent to $\mathbb{C}$ at 0, and let v have unit length with respect to the Poincaré metric in $\Omega(S)$. This means that $|v| = |\phi'(0)|/2$, so that $1/2 < |v| \le 2$. (Here $|v|$ denotes the euclidean norm of v.) With respect to the Poincaré metric $2ds/1 - r^2$ on D, we therefore have $1 < \|v\|_{\mathrm{hyp}} \le 4$. From the preceding two lemmas it follows that the image of v under $\sigma\colon \Omega(S) \to S_\varepsilon$ satisfies

$$\|v\|_{\mathrm{hyp}} \cosh \varepsilon \geq \|\sigma_* v\| \geq \sinh \varepsilon \|v\|_{\mathrm{hyp}}.$$

Hence,

$$4\cosh \varepsilon \geq \|\sigma_* v\| \geq \sinh \varepsilon.$$

We deduce that σ is $4\cosh \varepsilon$-lipschitz and that σ^{-1} is $1/\sinh \varepsilon$-lipschitz.

Finitely bent case

Proof. **Epsilon surface to infinity:** Let S_n be a standard sequence of outer approximations. Let r_n be the associated nearest point retraction. Let R_n be the unit vector field which, at a point x lying above S_n, points directly away from $r_n(x)$. Let R be the similarly defined unit vector field defined using the nearest point retraction r to S. Let $V \subset \mathbb{H}^3$ be the inverse image of r in the complement of $C(\Lambda)$, and let V_n be defined similarly using r_n. Given any compact subset C of V, we can make it lie above S_n by taking n large enough. We will choose C to be a compact subset of S_ε. Note that $S_{n\varepsilon}$ lies above S_ε (although parts of the two surfaces will coincide).

Since $r_n x$ lies on a support plane for $C(\Lambda)$, it is easy to see that if $x \in V_n$ then the angle between $R(x)$ and $R_n(x)$ is acute. It follows that the distance from S is strictly increasing along the trajectories of R_n. Therefore a trajectory of R_n meets S_ε at most once.

By following trajectories of R_n, we therefore define a map of C to $S_{n\varepsilon}$ if n is large, and this map will almost be a diffeomorphism on the interior of C, with derivative near to the identity, if n is large. More precisely, the map is a homeomorphism which is a diffeomorphism on pieces which we will now describe and the derivatives do not match up on the intersection of two pieces.

We consider the map $r_n\colon V_n \to S_n$, and the inverse image under r_n of the various facets of S_n. If the facet is a geodesic, the inverse image is a wedge of $\mathbb{H}^3$, bounded by two hyperbolic half-planes, which meet along that geodesic. In the upper half-space model, it is an infinite "slice of cake", bounded by two vertical quarter-planes. If the facet is a flat piece (recall that a flat piece does not contain its boundary edges), the closure of its inverse image is once again a 3-dimensional submanifold of $\mathbb{H}^3$, each of whose boundary components is a hyperbolic half plane.

From Lemma II.2.3.2 (*Uniform convergence of R*), we know that R_n converges uniformly to R on the compact subset C of S_ε. Let $\delta = \delta_n(x)$ be the angle between $R_n(x)$ and $R(x)$ for $x \in C$. Then δ_n converges uniformly to zero on C.

Since R_n is tangent to the boundary half planes referred to above, S_ε is transverse to these half spaces on and near C, and the intersection of the half planes with S_ε near C is a union of C^1 embedded curves near $S_{n\varepsilon}$.

There is a retraction $\alpha_n\colon V_n \to S_{n\varepsilon}$, defined by following the trajectory of R_n until it meets $S_{n\varepsilon}$. Each trajectory of R_n meets $S_{n\varepsilon}$ exactly once, so this map is well-defined. It is continuous and, on each of the 3-dimensional pieces referred to above (which are closed in the $\mathbb{H}^3 \setminus S_n$), it is analytic. However the derivatives do not match up along the boundary half planes.

Let us work out the derivatives explicitly. Let $x \in V_n$ be a a distance $t = t_n(x)$ from S_n. We extend $R_n(x)$ to an orthonormal basis $R_n(x)$, e_1, e_2 for the tangent space at x, as follows. If $r_n(x)$ is contained in a flat piece, there is no restriction on e_1 and e_2 (except that they provide an orthonormal basis). If $r_n(x)$ is contained in a facet which is a geodesic, we choose e_1 to map under r_n to a vector which is tangent to the geodesic. Then e_2 is chosen so as to complete the basis. We choose a basis for the tangent space to $S_{n\varepsilon}$ at any point y in the same way. With respect to such bases, the derivative of α_n at a point x is represented by the matrix

$$\begin{bmatrix} 0 & \dfrac{\cosh \varepsilon}{\cosh t} & 0 \\ 0 & 0 & \dfrac{\cosh \varepsilon}{\cosh t} \end{bmatrix}$$

if x is in the closure of the inverse image of a flat piece, and is represented by

$$\begin{bmatrix} 0 & \dfrac{\cosh \varepsilon}{\cosh t} & 0 \\ 0 & 0 & \dfrac{\sinh \varepsilon}{\sinh t} \end{bmatrix}$$

if x is in the inverse image of a face which is a geodesic. (These derivatives are worked out as follows. The terms involving cosh arise from the fact that, in the hyperbolic plane, if β is the curve at constant distance ε from the geodesic γ, then orthogonal projection of β onto γ reduces distances by a factor exactly equal to $\cosh \varepsilon$. The terms involving sinh arise from the fact that a circle in the hyperbolic plane of radius ε has circumference $2\pi\sinh \varepsilon$.)

We can use this information to estimate the quasiconformal and bilipschitz constants of $\alpha_n | S_\varepsilon \cap C$. The computations are performed by restricting in turn to each of the 3-dimensional pieces discussed above. A unit vector at x has the form

$$uR_n(x) + ve_1 + we_2$$

where $u^2 + v^2 + w^2 = 1$. The vector is tangent to S_ε if its inner product with $R(x)$ is zero. Now

$$R(x) = \cos \delta \, R_n(x) + be_1 + ce_2$$

where $b^2 + c^2 = \sin^2(\delta)$, and the condition that the unit vector be tangent is

$$u \cos \delta + bv + cw = 0.$$

It follows that $|u| \leq \tan(\delta)$. The image of the unit tangent vector under α_n is computed from the above matrices. We obtain either

$$v\frac{\cosh \varepsilon}{\cosh t}e_1 + w\frac{\cosh \varepsilon}{\cosh t}e_2 \quad \text{or} \quad v\frac{\cosh \varepsilon}{\cosh t}e_1 + w\frac{\sinh \varepsilon}{\sinh t}e_2.$$

Now, for $x \in C \subset S_\varepsilon$, $\delta = \delta_n(x)$ converges uniformly to zero, and $t = t_n(x)$ converges uniformly to ε as n tends to infinity. For large n, the above vectors will be very near to being unit vectors. Therefore the bilipschitz constant for $\alpha_n|S_\varepsilon \cap C$ tends to 1 as n tends to infinity, and so does the quasiconformal constant.

Let $\sigma_n: \Omega(S_n) \to S_{n\varepsilon}$ and $\sigma: \Omega \to S$ be the nearest point maps. We have already proved that σ_n is coth ε-quasiconformal and $4\cosh \varepsilon$-lipschitz and that σ^{-1} is cosech ε-lipschitz. We now prove that the same constants work for σ. Recall that, by Schwarz's lemma, any holomorphic map from the unit disk to the unit disk reduces the length with respect to the Poincaré metric of every vector, unless it is a Möbius transformation, in which case it preserves the length. In particular, the inclusion $\Omega(S_n) \to \Omega(S)$ is a distance decreasing map, if the Poincaré metric is used on each of these two open sets (i.e. each is given the metric resulting from an application of the Uniformization theorem).

Let D be the unit disk in the complex plane, and let $\phi: D \to \Omega(S)$ be a conformal isomorphism. Let $t < 1$ be very nearly equal to 1 and let D_t be the disk concentric to D, with radius t. The Poincaré metric on D_t is

$$\frac{2ds}{t(1 - r^2/t^2)}$$

which converges to the Poincaré metric on D as t tends to 1. If t is given, then, for large n

$$\phi(D_t) \subset \Omega(S_n) \subset \Omega(S).$$

Since holomorphic maps reduce the Poincaré metric, we see that the Poincaré metric on $\Omega(S_n)$ converges uniformly to the Poincaré metric on $\Omega(S)$ on compact subsets of $\Omega(S)$.

We claim that, for $x \in S_{n\varepsilon}$ and $\alpha_n^{-1}x \in S_\varepsilon$, $d(x, \alpha_n^{-1}x)$ tends to zero as n tends to infinity, and that this convergence is uniform on compact subsets of V. To see this, let $\delta > 0$ be a small number and recall from Lemma II.2.3.2 (*Uniform convergence of R*) that the angle between R and R_n becomes smaller than δ for n large. Moving along a trajectory of R_n a distance t therefore increases $d(x, rx)$ by at least $t\cos\delta$. For large n, and in a fixed compact region, $S_{n\varepsilon}$ will lie between S_ε and $S_{\varepsilon+\delta}$. Therefore the distance from $\alpha_n^{-1}x \in S_\varepsilon$ to $x \in S_{n\varepsilon}$ along a trajectory of R_n is at most $\delta/\cos\delta$. This proves the claim.

For a fixed compact subset X of $\Omega(S)$, we have proved above that the quasiconformal constant and the lipschitz constants of $\sigma: \Omega(S) \to S_\varepsilon$ on X and of σ^{-1} on $\sigma(X)$ can be estimated by using the Poincaré metric of $\Omega(S_n)$. Now on X, σ is the uniform limit of the composites

$$\Omega(S_n) \xrightarrow{\sigma_n} S_{n\varepsilon} \xrightarrow{\alpha_n^{-1}} S_\varepsilon.$$

This is because σ_n converges to σ and x is near $\alpha_n^{-1}(x)$. (We work within a fixed small compact neighbourhood of $\sigma(X)$ in $\mathbb{H}^3$.) For fixed distinct $z_1, z_2 \in \Omega$, we deduce that

$$\cosh\varepsilon \geq \frac{d_\varepsilon(\sigma z_1, \sigma z_2)}{d_\Omega(z_1, z_2)} \geq \sinh\varepsilon.$$

This is exactly what is claimed in the statement of the theorem.

> **Epsilon surface to infinity**

We will have to apply the above result when the underlying manifold is not a smooth Riemannian manifold. To avoid unnecessary complication, we give the result only in the situation we will actually meet.

II.2.4. Extending a lamination to a pair of orthogonal foliations

In the previous section, we have described how the nearest point mapping gives us universal quasiconformal and bilipschitz constants for the nearest point mapping going from infinity to S_ε. In order to construct a natural homeomorphism between S_ε and S, we will extend the bending lamination to a foliation. This foliation has an orthogonal foliation and the map will be defined by averaging the nearest point mapping to S over the orthogonal leaves. We will need very precise information about these foliations. Since S is isometric to the hyperbolic plane (Corollary I.1.12.7. (*Isometry to the hyperbolic plane*)) we find it convenient to work in $\mathbb{H}^2$ rather than in $\mathbb{H}^3$.

II.2.4.1. Definition. We recall Thurston's definition of a *geodesic lamination* on a complete hyperbolic 2-dimensional manifold. This is a set of disjoint geodesics, whose union is a closed subset.

On a compact manifold, the support of a geodesic lamination has Lebesgue measure zero. However, our laminations are mostly on the hyperbolic plane, and there is no restriction on the measure. One example is to take a single geodesic α, and define the lamination to be all geodesics orthogonal to α. We recall that we have, in effect, discussed how laminations arise naturally from convex hulls of closed subsets of S^2.

II.2.4.2. Definition. If $h\colon \mathbb{H}^2 \to S$ is the covering map which is a local isometry, described in Section I.1.12 (*The boundary is a complete hyperbolic manifold*), then we define the *bending lamination* on $\mathbb{H}^2$ to be the set of geodesics which are inverse images under h of bending lines.

In order to prove Sullivan's theorem, we need to investigate arbitrary geodesic laminations. (In fact any geodesic lamination is a bending lamination for some convex hull.) Let $\mu < (\log 3)/2$. (This is the distance of the centre of an ideal triangle to one of its sides.) Let $\mathcal{L}$ be a geodesic lamination in a complete hyperbolic surface. We will produce a canonical extension of this lamination to $\mathcal{L}_\mu$, which will be a foliation of U_μ, the μ-neighbourhood of $\mathcal{L}$. Here we mean by "canonical" that the construction is local. In particular, if there is a group of isometries, which preserves the lamination, then the foliation will also be preserved. The foliation $\mathcal{L}_\mu$ and the orthogonal foliation $\mathcal{F}_\mu$ will be fundamental tools in this paper.

We work in the universal cover of the surface. Since the construction is canonical, it will descend to the original surface. So from now on we assume that we have a lamination $\mathcal{L}$ of $\mathbb{H}^2$.

Before extending $\mathcal{L}$, we state a corollary to a lemma proved in the Appendix (Lemma II.A.5.1 (*Nearest geodesics*)). Let C be a component of the complement of $\mathcal{L}$ and let l_1, l_2, l_3 be three boundary components of C.

II.2.4.3. Corollary: Nearest geodesics corollary. *Let $\mu < (\log 3)/2 =$ arcsinh $(1/\sqrt{3})$, and suppose we have a point $z \in C$, which is at a distance less than or equal to arcsinh $(e^{-\mu})$ from both l_2 and l_3. Then its distance from l_1 is greater than μ.*

We now fix $\mu < (\log 3)/2$ for the rest of the chapter. Towards the end of the chapter, we will give the result of a computer program showing what the best value is for the purpose of computing Sullivan's constants.

Let C be a component of $\mathbb{H}^2 \backslash \mathcal{L}$. We have to define a foliation on $U_\mu \cap C$, such that the boundary geodesics of C are leaves of the foliation. To carry out this extension, we express $U_\mu \cap C$ as the union of three different types of sets, which are closed subsets of U_μ. Let β and γ be typical geodesic boundary components of C. We write

$$U_1(\beta, \gamma) = \{z \in C \,|\, d(z, \beta) < \mu \quad \text{and} \quad d(z, \gamma) \le \mu\}$$

$$U_2(\beta) = \{z \in C \,|\, d(z, \beta) < \mu \quad \text{and} \quad d(z, \mathcal{L} \backslash \beta) \ge \text{arcsinh}(e^{-\mu})\}$$

$$U_3(\beta, \gamma) = \{z \in C \,|\, d(z, \beta) < \mu \quad \text{and} \quad \mu \le d(z, \gamma) \le \text{arcsinh}(e^{-\mu})\}.$$

Roughly, $U_1(\beta, \gamma)$ consists of all points which are very near both β and γ, $U_2(\beta)$ consists of points which are very near β and are far from any other geodesic of $\mathcal{L}$, and $U_3(\beta, \gamma)$ consists of points which are very near β and are moderately, but not very, near γ.

II.2.4.4.

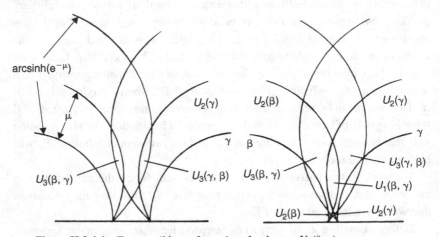

Figure II.2.4.4 *Two possible configurations for the sets $U_1(\beta, \gamma)$, etc.*

As β and γ vary, we get various sets, some of which may be empty. By Lemma II.2.4.3 (*Nearest geodesics corollary*), we see that these sets can intersect each other only in the following ways. $U_1(\beta, \gamma)$ can meet $U_1(\gamma, \beta)$; $U_1(\beta, \gamma)$ can meet $U_3(\beta, \gamma)$; and $U_2(\beta)$ can meet $U_3(\beta, \gamma)$.

In $U_1(\beta, \gamma)$, we extend the lamination by putting in extra geodesics as follows. Let α be the common orthogonal to β and γ or their common end-point at infinity. We foliate $U_1(\beta, \gamma)$ by $G(\alpha)$, the foliation consisting of all geodesics orthogonal to α, if α is a geodesic, and consisting of all geodesics asymptotic to α, if α is a point at infinity.

We foliate $U_2(\beta)$ by $E(\beta)$, the foliation by curves equidistant from β.

II.2.4.5.

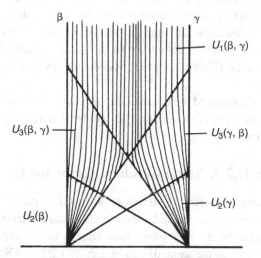

Figure II.2.4.5 *The foliation of $U_2(\beta)$ and $U_2(\gamma)$ is by equidistant curves. The foliation of $U_1(\beta, \gamma)$ is by geodesics. The foliation of $U_3(\beta, \gamma)$ is by a compromise between the two extreme situations. This figure shows how to join the foliation by equidistant curves to the foliation by geodesics.*

It remains to foliate $U_3(\beta, \gamma)$ in such a way that the foliation matches up with the foliations already defined on sets of the form U_1 and U_2. From the discussion above, the only matching up necessary will be on $U_1(\beta, \gamma) \cap U_3(\beta, \gamma)$ and on $U_2(\beta) \cap U_3(\beta, \gamma)$. Let $\phi \colon \mathbb{R} \to \mathbb{R}$ be the function, which is linear on $[\mu, \operatorname{arcsinh}(e^{-\mu})]$ and such that $\phi(t) = 0$ for $t \geq \operatorname{arcsinh}(e^{-\mu})$ and $\phi(t) = 1$ for $t < \mu$. We fix an orientation of γ and orient β consistently. Let α be the common orthogonal or the common endpoint of β and γ. Let $G = G(\alpha)$ be the (real analytic) unit vector field described above and which points in the same direction as γ and β. Let $E = E(\gamma)$ be the (real analytic) unit vector field, which is tangent to the curves which are equidistant from γ, and which points in the same direction as γ. In $U_3(\beta, \gamma)$ the angle between E and G is small, so one gets a non-zero vector field from a convex combination of them. The particular combination we take at a point z is

$$\phi(d(z, \gamma))\, G(z) + (1 - \phi(d(z, \gamma)))\, E(z).$$

We have to check that this gives a well-defined line field on a μ-neighbourhood of $\mathcal{L}$ when this construction is applied simultaneously to all components C of the complement of $\mathcal{L}$ and to all components of the boundary of C. First of all we have made a choice of orientation of γ, in order to be able to orient and then add G and E. But changing the orientation of γ will change the sign for G and for E, and hence for our convex combination. So the line field is independent of the choice of orientation although the vector field is not. Second we check immediately that the various definitions of the foliation match up to give us a lipschitz line field on C, and that on U_μ it is defined by a linefield which is at least continuous.

We have now completed the construction of a line field and foliation in U_μ the μ-neighbourhood of the lamination $\mathcal{L}$, and we denote by $\mathcal{L}_\mu$ the foliation of U_μ thus constructed.

II.2.5. Some standard vector fields

Before continuing with our investigation, we need to pause in order to do some computations on the standard vector fields G and E. Let α be an oriented geodesic. Let $E(\alpha)$ be the unit vector field tangent to the curves equidistant from α and pointing in the same direction. Let $R\colon T\mathbb{H}^2 \to T\mathbb{H}^2$ be the vector bundle map over the identity on $\mathbb{H}^2$, which is rotation through $\pi/2$ in a clockwise direction on each tangent space. We define $G(\alpha)$ so that $R(G(\alpha)) = E(\alpha)$. Then $G(\alpha)$ is one of the two unit vector fields tangent to the foliation by geodesics which are orthogonal to α. Let t be the signed distance from α, with the sign chosen so that $G(\alpha) \cdot t = 1$.

We will discuss at the same time the limiting case, where α is infinitely distant. In that case α is interpreted as a point on the boundary of hyperbolic space, $G(\alpha)$ is tangent to the foliation by geodesics asymptotic to α and pointing away from α, $E(\alpha)$ is tangent to the horocyclic foliation centred at α. We will fix some definite horocycle, and define x to be the signed distance from that horocycle, so that $G(\alpha) \cdot x = 1$.

II.2.5.1. Proposition: Covariant derivatives of G and E. *We abbreviate to $E = E(\alpha)$ and $G = G(\alpha)$:*

(1) $\nabla_G G = \nabla_G E = 0$,
(2) *G and $\cosh t \cdot E$ are commuting vector fields. (In the limiting case, G and $e^x E$ are commuting vector fields.)*
(3) *$\nabla_E E = -\tanh t\, G$ and $\nabla_E G = \tanh t \cdot E$ (In the limiting case, $\tanh t$ is replaced by 1.)*
(4) *The classical lipschitz constant of E and G is $|\tanh t|$, within a $|t|$-neighbourhood of α. (In the limiting case, the classical lipschitz constant is 1.)*
(5) *The classical lipschitz constant of $\cosh t \cdot E$ is $|\sinh t|$ within a $|t|$-neighbourhood of α.*

(6) *Let v be a tangent vector in the hyperbolic plane and let a and b be*
constants. Then $\nabla_v (aE + bG) = \langle b, E \rangle \tanh t(-aG + bE)$.

Proof. **Covariant derivatives of G and E:** We have $\nabla_G G = 0$, since this is
one of the defining properties of a geodesic. Therefore

$$\nabla_G E = \nabla_G R(G) = R(\nabla_G G) = 0$$

since R commutes with covariant differentiation.

To see that G and $\cosh tE$ are commuting vector fields, note that they are the
coordinate vector fields for coordinates on $\mathbb{H}^2$. (These coordinates are called
Fermi coordinates by differential geometers.) The coordinates of a point z are
(s, t), where t is the signed distance of z from α and s is the signed distance
along α of a fixed origin on α to the orthogonal projection of z into α.

For any smooth vector fields X and Y, we have

$$\nabla_X Y - \nabla_Y X = [X, Y].$$

From this we deduce

$$\cosh t \cdot \nabla_E G = \nabla_G(\cosh t \cdot E) = \sinh t \cdot E$$

since $\nabla_G E = 0$. We find

$$\nabla_E G = \tanh t \cdot E.$$

Also,

$$\nabla_E E = \nabla_E(RG) = R(\nabla_E G) = -\tanh t \cdot G.$$

This is sufficient to prove all the results claimed. The lipschitz constant for
$\cosh t \cdot E$ results from the computations

$$\nabla_E \cosh t \cdot E = \sinh t \cdot G \quad \text{and} \quad \nabla_G \cosh t \cdot E = \sinh t \cdot E.$$

> **Covariant derivatives of G and E**

II.2.5.2. Corollary: Disjoint geodesics mark 2. *Let β and γ be disjoint*
geodesics in $\mathbb{H}^2$ and let them "point in the same direction" (i.e. if an oriented
geodesic cuts both β and γ then the two intersections have the same sign). Let
v be a unit tangent vector to β at x and let w be a unit tangent vector to γ at y.
Let v' be the parallel translate of v along the geodesic from x to y. Then

$$\|v' - w\| \le d(x, y).$$

Proof. It is easy to give a direct proof using hyperbolic trigonometry. However, it is easier for us to rely on what we have just proved. Since β and γ are disjoint, they either have a common orthogonal geodesic α or they are asymptotic to the same point α of the boundary of $\mathbb{B}^2$. Then β and γ are both trajectories of $G(\alpha)$, and Proposition II.2.5.1 (*Covariant derivatives of G and E*) applies. The classical lipschitz constant of $G(\alpha)$ is at most 1, and the desired result follows.

$\square$

II.2.6. Lipschitz line fields in the hyperbolic plane

We now show that the line fields constructed in Section 2.4 (*Extending a lamination to a pair of orthogonal foliations*) are lipschitz. This is one of the few places where the search for good constants, as opposed to any constants, in Sullivan's theorem causes additional complications. The formula $k(t) = 1 + 4t$, in place of the formula in the statement of the next proposition, is easy to prove. Any bound for the lipschitz constant of the vector field $\mathcal{L}_\mu$ is good enough to prove a version of Sullivan's theorem. Since this constant appears as an exponent, its size makes a big difference to the numerical estimates.

We recall from Section 2.4 (*Extending a lamination to a pair of orthogonal foliations*) that $\mathcal{L}_\mu$ was defined using a piecewise linear map $\phi \colon \mathbb{R} \to \mathbb{R}$. This map is constant, except on a certain interval, where the slope A is constant. We have

$$A = \frac{1}{\operatorname{arcsinh}(e^{-\mu}) - \mu}.$$

We recall that μ has been fixed with $\mu < (\log 3)/2$.

II.2.6.1. Proposition: Lipschitz (H2). *The line field tangent to $\mathcal{L}_\mu$ is lipschitz, with classical lipschitz constant at most $k(t)$ at a point a distance t from the nearest geodesic of $\mathcal{L}$, where*

$$k(t) = \frac{2 + 4\tanh t + 4A\sin(t/2)(1 + \sin(t/2)) + \dfrac{\sin t}{1 + e^{-2t}}}{1 + \cos t}.$$

By substituting $t = \mu$ we get an upper bound k for the classical lipschitz constant in U_μ. In regions where the distance to each of two different boundary geodesics is less than μ, the classical lipschitz constant is 1.

We have not yet said what is meant by a lipschitz line field. We will mean simply that one of the two unit tangent vector fields (and hence the other) is lipschitz. (There could be an objection that the line field will in general not be orientable and so there is no tangent vector field. The line field is definitely not orientable if, for example, the lamination $\mathcal{L}$ contains an ideal triangle. However, because our concept of "lipschitz" is relative to paths, we only need to define the concept in small open sets where there is no problem about the orientation.)

Proof. Lipschitz (H2): We first compute the lipschitz constant on the complement of $\mathcal{L}$. Here the field only fails to be C^∞ along curves where the distance to some geodesic in the boundary of a component is either μ or arcsinh $(e^{-\mu})$. Such curves do not intersect each other in U_μ. Therefore we can compute the classical lipschitz constant as the covariant derivative. Let $E = E(\gamma)$ and $G = G(\alpha)$, where α is a geodesic orthogonal to the geodesic γ, (or α is a point on the boundary of $\mathbb{B}^2$ and γ is a geodesic through α), and E and G are the fields which appear in the construction of $\mathcal{L}_\mu$. Let v be a unit vector tangent at a point $z \in \mathbb{H}^2$ at a distance t from γ. From Proposition II.2.5.1 (*Covariant derivatives of G and E*), we know that $\nabla_v E$ has norm at most $\tanh t$ and that $\nabla_v G$ has norm at most 1.

So we may assume that we are in a set of the form $U_3(\beta, \gamma)$, defined in Section II.2.4 (*Extending a lamination to a pair of orthogonal foliations*). By abuse of notation, we denote by $\phi(z)$ the quantity $\phi(d(z, \gamma))$. Let X be the vector field $\phi G + (1 - \phi)E$. This is a tangent field to the foliation $\mathcal{L}_\mu$, but is not a unit field in general. Let v be a unit vector at a point $z \in U_3(\beta, \gamma)$. We have

$$\nabla_v X = \nabla_v E + \nabla_v(\phi)(G - E) + \phi . \nabla_v(G - E).$$

We will compute the covariant derivative for the corresponding unit field. We have

$$\nabla_v(X/\|X\|) = \frac{\nabla_v X}{\|X\|} - \frac{\langle \nabla_v X, X \rangle X}{\|X\|^3}.$$

In order to make the estimate, we first estimate $\langle E_\gamma, G_\alpha \rangle$. By Lemma II.2.5.1 (*Covariant derivatives of G and E*), we see that if $v = G_\gamma(z)$ for some z, then

$$|\nabla_v \langle E_\alpha, E_\gamma \rangle| \leq |\langle G_\alpha, E_\gamma \rangle| = \sqrt{1 - \langle E_\alpha, E_\gamma \rangle^2}.$$

If we write this as a differential inequality, we find that

$$\left| \frac{du}{dt} \right| \leq \sqrt{1 - u^2}$$

where t is the distance from γ along one of the solutions of G_γ, and $u(t) = \langle E_\alpha, G_\gamma \rangle(t)$. This implies that $|u(t)| \leq \sin t$. (Note that the theorem about differential inequalities only holds for lipschitz vector fields. Therefore the above inequality will only hold for $t \leq \pi/2$. After that a solution could continue with the constant value 1. In our case $t < \mu < (\log 3)/2 < \pi/2$.) We deduce that $\langle E_\alpha, E_\gamma \rangle = \langle G_\alpha, G_\gamma \rangle$ is less than or equal to $\sin t$ in absolute value, and that $\langle G_\alpha, E_\gamma \rangle = \langle E_\alpha, G_\gamma \rangle$ is greater than $\cos t$ (provided the orientations of α and γ are properly chosen, otherwise these terms are less than $\cos t$). We also deduce that,

$$|\nabla_v \langle E_\gamma, G_\alpha \rangle| \leq \tanh t \cdot \sin t + \sin t$$

for any unit vector v. Also the norm of $\nabla_v G_\alpha$ is bounded by 1 and the norm of $\nabla_v E_\gamma$ is bounded by $\tanh t$.

We are now able to give a good estimate for $\langle X, X \rangle$, We have

$$\langle X, X \rangle = 1 - 2\phi(1 - \phi)(1 - \langle E_\gamma, G_\alpha \rangle).$$

The maximum value of $\phi(1 - \phi)$ is 1/4 and its maximum slope is A. It follows that

$$\langle X, X \rangle \geq (1 + \cos t)/2.$$

We also have

$$2\langle \nabla_v X, X \rangle = \nabla_v \langle X, X \rangle$$
$$= 2(1 - \langle E_\gamma, G_\alpha \rangle) \nabla_v (\phi(1 - \phi)) + 2\phi(1 - \phi) \nabla_v \langle E_\gamma, G_\alpha \rangle.$$

This implies that

$$|\langle \nabla_v X, X \rangle| \leq A(1 - \cos t) + \frac{(1 + \tanh t) \sin t}{4}$$
$$\leq 2A \sin^2(t/2) + \frac{\sin t}{2(1 + e^{-2t})}.$$

Next we have to estimate $||\nabla_v X||$. We have

$$\nabla_v X = \nabla_v E + (\nabla_v \phi)(G - E) + \phi \nabla_v (G - E).$$

We have

$$||G - E||^2 = 2(1 - \langle G_\alpha, E_\gamma \rangle) \leq 4 \sin^2(t/2)$$

and so

$$||G - E|| \leq 2\sin(t/2).$$

Hence

$$||\nabla_v X|| \leq \tanh t + A||G_\alpha - E_\gamma|| + (1 + \tanh t) \leq 1 + 2\tanh t + 2A \sin(t/2).$$

Putting all these inequalities together, we see that the result for a single complementary region is proved.

Since some points have neighbourhoods which meet a countable number of distinct complementary regions further argument is necessary in general. By Lemma II.A.2.4 (*Lipschitz constant*), what we need to do is to show that if a geodesic arc xy is entirely contained within U_μ, then

$$||\tau\mathcal{L}_\mu(x) - \mathcal{L}_\mu(y)|| \leq kd(x,y)$$

where τ is parallel translation along xy. By breaking xy up into at most three shorter arcs, it is easy to see that we may assume one of the following two conditions holds:

(1) The points x and y both lie on geodesics of the lamination $\mathcal{L}$. In this case the proposition follows immediately from Lemma II.2.5.2 (*Disjoint geodesics mark 2*).

(2) The interior of the geodesic arc xy is disjoint from $\mathcal{L}$. In this case the result has been proved above, since we are working in a single complementary region.

II.2.6.2. Lemma: Continuous dependence (H2). *If the sequence of laminations $\mathcal{L}_n$ converges to $\mathcal{L}$ on $\mathbb{H}^2$, then the tangent field to the foliation $\mathcal{L}_{n,\mu}$ converges uniformly to the tangent field to $\mathcal{L}_\mu$ on compact subsets of U_μ. (Any compact subset of U_μ will eventually be contained in each $U_{n,\mu}$.) Similarly for the line fields $\mathcal{F}_\mu$, orthogonal to $\mathcal{L}_\mu$.*

II.2.6.3. Remark: If we work with the wrong topology on the space of laminations, it will not necessarily be true that $U_{n,\mu}$ converges to U_μ, even if the laminations arise from finite approximations. Nor will it necessarily be true that $\mathcal{L}_{n,\mu}$ converges to $\mathcal{L}_\mu$ on U_μ. For example, a half space can be approximated by a bent surface, with the bending taking place along an arbitrary lamination, provided the amount of bending is sufficiently small. The lamination corresponding to the half space is the empty lamination, but the finite approximations are all equal to the same non-trivial lamination. An important property of the standard sequence of outer approximations – see Section I.1.13 (*Finite approximations to the convex hull boundary*) – is that it does give a sequence of laminations in the hyperbolic plane which converges to the bending lamination in the relevant topology on the space of laminations.

Proof. Continuous dependence (H2): If x is in the interior of a flat piece F of $\mathcal{L}$, then, for large n it will be in the interior of a flat piece F_n of $\mathcal{L}_n$. Also, if $x \in U_\mu$, then $x \in U_{n,\mu}$ for n large. By Lemma II.A.5.1 (*Nearest geodesics*), if $x \in F \cap U_{n,\mu}$ ($\mu < (\log 3)/2$), then we can pick out either the nearest geodesic or the nearest two geodesics.

We refer to the notation of Section II.2.4 (*Extending a lamination to a pair of orthogonal foliations*). Let β and γ be geodesics in $\mathcal{L}$. Then for n large, there will be geodesics $\beta_n, \gamma_n \in \mathcal{L}_n$, such that β_n is near β and γ_n is near γ. If $x \in U_1(\beta, \gamma)$, then $x \in U_1(\beta_n, \gamma_n)$ for n large. Similarly for $U_2(\beta)$ and for $U_3(\beta, \gamma)$.

If α is the common orthogonal or the common endpoint of β and γ, then α_n converges to α. Then $G(\alpha_n)$ converges to $G(\alpha)$, uniformly on compact subsets of $\mathbb{H}^2$. Also $E(\beta_n)$ converges uniformly on compact subsets to $E(\beta)$. It follows that in $U_1(\beta, \gamma)$ and in $U_2(\beta)$, $\mathcal{L}_{n,\mu}$ converges pointwise to $\mathcal{L}_\mu$. Recall that, in $U_3(\beta, \gamma)$, the function of z given by $\phi(d(z, \gamma))$ is used to define $\mathcal{L}_\mu$. Since $\phi(d(z, \gamma_n))$ converges pointwise to $\phi(d(z, \gamma))$, we obtain pointwise convergence of $\mathcal{L}_{n,\mu}$ to $\mathcal{L}_\mu$ on $U_\mu \setminus \mathcal{L}$.

If $x \in \mathcal{L}$, then, for large values of n, a geodesic of $\mathcal{L}_n$ will pass near to x, nearly parallel to the $\mathcal{L}$-geodesic through x (see Lemma I.1.14.3 (*Convergence of bending laminations*)). Let k be given by Lemma II.2.6.1 (*Lipschitz* (H2)). Since $\mathcal{L}_{n,\mu}$ is k-lipschitz, pointwise convergence at x follows. The lipschitz property (*cf.* Lemma II.2.6.1 (*Lipschitz* (H2))) now enables us to prove uniform convergence on compact subsets.

> **Continuous dependence (H2)**

II.2.7. Foliation coordinates

The line field $\mathcal{L}_\mu$ and its orthogonal line field $\mathcal{F}_\mu$ can be used to construct coordinates on small open subsets of U_μ. These coordinates are particularly important for the averaging procedure we will use to go from the surface S_ε, at a distance ε from the convex hull boundary, to S, a component of the convex hull boundary. For that reason we will investigate the properties of such coordinates at length.

II.2.7.1. Definition. Let M be a 2-dimensional manifold with a path metric, and suppose that M has two transverse foliations X and Y. We assume that the leaves are rectifiable. Let $z_0 \in M$ be a fixed point. Let $a \leq 0 \leq b$ and $c \leq 0 \leq d$ and suppose that $a < b$ and $c < d$. We say that $\phi \colon (a, b) \times (c, d) \to M$ is a *foliation chart* giving *foliation coordinates relative* to X and Y, and *centred* on z_0, if the following conditions are satisfied.

(1) $\phi(0,0) = z_0$.

(2) ϕ maps horizontal lines to leaves of X and vertical lines to leaves of Y.

(3) ϕ preserves path distances along the horizontal line through the origin and along the vertical line through the origin.

We will say that a mapping ϕ, whose domain is the product of two closed intervals, is a foliation chart, if it is the restriction of a foliation chart defined on the product of two larger open intervals.

II.2.7.2.

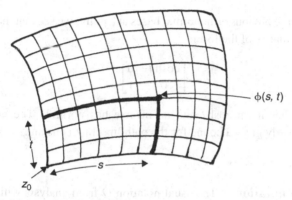

$\phi(s, t)$

Figure II.2.7.2 *A foliation chart.*

We will use this concept when M is a submanifold of $\mathbb{H}^3$. In particular, we will use it when M is the boundary of a convex hull. It is for this reason that we cannot assume that M is differentiable. Usually, but not always, the foliations X and Y will arise from integrating lipschitz vector fields. By abuse of notation, the vector fields will bear the same names as the corresponding foliations.

We investigate foliation coordinates relative to X and Y, by comparing with a very special type of foliation coordinates, which we understand completely.

II.2.7.3. Definition. Let α be an oriented geodesic and let $G(\alpha)$ and $E(\alpha)$ be the associated unit vector fields (see Section II.2.5 (*Some standard vector fields*)). Let $z_0 \in \alpha$. We define *standard foliation* coordinates relative to α, as the coordinates defined for the whole of $\mathbb{H}^2$ relative to $G(\alpha)$ and $E(\alpha)$, centred at z_0. Given a vector field X, defined near z_0, we will also talk of standard foliation coordinates relative to X. These are the same as standard foliation coordinates relative to α, where α is the geodesic with tangent $X(z_0)$.

Note that standard foliation coordinates are just Fermi coordinates, already discussed in Proposition II.2.5.1 (*Covariant derivatives of G and E*). The associated coordinate vector fields are G and $\cosh t \cdot E$, where t is the (signed) distance from α. We abbreviate "standard foliation coordinates" to "sfc", and we talk of the sfc metric, etc.

II.2.7.4. Lemma: Lipschitz constant for standard foliation coordinates.   *Let $\phi_{\mathrm{sfc}} \colon \mathbb{R}^2 \to \mathbb{H}^2$ denote standard foliation coordinates relative to an oriented geodesic α and centred on z_0. We give $\mathbb{R}^2$ its usual euclidean metric. Then, in a t-neighbourhood of α, ϕ_{sfc} has lipschitz constant $\cosh t$ and ϕ_{sfc}^{-1} has lipschitz constant 1.*

Proof.   Using obvious orthonormal bases at a point, we see that the derivative of ϕ_{sfc} has a matrix of the form

$$\begin{bmatrix} \cosh t & 0 \\ 0 & 1 \end{bmatrix}.$$

This matrix has norm $\cosh t$ and its inverse has norm 1. The norm of the derivative clearly gives a bound for the path lipschitz constant.

$\square$

II.2.7.5. O notation.   The usual notation O from analysis will be used in this paper in such a way that $O(x)/x$ is bounded by some universal constant (i.e. a definite number like 1.791). When we want to display dependence on some other constant k, we will write $O(x; k)$, etc. In that case, the rate of convergence as well as the actual lim sup may depend on k.

We denote by $B(z, r)$ the open ball in the hyperbolic plane of radius r.

II.2.7.6. Lemma: Angle estimate.   *Let X and Y be orthogonal lipschitz vector fields with classical lipschitz constant k on $B(z_0, r)$ where r is small. Let $\phi_{\mathrm{sfc}} \colon \mathbb{R}^2 \to \mathbb{H}^2$ be standard foliation coordinates relative to X and centred at z_0. Then the angle between X and $\partial/\partial s$ and the angle between Y and $\partial/\partial t$ are bounded by $kr + O(r^2; k)$, where the measurement of angle is done with respect to the sfc metric (and the measurement of r is hyperbolic).*

Proof.   Let d be the signed distance from α. Then $d < r$. The classical lipschitz constants of $X - \cosh d \cdot E$ and $Y - G$ are bounded by $k + \sinh d$, as we see from Proposition II.2.5.1 (*Covariant derivatives of G and E*). Since the two fields are equal to zero at z_0, $\|X - \cosh d \cdot E\|$ and $\|Y - G\|$ are bounded by $(k + \sinh r)r$. Since ϕ_{sfc}^{-1} has lipschitz constant 1 (see Lemma II.2.7.4 (*Lipschitz constant for standard foliation coordinates*)), we have $\|X - \partial/\partial s\|_{\mathrm{sfc}} \leq (k + \sinh r)r$

and $||Y - \partial/\partial t||_{\text{sfc}} \le (k + \sinh r)r$. Now, $\arcsin u = u + O(u^3)$ if u is small. Also $\partial/\partial s$ and $\partial/\partial t$ are unit vectors in foliation coordinates. It follows that the angle between X and $\partial/\partial s$ and the angle between Y and $\partial/\partial t$ are bounded by $kr + O(r^2; k)$, where this angle is measured in standard foliation coordinates.

$\square$

II.2.7.7. Lemma: Leaves meet.

Let X and Y be orthogonal lipschitz vector fields with classical lipschitz constant k on $B(z_0, 2r)$ where r is small. Let A be an X-leaf which meets $B(z_0, r)$ and let B be the Y-leaf through z_0. Then A meets B exactly once, at a point which we denote by u. We have $d(u, z_0) \le r + O(r^2; k)$.

Proof. The idea of the proof is that the result is reasonably obvious in euclidean geometry. Here we want to prove the result in hyperbolic geometry, so we use standard foliation coordinates relative to X, and these induce euclidean geometry. But then we need to know to what extent standard foliation coordinates enable us to accurately estimate hyperbolic distances. With these explanations, we start the formal proof.

Let $z \in A \cap B(z_0, r)$. We work with euclidean geometry on $\mathbb{R}^2$, using standard foliation coordinates relative to X and centred on z_0. We denote by $B_{\text{sfc}}(z_0, r)$ the disk centred at z_0 and radius r, as measured in sfc coordinates. By Lemma II.2.7.4 (*Lipschitz constant for standard foliation coordinates*) $B(z_0, r)$ is contained within $B_{\text{sfc}}(z_0, r)$. Lemma II.2.7.6 (*Angle estimate*) tells us that, within $B_{\text{sfc}}(z_0, 2r) \subset B(z_0, 2r\cosh r)$, the sfc angle of X to the horizontal is bounded by

$$2kr + O(r^2; k)$$

and the sfc angle of Y to the vertical has the same bound.

Let θ be a small angle. Consider the euclidean triangle xyz_0, where x lies on the circle with centre z_0 and radius r so that $z_0 x$ is at an angle θ to the vertical ray pointing upwards from z_0, xy is tangent to the circle, and y lies on the other side of the vertical so that $z_0 y$ is also at an angle θ to the vertical ray. Then $z_0 y$ has length $r \sec 2\theta$.

We apply this in our situation with $\theta = 2kr + O(r^2; k)$. Let l be the horizontal line through z_0. Then l is straight in sfc coordinates, and is a geodesic in hyperbolic coordinates. Let v be any point in the sfc triangle xyz_0. Then

$$d_{\text{hyp}}(v, l) = d_{\text{sfc}}(v, l).$$

II.2.7.8.

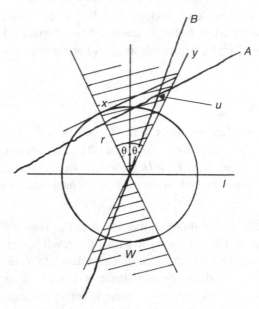

Figure II.2.7.8 *The leaf A meets the leaf B inside the triangle xyz_0. The angle θ shown satisfies $\theta = 2kr + O(r^2; k)$. The picture is drawn in sfc coordinates.*

by the definition of sfc coordinates, and

$$d_{\text{sfc}}(v, l) \leq d_{\text{sfc}}(v, z_0) \leq r \sec(4kr + O(r^2; k)) = r + O(r^2; k).$$

Now the sfc distance from v to l is equal to the hyperbolic distance from v to l. Therefore, from Lemma II.2.7.4 (*Lipschitz constant for standard foliation coordinates*), we see that

$$d_{\text{hyp}}(v, z_0) \leq \cosh(r + O(r^2; k)) \cdot d_{\text{sfc}}(v, z_0) \leq r + O(r^2; k).$$

Hence, for small values of r (how small depends only on k), the triangle xyz_0 is contained in a hyperbolic disk of hyperbolic radius $2r$ and the vector field X is defined on it.

Let W be the closed subset consisting of z_0 and of all the points p in $\mathbb{R}^2$, such that the sfc angle of the sfc straight line pz_0 to the vertical is less than or equal to θ. Then A must intersect this wedge, since the vector field X is defined on a disk whose sfc radius is $2r - O(r^2)$ (in fact the radius is greater than $2r/\cosh 2r$), the angle of A to the horizontal is less than $4kr + O(r^2; k)$, and A meets the sfc disk of radius r. Moreover, it will meet B inside this wedge and inside the triangle xyz_0 (or a triangle like xyz_0, obtained by reflection in the horizontal line through z_0).

Let u be the intersection point of A and B. The sfc distance of u to z_0 is bounded by the sfc length of yz_0. For the hyperbolic distance we therefore have (by Lemma II.2.7.4 (*Lipschitz constant for standard foliation coordinates*))

$$d(u, z_0) \leq \cosh(r + O(r^2; k)) \sec(4kr + O(r^2; k))r = r + O(r^2; k).$$

This is the required result.

$\square$

II.2.7.9. Lemma: Comparison with intrinsic metric. *Let X be a unit orthogonal lipschitz vector field with classical lipschitz constant k on a convex open subset U of a smooth surface. (The result also applies to surfaces like those discussed in Lemma II.2.2.1* (Geometry of a piecewise smooth manifold), *which are obtained by gluing together a finite number of pieces of constant non-negative curvature.) Let γ be the geodesic such that $\gamma(0) = z_0 \in U$ and $\gamma'(0) = X(z_0)$. Let $t > 0$ and let*

$$k_1 = k_1(t) = \frac{e^{kt} - kt - 1}{kt^2}.$$

Then

$$d(\gamma(s), x_s(z_0)) \leq k_1 |s|^2$$

where x_s denotes the solution to X, provided the two curves do not exit from U and provided $|s| < t$.

Proof. We treat the geodesic γ as an approximate solution of the vector field X. There is no loss of generality in taking $s > 0$. We have the estimate $\varepsilon(s) \leq ks$, where ε is the error term. This follows using parallel translation along γ. We now apply the inequality of Theorem II.A.3.2 (*Approximation inequality.*) The integral in the statement of that Theorem is

$$(e^{kt} - kt - 1)/k.$$

By expanding in a power series, we see that

$$(e^{ks} - ks - 1)/ks^2$$

is an increasing function of s. The result follows.

$\square$

II.2.7.10. Lemma: Length continuous.   *Let X and Y be unit orthogonal lipschitz vector fields, defined on some convex open subset U of the hyperbolic plane. Let $a \leq 0 \leq b$ and $c \leq 0 \leq d$. Let $\phi: [a, b] \times [c, d] \to U$ be foliation coordinates relative to X and Y. Then the path length of $\phi([a, b] \times \{t\})$ is a continuous function of t.*

Proof.   Without loss of generality $a = 0$. We need only prove continuity at $t = 0$. By the definition of foliation coordinates, we have to prove that the lengths of the curves are near b, provided that $|t|$ is near zero.

Let x_s be the 1-parameter group of homeomorphisms associated with the unit vector field X. By Theorem II.A.3.2 (*Approximation inequality*), we have

$$d(\phi(s, 0), x_s(\phi(0, t))) = d(x_s(\phi(0, 0)), x_s(\phi(0, t))) \leq e^{ks}|t|.$$

Hence $x_s(\phi(0, t))$ tends to $\phi(s, 0)$ as $|t|$ tends to zero. Taking $s = b - \varepsilon$ and $s = b + \varepsilon$, we find that, for small values of $|t|$, the curve $x_s(\phi(0, t))$ passes through $\phi(b, t)$ for some s with $b - \varepsilon < s < b + \varepsilon$.

$\square$

II.2.8. Formulas in the hyperbolic plane

II.2.8.1. Parallel proofs.   We now prove three lemmas containing the main estimates we need. Analogous lemmas with practically the same statement will be proved in this section on the hyperbolic plane, in Section II.2.9 (*Flat equidistant surfaces*), in Section II.2.12 (*Finitely bent equidistant*) and in lemma under Section II.2.13.1 (*General equidistant*). We attempt to help the reader find his/her way by naming these lemmas with similar names in the different contexts. Each time, the proof in the later sections consists of referring back to the preceding version.

II.2.8.2. Lemma: Length lipschitz.   *Let U be a convex open set in the hyperbolic plane, and let X and Y be unit orthogonal vector fields on U with classical lipschitz constant k. Let $a \leq 0 \leq b$ and $c \leq 0 \leq d$ and let ϕ: $[a, b] \times [c, d] \to U$ be a foliation chart relative to X and Y. By the previous lemma, there is an upper bound, m say, for the lengths of the horizontal curves $\phi([a, b] \times t)$. Let $u_{a,b}$: $\phi(a \times [c, d]) \to \phi(b \times [c, d])$ be the homeomorphism which maps $\phi(a, t)$ to $\phi(b, t)$. Then $u_{a,b}$ is e^{km} path-lipschitz.*

Proof.   The two cases $a = 0 < b$ and $a < b = 0$ are really the same. By combining these, we see that there is no loss of generality in assuming that $a = 0$.

Let $s =$ length $\phi|[0, b] \times t'$. Then $s \leq m$ and $\phi(b, t') = x_x \phi(0, t')$. By Theorem II.A.3.2 (*Approximation inequality*) we have

$$d(\phi(b, t'), x_s \phi(0, t)) \leq |t - t'|e^{km}.$$

Let $\varepsilon > 0$ be chosen arbitrarily. If $|t - t'|$ is small enough, we can apply Lemma II.2.7.7 (*Leaves meet*) to deduce that

$$d(\phi(b, t), \phi(b, t') \leq |t - t'|e^{km}(1 + \varepsilon).$$

Taking a sequence of finer and finer partitions of the domain and range of $u_{a,b}$, we find that

$$\text{length } \phi|b \times [t', t] \leq (1 + \varepsilon)e^{km}\text{length } \phi|0 \times [t', t].$$

Since ε is arbitrary, the lemma is proved.

$\square$

II.2.8.3. Lemma: Estimates for psi.   *Let U be a convex open set in the hyperbolic plane, and let X and Y be unit orthogonal vector fields on U with classical lipschitz constant k. Let ϕ: $[0, s] \times [0, v] \to U$ be a foliation chart, relative to X and Y, with v small. Let ψ be defined by*

$$\psi(t) = \text{length } \phi|[0, t] \times v.$$

Then ψ is a homeomorphism from one interval to another. It is differentiable almost everywhere, and its Radon–Nikodym derivative is equal to its derivative in the usual sense. (This is a fancy way of saying that the integral of the derivative is equal to the function.) We have the following estimates

$$1 - ke^{kt}v - O(v^2; k, s) \leq \psi'(t) \leq 1 + ke^{kt}v + O(v^2; k, s)$$

for any value of t such that $\psi'(t)$ exists, and

$$t - v(e^{kt} + 1) - O(v^2; k, s) \leq \psi(t) \leq t + v(c^{kt} + 1) + O(v^2; k, s).$$

II.2.8.4. Remark: The proof of this result will use only the estimate in Lemma II.2.8.2 (*Length lipschitz*). Therefore the results will be applicable whenever we can prove the estimate of that lemma. We will in fact want to apply the result in other situations, in particular on the δ-distant surface from the convex hull boundary.

Proof. The previous lemma shows that ψ is lipschitz. Also ψ is strictly mono-tonic. It is a standard fact from Analysis (see [Hewitt–Stromberg, 1965, p. 344]) that ψ is then differentiable almost everywhere, and that its derivative is equal to its Radon–Nikodym derivative.

Let

$$\varepsilon(t,v) = \max_{0 \le w \le v} \{\text{length } \phi | [0,t] \times w - t, 0\}.$$

Let $v(t) = \text{length } \phi | t \times [0,v]$. We abbreviate $\varepsilon(t,v)$ to ε. Then by Lemma II.2.8.2 (*Length lipschitz*)

$$v(t) \le \exp(kt) \exp(k\varepsilon)v.$$

Applying the same lemma again, but with vertical and horizontal interchanged, we find:

$$t + \varepsilon \le \exp(k \max_{0 \le u \le t} v(u))t.$$

We deduce that

$$\varepsilon \le [\exp(kv \exp(k\varepsilon) \exp(kt)) - 1]t \le [\exp(kv \exp(k\varepsilon) \exp(ks)) - 1]s.$$

For a fixed value of v, we regard $[\exp(kv \exp(k\varepsilon) \exp(ks)) - 1]s$ as a function of ε, forgetting for the moment about the dependence of ε on t and v. Its graph crosses the graph of the identity function exactly twice, provided v is small enough, and, as v tends to zero, one of the crossing points tends to the origin and the other tends to infinity. By the Implicit Function theorem, there is an analytic function $\alpha(v)$, corresponding to the crossing point near the origin, and such that $\alpha(0) = 0$. Since $\varepsilon(0,v) = 0$ and $\varepsilon(t,v) \ge 0$ for all $t \le s$, we deduce from continuity in $w(0 \le w \le v)$ that $\varepsilon(t,w) \le \alpha(w) = O(w; k, s)$. Therefore

$$\varepsilon \le [\exp\{kv(1 + O(v; k,s)) \exp(kt)\} - 1]t \le kvte^{kt} + O(v^2; k, s).$$

It now follows that

$$v(t) \le \exp(kt)v + O(v^2; k, s).$$

The derivative of ψ at t, if it exists, is bounded above by $e^{kv(t)}$ and below by $e^{-kv(t)}$, because the derivative must be smaller than the lipschitz constant, and we know a bound for the lipschitz constant from Lemma II.2.8.2 (*Length lipschitz*). The inequalities in the statement of the lemma for the derivative of ψ now follow. The inequalities for ψ itself follow by integration.

II.2.8.5. Lemma: Infinitesimal distances. *Let X and Y be orthogonal unit vector fields on a convex open subset U of the hyperbolic plane, and suppose they have classical lipschitz constant k. Let z_1 and z_2 be nearby points of U. Let $\phi: [0, u] \times [0, v] \to U$ be a foliation chart relative to X and Y, centred on z_1 and such that $\phi(u, v) = z_2$. Let $d = d(z_1, z_2)$ and let $r^2 = u^2 + v^2$. Then*

$$\frac{d}{r} - 1 = O(d; k) = O(r; k).$$

II.2.8.6.

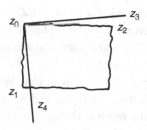

Figure II.2.8.6 *Relation of foliation coordinates to geodesic.*

Proof. Let $z_0 = \phi(0, v)$. Let $u_1 = \text{length } \phi|[0, u] \times v$. By Lemma II.2.8.3 *(Estimates for psi)*, $u_1 \le u + O(r^2; k)$.

Let α and β be the geodesics through z_0, such that $\alpha(0) = \beta(0) = z_0$, $\alpha'(0) = X(z_0)$ and $\beta'(0) = -Y(z_0)$. Let $z_3 = \alpha(u)$ and let $z_4 = \beta(v)$. By Lemma II.2.7.9 *(Comparison with intrinsic metric)*,

$$d(z_1, z_4) = O(v^2; k) \text{ and}$$
$$d(z_2, z_3) = O(u_1^2; k) = O(u^2; k) + O(r^3; k).$$

We deduce that

$$\left| \frac{d(z_1, z_2)}{r} - 1 \right| \le \left| \frac{d(z_3, z_4)}{r} - 1 \right| + \left| \frac{d(z_1, z_4) + d(z_2, z_3)}{r} \right|$$

$$\le \left| \frac{\text{arccosh} (\cosh u \cosh v)}{r} - 1 \right| + O(r; k) = O(r; k).$$

It follows that anything which is $O(r; k)$ is also $O(d; k)$ and the lemma is proved.

◻

II.2.9. Flat equidistant surfaces

We now repeat the lemmas of Section II.2.8 (*Formulas in the hyperbolic plane*), in the case of a flat equidistant surface. The reader is referred to Section II.2.8.1 (*Parallel proofs*) for an explanation.

We fix a hyperbolic plane S in hyperbolic 3-space, and an equidistant surface S_δ, at a distance δ from S. We measure distances on S_δ by using the path metric in S_δ. The nearest point map $r: S_\delta \to S$ reduces distances by a factor $\cosh \delta$, and is a diffeomorphism. Given a lamination $\mathcal{L}$ on S, we obtain an induced lamination $\mathcal{L}_\delta$ on S_δ. We recall from Section II.2.4 (*Extending a lamination to a pair of orthogonal foliations*), that μ is a fixed number with $\mu < (\log 3)/2$, that U_μ is the open μ-neighbourhood of $\mathcal{L}$, and that $\mathcal{L}_\mu$ and $\mathcal{F}_\mu$ are the orthogonal foliations defined in that section. Let k be the lipschitz constant of the line fields $\mathcal{L}_\mu$ and $\mathcal{F}_\mu$. The foliations $\mathcal{L}_\mu$ and $\mathcal{F}_\mu$ are also mapped by r^{-1} to foliations on S_δ, which we denote by $\mathcal{L}_{\delta,\mu}$ and $\mathcal{F}_{\delta,\mu}$.

Let U be an open subset of $U_\mu \subset S$, and let $U_\delta \subset S_\delta = r^{-1}(U)$. The results are deduced immediately from their analogues and we omit the proofs, confining ourselves to formulating the results. The only point we need to watch for is to check that the change of scale $\cosh \delta$ acts in our favour and not against us. (So the estimates could be very slightly improved. However, when we come to apply these results, δ is so small that the factor $\cosh \delta$ makes no practical difference.)

Let $a \leq 0 \leq b$ and $c \leq 0 \leq d$ and let $\phi: (a, b) \times (c, d) \to U_\mu$ be a foliation chart relative to $\mathcal{L}_\mu$ and $\mathcal{F}_\mu$. Let R be the image of ϕ. We suppose that R is contained in a convex subset of U_μ. Let R_δ be the inverse image in S_δ of R under r. In order to retain the symmetry of the results, we will use X and Y to denote the unit vector fields associated to $\mathcal{L}_\mu$ and $\mathcal{F}_\mu$, though not necessarily in that order. Let k be the lipschitz constant for X and Y. We use X_δ and Y_δ to similarly denote unit vector fields associated to $\mathcal{L}_{\delta,\mu}$ and $F_{\delta,\mu}$. The classical lipschitz constant of X_δ and Y_δ with respect to the intrinsic metric on S_δ is $k/\cosh \delta$. We replace this by the larger constant k.

II.2.9.1. Lemma: Length lipschitz (flat equidistant).   *Let ABDC be a subrectangle of R_δ, such that AB and CD are X_δ-curves and AC and BD are Y_δ-curves. Let m be the maximum length of any of the Y_δ-leaves in the rectangle R_δ. Then*

$$\text{length}(AB) \leq \exp(km) \times \text{length}(CD).$$

II.2.9.2. Lemma: Estimates for psi (flat equidistant).   *Let $U \subset U_\mu$ be a convex open set in the hyperbolic plane, and let X and Y be the unit orthogonal* *vector fields on U, with classical lipschitz constant k, described above. Let* $\phi_\delta \colon [0, s] \times [0, v] \to U_\delta$ *be a foliation chart, relative to* X_δ *and* Y_δ, *with v small. Let*

$$\psi_\delta(t) = \text{length } \phi_\delta | [0, t] \times v.$$

Then

$$1 - ke^{kt}v - O(v^2; k, s) \le \psi'_\delta(t) \le 1 + ke^{kt}v + O(v^2; k, s)$$

for any value of t such that $\psi'_\delta(t)$ *exists, and*

$$t - v(e^{kt} - 1) - O(v^2; k, s) \le \psi_\delta(t) \le t + v(e^{kt} + 1) + O(v^2; k, s).$$

II.2.9.3. Lemma: Infinitesimal distances (flat equidistant). *Let X and Y be the orthogonal unit vector fields described above on a convex open subset* $U \subset U_\mu$. *Let* z_1 *and* z_2 *be nearby points of* U_δ. *Let* $\phi_\delta \colon [0, u] \times [0, v] \to U_\delta$ *be a foliation chart relative to* X_δ *and* Y_δ, *centred on* z_1 *and such that* $\phi_\delta(u, v) = z_2$. *Let* $d = d_\delta(z_1, z_2)$, *where* d_δ *is the path distance in* U_δ, *and let* $r^2 = u^2 + v^2$. *Then*

$$d/r - 1 = O(d; k) = O(r; k).$$

II.2.10. Foliations on the convex hull

We now turn our attention to a convex hull boundary, which we denote by S. We continue to work with the fixed number μ, where $0 < \mu < (\log 3)/2$. For simplicity, we will assume that S is simply connected, although some of the proofs in this section will go through, even if S is not simply connected. (However Sullivan's theorem fails for general non-simply-connected S. See Section II.2.16 (*Counterexample*).) Thus, we have a compact subset X of the 2-sphere, whose complement is a topological open disk, and S is the boundary of the hyperbolic convex hull of X. Equivalently, X is a connected compact subset.

Let $h \colon \mathbb{H}^2 \to S$ be the isometry defined in Corollary I.1.12.7 (*Isometry to the hyperbolic plane*). Let $\mathcal{L}$ be the corresponding bending lamination on $\mathbb{H}^2$. We obtain a well-defined line field and foliation $\mathcal{L}_\mu$ on $U_\mu \subset \mathbb{H}^2$, as described in Section II.2.4 (*Extending a lamination to a pair of orthogonal foliations*). We transfer $\mathcal{L}$ and $\mathcal{L}_\mu$ to S by using h, and we denote the corresponding lamination and line field by $h\mathcal{L}$ and $h\mathcal{L}_\mu$, respectively. The line field $h\mathcal{L}_\mu$ is not in general lipschitz in the metric of $\mathbb{H}^3$, because parts of S which have nothing to do with each other may be near in $\mathbb{H}^3$ (see Figure II.2.10.1). In Section II.A.1 (*The lipschitz property*) we defined the concept of "path-lipschitz", and it is in this sense that the line field defined on S is lipschitz. We have to restrict our attention to how the line field varies along rectifiable paths in S. Moreover, the metric in the range space is the metric which comes from the tangent bundle of $\mathbb{H}^3$ (*cf.* Section II.A.2.1 (*Metric on tangent bundle*)).

II.2.10.1.

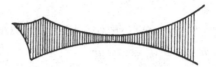

Figure II.2.10.1   *The convex hull boundary may come back near itself.*

II.2.10.2. Lemma: Lipschitz on convex hull boundary. *The line field $h \mathcal{L}_\mu$ is k_0-path-lipschitz with respect to the path metric on S, where $k_0 = \sqrt{1 + k^2}$.*

The expression for the lipschitz constant comes from Section II.2.6.1 (*Lipschitz (H2)*) and from Lemma II.A.2.4 (*Lipschitz constant*).

Proof.   **Lipschitz on convex hull boundary:** Let S_n be a standard sequence of outer approximations, and let $h_n \colon \mathbb{H}^2 \to S_n$ be the isometry of Proposition II.1.12.6 (*Covering maps from the hyperbolic plane*). Let $\mathcal{L}_n$ be the corresponding bending lamination in $\mathbb{H}^2$. Let $U_{n,\mu}$ be the open μ-neighbourhood of $\mathcal{L}_n$ in $\mathbb{H}^2$ and let $\mathcal{L}_{n,\mu}$ and $\mathcal{F}_{n,\mu}$ be the pair of orthogonal foliations on $U_{n,\mu}$.

We will first prove that $h_n \mathcal{L}_{n,\mu}$ is k_0-path-lipschitz. To see this, we need to check along paths in S_n, and, as usual, we can reduce first to piecewise geodesic paths, and then to short paths which are geodesic in $\mathbb{H}^3$. The lipschitz property in the case of such finitely bent surfaces then follows from Lemma II.2.6.1 (*Lipschitz (H2)*).

We now prove that the line field $h \mathcal{L}_\mu$ is k_0-lipschitz. We want to prove this by limiting arguments, and here a little care must be taken. Note that it is quite possible that each bending line of S_n fails to be a bending line of, or even lie on, S. However, many geodesics in the lamination of S do lie on S_n – more and more of them, more and more densely packed as n increases. These geodesics may lie in the interior of flat pieces of S_n.

Using an argument by contradiction, we now show that $h \mathcal{L}_\mu$ is k_0-path-lipschitz. Let γ be a rectifiable path in S, and suppose that

$$\operatorname{length}(h \mathcal{L}_\mu {}^\circ \gamma) > k_0 \operatorname{length}(\gamma).$$

We recall that the length of a rectifiable path is approximated by taking a very finely spaced sequence of points along the path, and taking the geodesic distances between these points. It follows from the last sentence of Lemma II.1.5.7 (*Intrinsic metric lipschitz*) that we may assume that there are points $x_1, x_2 \in hU_\mu$ such that

$$d_T(h\mathcal{L}_\mu(x_1), h\mathcal{L}_\mu(x_2)) > k_0 d_S(x_1, x_2) \qquad \text{Inequality D}$$

where d_T is the distance in the tangent bundle of $\mathbb{H}^3$ (*cf.* Section II.A.2.1 (*Metric on tangent bundle*)) and d_S is the path distance in S. We will also assume that x_1 and x_2 lie in some small shell neighbourhood and that the intersection of the closure of the shell with S is contained in hU_μ.

As pointed out above, $h\mathcal{L}_\mu$ and $h_n\mathcal{L}_{n,\mu}$, need not agree at a special point. However, if x lies in the interior of a flat piece F of S, then the flat piece F_n of S_n containing x will converge towards F. The continuity properties of $\mathcal{L}_\mu$ discussed in Section II.2.6.2 (*Continuous dependence (H2)*) show that $h_n\mathcal{L}_{n,\mu}(x)$ converges towards $h\mathcal{L}_\mu(x)$ in this case.

If x is a special point in a bending line of $h\mathcal{L}$, let F_n be a flat piece of S_n containing x (possibly in the boundary of F_n). Let α be a nearest boundary component of F_n to x, and let y be the nearest point on α to x. Then y (which depends on n, although the notation obscures this fact) converges to x as n tends to infinity. The k_0-path-lipschitz property of $h_n\mathcal{L}_{n,\mu}$ shows that $h_n\mathcal{L}_{n,\mu}(x)$ is near $h_n\mathcal{L}_{n,\mu}(y)$. The geodesic of $h\mathcal{L}$, on which x lies, lies in F_n and is disjoint from or equal to α. It follows from Lemma I.1.8.2 (*Disjoint geodesics*) that $h\mathcal{L}_\mu(x)$ is near $h_n\mathcal{L}_{n,\mu}(y)$. Therefore $h_n\mathcal{L}_{n,\mu}(x)$ converges to $h\mathcal{L}_\mu(x)$.

Since special points are dense in S, we may assume that in Section II.2.10.2 (*Inequality D*) x_1 and x_2 are such points. If d_n is the path distance in S_n, then $d_n(x_1, x_2)$ converges to $d_S(x_1, x_2)$. Using the results of the preceding two paragraphs, we may assume that

$$d_T(h_n\mathcal{L}_{n,\mu}(x_1), h_n\mathcal{L}_{n,\mu}(x_2)) > k_0 d_n(x_1, x_2).$$

Since we are working in a shell, we may assume that $d_n(x_1, x_2)$ is given by a path in hU_μ (see the last sentence of the statement of Lemma I.1.5.7 (*Intrinsic metric lipschitz*)). But this contradicts what we have already proved, namely that $h_n\mathcal{L}_{n,\mu}$ is k_0-path-bilipschitz on S_n.

Lipschitz on convex hull boundary

II.2.10.3. Lemma: Continuous dependence (convex hull boundary).

Let S_n be a standard sequence of finite outer approximations to S. The line fields

$h_n \mathcal{L}_{n,\mu}$ converge to $h\mathcal{L}_\mu$, uniformly on compact subsets of $S \cap hU_\mu$, in the sense that, given a compact subset $K \subset S \cap hU_\mu$ and a small positive number δ, we have for n large, depending only on δ and K, if $x_n \in S_n$ is in the δ-neighbourhood of $x \in K$, then $x \in h_n U_{n,\mu}$ and the tangent line to $h\, \mathcal{L}_{n,\mu}$ at x_n is near the tangent line to $h\, \mathcal{L}_\mu$ at x.

Proof.   Continuous dependence (convex hull boundary): Let K be a compact subset of $S \cap hU_\mu$, and suppose the convergence statement is false for K. Then, for each n, we have points $x_n \in S_n$ and $y_n \in K$, such that the distance of x_n to y_n converges to zero, and $h_n \mathcal{L}_{n,\mu}(x_n)$ is not near $h\mathcal{L}_\mu(y_n)$. Without loss of generality, we may assume that x_n and y_n converge to $x \in S$.

If x lies in a flat piece of S, then the plane of this flat piece is eventually one of the defining planes for S_n. Therefore, by the convergence properties of $h_n \mathcal{L}_{n,\mu}$ which follow from Section II.2.6.2 (*Continuous dependence (H2)*), x cannot lie in the interior of a flat piece of S.

Let $z_n \in S_n \cap S$ be a nearest point (in the metric of $\mathbb{H}^3$) to x_n, which lies in a bending line B_n of $h\mathcal{L}$. Since x is not contained in a flat piece of S, z_n will converge to x (special points are dense). Either B_n is also a bending line for S_n, or it lies in a flat piece of S_n, near the boundary of the flat piece if n is large. In the first case, $h_n(\mathcal{L}_{n,\mu}(z_n))$ is equal to $h\mathcal{L}_\mu(z_n)$. In the second case, $h_n \mathcal{L}_{n,\mu}(z_n)$ is near $h\mathcal{L}_\mu(z_n)$ as we see by applying Lemma II.1.8.2 (*Disjoint geodesics*) to B_n and the nearby bounding geodesic of the flat piece, and also by noting that, by Lemma II.2.10.2 (*Lipschitz on convex hull boundary*), $\mathcal{L}_{n,\mu}$ is lipschitz.

Now z_n is near x_n in the metric of $\mathbb{H}^3$. Since y_n is near x_n in the metric of $\mathbb{H}^3$, it is also near z_n in the metric of $\mathbb{H}^3$. It follows from Lemma II.1.5.7 (*Intrinsic metric lipschitz*) that z_n is near x_n in the intrinsic metric on S and that z_n is near y_n in the intrinsic metric on S_n. Since $h_n \mathcal{L}_{n,\mu}$ and $h\mathcal{L}_\mu$ are both k_0-path-lipschitz, we obtain a contradiction.

> **Continuous dependence (convex hull boundary)**

II.2.11.   Three orthogonal fields

We continue to work with a fixed number $\mu > 0$, where $\mu < (\log 3)/2$. Let V be the component of the complement of the convex hull, which retracts onto a fixed simply connected boundary component, S, using the nearest point retraction r. Let $V_\mu = r^{-1}(U_\mu)$. We construct three orthogonal line fields on V_μ.

We will fix $\delta_1 < \delta_2$, and work in a region where the distance from S lies between δ_1 and δ_2:

(1) The vector field R on V is the unit vector at x pointing away from rx. This vector is the composite

$$V \rightarrow \mathbb{H}^3 \times \mathbb{H}^3 \backslash \Delta \rightarrow T\mathbb{H}^3 \backslash 0 \rightarrow T_1 \mathbb{H}^3$$

where the first map is $x \mapsto (x, rx)$, the second map is the inverse of $(x, v) \mapsto (x, \exp_x v)$ and the third map is $(x, v) \mapsto (x, v)/||v||$. The first map is 2-path-lipschitz since r is distance decreasing. The second and third maps are real analytic, and do not depend in any way on the particular situation. Therefore the composite is k_1-path-lipschitz, where k_1 depends only on δ_1 and δ_2.

(2) On $V_\mu = r^{-1}(U_\mu)$, we define a field $\tau(\mathcal{L}, V_\mu)$, orthogonal to R, as follows. Given $x \in V_\mu$, consider the plane containing x, rx and the tangent to the leaf of $\mathcal{L}_\mu$ through rx. Then $\tau(\mathcal{L}, V_\mu)(x)$ is the unique line in this plane orthogonal to R and containing x.

The line field $\tau(\mathcal{L}, V_\mu)$ is tangent to the surfaces equidistant from S, since it is orthogonal to R. The foliation on a single equidistant surface S_δ is the inverse image under $r: S_\delta \rightarrow S$ of the foliation $\mathcal{L}_\mu$ on S. This is seen by restricting to single facets of S.

If we are dealing with a finite lamination on S, then the field $\tau(\mathcal{L}, V_\mu)$ is constructed as follows. If $x \in V_\mu$ and rx is contained in a flat piece of S, then we have the equidistant surface from the plane of the flat piece which passes through x. The vector at x is tangent to the equidistant surface and comes from the tangent vector at rx by parallel translation. If $x \in V_\mu$ and rx is contained in a bending line, then the equidistant surface to the bending line is a cylinder, in fact isometric to a euclidean cylinder. In this case the tangent vector at x is along a generator of the cylinder. So the vector field, restricted to an equidistant surface, in the case of a finite lamination, consists of pieces on which it is a copy of what is happening on S, and pieces of cylinders on which it is a foliation by generators of the cylinder.

(3) The line field $\tau(\mathcal{F}, V_\mu)$ is defined to be orthogonal to the preceding two fields. On an equidistant surface, the foliation is the inverse image under r of the foliation $h\mathcal{F}_\mu$. The vector field can be described in the case of a finite lamination, just as in the previous case. Over flat pieces, it is a copy of the orthogonal foliation. On the cylindrical part, it consists of vectors orthogonal to the generators.

II.2.11.1. Lemma: Continuous dependence (H3).   *If S_n is a standard sequence of finite outer approximations to S, then we have uniform convergence $R_n \to R$ on compact subsets of V, and uniform convergence*

$$\tau(\mathcal{L}_n, V_{n,\mu}) \to \tau(\mathcal{L}, V_\mu), \tau(\mathcal{F}_n, V_{n,\mu}) \to \tau(\mathcal{F}, V_\mu)$$

on compact subsets of V_μ.

Proof.   Continuous dependence (H3): The first statement follows from the factorization of R as

$$V \to \mathbb{H}^3 \times \mathbb{H}^3 \backslash \Delta \to T\mathbb{H}^3 \backslash 0 \to T_1 \mathbb{H}^3$$

since only the first map changes and r_n converges to r. To prove the second statement, we apply Lemma II.2.10.3 (*Continuous dependence (convex hull boundary)*). The third statement follows from the first two.

> **Continuous dependence (H3)**

II.2.11.2.   We now show that the three fields are lipschitz. In fact, we have already seen that R is lipschitz, with constants depending only on δ_1 and δ_2. Suppose we can show that $\tau(\mathcal{L}, V_\mu)$ is k_0-path-lipschitz, for some universal constant k_0 (we can take $k_0 = \sqrt{1 + k^2}$). It will then also follow that $\tau(\mathcal{F}, V_\mu)$ is also lipschitz, with constant depending only on δ_1 and on δ_2. To see this, note that the function which constructs from two orthogonal unit vectors a third unit orthogonal vector, which is uniquely determined up to sign, is an analytic function on a compact space. It is therefore path-lipschitz, with a universal lipschitz constant. So the result required about $\tau(\mathcal{F}, V_\mu)$ follows.

II.2.11.3. Lemma: Lipschitz L (H3).   *The vector field $\tau(\mathcal{L}, V_\mu)$ is k_0-path-lipschitz, where $k_0 = \sqrt{1 + k^2}$.*

Proof.   Lipschitz L (H3): We prove the result first for a finitely bent surface S. Here the field is built up from a finite number of pieces, namely the inverse images under the nearest point retraction r of the facets of S. Over the flat pieces of S, the field is k_0-path-lipschitz, by Section II.2.9 (*Flat equidistant surfaces*). Over a bending line B the path-lipschitz constant is $\sqrt{2}$. This is because parallel translation along any hyperbolic plane orthogonal to a bending line of S preserves the vector field tangent to the foliation by lines equidistant from the bending line. Hence covariant differentiation in a direction orthogonal to the field is zero. Covariant differentiation in the direction of the field can be computed in the plane which contains B and a single tangent vector to the field. In this plane, the field is equal to E, and the path lipschitz constant can be computed from the classical lipschitz constant, which is given by Lemma II.2.5.1 (*Covariant derivatives of G and E*).

The above argument proves the result for a finitely bent surface. The general result follows since $\tau(\mathcal{L}_n, V_{n,\mu})$ converges uniformly to $\tau(\mathcal{L}, V_\mu)$ on compact subsets.

> **Lipschitz L (H3)**

II.2.11.4. Remark. Unbounded lipschitz constant.

The vector field $\tau(\mathcal{L}, v_\mu)$ extends continuously to the field $h\mathcal{L}_\mu$ on S. In contrast, while $h\mathcal{F}_\mu$ exists as a foliation, it does not correspond to a continuous line field in general. At a bending line with non-zero bending angle, we will have a discontinuity. Thus, as δ_1 becomes smaller, the lipschitz constant of $\tau(\mathcal{F}, V_\mu)$ will, in most cases, tend to infinity.

II.2.12. The equidistant surface from a finitely bent convex hull boundary

In this section, S will be a simply connected finitely bent convex hull boundary, and we will show how to use the pair of orthogonal foliations on the equidistant surface S_δ, constructed from the bending lamination, to measure distances on S_δ. We will prove analogues of the results in Section II.2.8 (*Formulas in the hyperbolic plane*), and we refer readers to Section II.2.8.1 (*Parallel proofs*) for an explanation.

The vector fields X_δ and Y_δ are the fields $\tau(\mathcal{F}, V_\mu)$ and $\tau(\mathcal{L}, V_\mu)$, restricted to the surface S_δ. (We may wish to interchange X_δ and Y_δ, but we will not allow arbitrary fields.) Let k be the lipschitz constant of $\mathcal{L}_\mu$ and $\mathcal{F}_\mu$ on the open μ-neighbourhood U_μ of the bending lamination $\mathcal{L}$ in $\mathbb{H}^2$. X_δ and Y_δ have classical lipschitz constant k with respect to the intrinsic metric on S_δ.

The surface S_δ is a finite union of pieces, each piece being the inverse image under r of a facet of S. There are two different types of pieces. We have *negatively curved pieces*, which are at distance δ from a flat piece of S. These have curvature $-1/\cosh^2(\delta)$, as pointed out at the beginning of Section II.2.2 (*The epsilon distant surface*). We also have *cylindrical pieces*, which are at a distance δ from a bending line of S. These have a euclidean metric. We call the boundary geodesics of a cylindrical piece *boundary geodesics*.

On S_δ, any two points are joined by a unique geodesic. We have seen this in Lemma II.2.2.1 (*Geometry of a piecewise smooth manifold*).

Let $a \leq 0 \leq b$ and $c \leq 0 \leq d$ and let $\phi : (a,b) \times (c,d) \to U_\mu$ be a foliation chart relative to $h\mathcal{L}_\mu$ and $h\mathcal{F}_\mu$. Let R be the image of ϕ. We suppose that R is contained in a convex subset of U_μ (i.e. convex with respect to the intrinsic metric which makes S isometric to the hyperbolic plane). Let R_δ be the inverse image in S_δ of R under r. We use X_δ and Y_δ to denote the unit orthogonal vector fields $h\mathcal{L}_{\delta,\mu}$ and $h\mathcal{F}_{\delta,\mu}$, though not necessarily in that order.

II.2.12.1. Lemma: Length lipschitz (finitely bent equidistant). *Let ABDC be a subrectangle of R_δ, such that AB and CD are X_δ-curves and AC and BD are Y_δ-curves. Let m be the maximum length of any of the Y_δ-leaves in the rectangle R_δ. Then*

$$\text{length}(AB) \leq \exp(km) \cdot \text{length}(CD).$$

Proof. Suppose first that AC is on an $\mathcal{L}$-curve. By breaking up the rectangle into negatively curved and flat pieces, we see that we may assume that the rectangle consists of only one piece. An appeal to Lemma II.2.9.1 (*Length lipschitz (flat equidistant)*) proves the desired inequalities for a negatively curved piece. For a cylindrical piece the length of the segment AB is equal to the length of the corresponding segment on CD, because on the cylindrical pieces the two foliations are orthogonal foliations by lines which are straight in the euclidean metric. This completes the proof if AC is an $\mathcal{L}$-curve.

II.2.12.2.

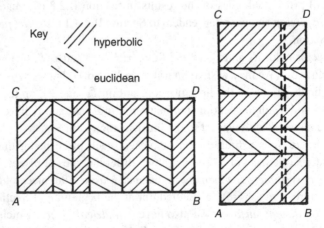

Figure II.2.12.2 *The different shadings represent negatively curved and flat pieces. The two cases illustrated are discussed separately in the proof.*

If AC is an $\mathcal{F}$-curve, then we break the rectangle up into a finite number of extremely thin pieces, as illustrated by the dotted lines in Figure II.2.12.2. That is, we may assume that AB is extremely small, so small in fact, that the length of the Y_δ-curves, in each of the subrectangles of constant curvature, hardly changes. Each thin subrectangle of constant curvature gives rise to one inequality, and, by composing the inequalities, we get the required result.

$\square$

II.2.12.3. Lemma: Estimates for psi (finitely bent equidistant). *Let $U \subset U_\mu$ be a convex open set in the hyperbolic plane, and let X and Y be the unit orthogonal vector fields on U described above. Let $\phi_\delta \colon [0, s] \times [0, v] \to U_\delta$ be a foliation chart, relative to X_δ and Y_δ, with v small. Let $\psi_\delta(t)$ be defined by*

$$\psi_\delta(t) = \text{length } \phi_\delta | [0, t] \times v.$$

Then

$$1 - ke^{kt} - O(v^2; k, s) \le \psi_\delta'(t) \le 1 + ke^{kt}v + O(v^2; k, s)$$

for any value of t such that $\psi_\delta'(t)$ exists, and

$$t - v(e^{kt} - 1) - O(v^2; k, s) \le \psi_\delta(t) \le t + v(e^{kt} - 1) + O(v^2; k, s).$$

As pointed out in the remark after Lemma II.2.8.3 (*Estimates for psi*), this follows from the preceding lemma. The following lemma is then a consequence, just as in the flat case.

II.2.12.4. Lemma: Infinitesimal distances (finitely bent equidistant).
Let X and Y be the orthogonal unit vector fields described above on a convex open subset $U \subset U_\mu$. Let z_1 and z_2 be nearby points of U_δ. Let $\phi_\delta \colon [0, u] \times [0, v] \to U_\delta$ be a foliation chart relative to X_δ and Y_δ, centred on z_1 and such that $\phi_\delta(u, v) = z_2$. Let $d = d_\delta(z_1, z_2)$, where d_δ is the path distance in U_δ, and let $r^2 = u^2 + v^2$. Then

$$\frac{d}{r} - 1 = O(d; k) = O(r; k).$$

II.2.13. Surfaces equidistant from a general convex hull boundary

In this section S is a general simply connected convex hull boundary, obtained from the convex hull of a closed subset X of the 2-sphere. We will prove results analogous to those in Section II.2.8 (*Formulas in the hyperbolic plane*), and we refer readers to Section II.2.8.1 (*Parallel proofs*) for an explanation.

As before, we have the vector fields $\tau(\mathcal{F}, V_\mu)$ and $\tau(\mathcal{L}, V_\mu)$, which are the tangent fields to foliations on $V_\mu \cap S_\delta$. We wish to use the foliations to measure distances on S_δ.

II.2.13.1. Lemma: General equidistant. *The three Lemmas II.2.12.1 (Length lipschitz (finitely bent equidistant)), II.2.12.3 (Estimates for psi (finitely bent equidistant)), and II.2.12.4 (Infinitesimal distances (finitely bent equidistant)) are true in the case of a surface equidistant from a general simply connected convex hull boundary. The same constants are valid.*

Proof. We will define a standard sequence of finite outer approximations S_n. We remind the reader that such a sequence is defined by means of special pairs. If x is a point outside the convex hull of X, we denote by P_x the plane through rx orthogonal to the geodesic from x to rx.

We start by proving the analogue of the first of the three lemmas. We will assume that the special pairs include (rA, P_A), (rB, P_B), (rC, P_C), and (rD, P_D). This means that we may assume that A, B, C, and D lie on $S_{n,\delta}$ for large enough n. We define B_n so that the length of the foliation curve on $S_{n,\delta}$ of AB_n is equal to the length on S_δ of AB. We define C_n similarly. Finally, we define D_n by completing the rectangle on $S_{n,\delta}$.

By Lemma II.2.11.1 (*Continuous dependence (H3)*), the vector field $\tau(\mathcal{F}_n, V_{n,\mu})$ converges uniformly on compact sets to $\tau(\mathcal{F}, V_\mu)$, and the vector field $\tau(\mathcal{L}_n, V_\mu)$ converges uniformly to $\tau(\mathcal{L}, V_\mu)$. It follows from Theorem II.A.3.2 (*Approximation inequality*) that the sides of the rectangle $A_n B_n D_n C_n$ will converge uniformly as paths parametrized by arclength to the sides of $ABDC$. The first of the three lemmas now follows by taking limits.

The second lemma now follows formally, as pointed out in the remark after Lemma II.2.8.3 (*Estimates for psi*). The third lemma also follows because path distance in S_n converges to path distance in S (provided the points, whose distance apart is being measured, lie in both surfaces), and by using Theorem II.A.3.2 (*Approximation inequality*) to keep control over lengths of leaves.

II.2.14. The map ρ

We have foliated a μ-neighbourhood of the lamination $\mathcal{L}$ with orthogonal foliations, denoted $\mathcal{L}_\mu$ and $\mathcal{F}_\mu$. We have fixed $\mu < (\log 3)/2$. Then k is determined by Lemma II.2.6.1 (*Lipschitz (H2)*). Let us fix $s \leq \mu/2$ and such that $k_1 s < 1$, where k_1 is defined in Lemma II.2.7.9 (*Comparison with intrinsic metric*). Let $\delta > 0$ be some fixed number. (Some limitations on the possible choices of δ in terms of μ will appear in due course.) Let S_δ be the surface in V at distance δ from S. By Lemma II.1.3.6 (*Smooth epsilon surface*), S_δ is a C^1-surface. We will define a bilipschitz map $\rho \colon S_\delta \to S$, and determine its bilipschitz constant.

Let $u \in S_\delta$. If $d_\delta(u, r^{-1}\mathcal{L}) > s$, we define $\rho(u) = r(u)$. If $d_\delta(u, r^{-1}\mathcal{L}) < s \cosh \delta$, we define $\rho(u)$ as follows. We parametrize the $\mathcal{F}$ leaf A through u and the $\mathcal{F}$-leaf B through ru by pathlength, making u and ru correspond to zero. Then r sends A to B (weakly) monotonically. The parametrization also involves choosing a direction on A, and we choose the direction on B so that r is monotonic increasing. Note that r is constant over an $\mathcal{F}$-subinterval on which the nearest point in S is a fixed point of a fixed bending line of $\mathcal{L}$. By abuse of notation, we regard A and B as subintervals of the real numbers containing zero, and we have $r \colon A \to B$, a monotonically increasing 1-lipschitz real valued function of a real variable. In this notation, we define

$$\rho(u) = \frac{1}{2s} \int_{-s}^{s} r(t)\mathrm{d}t \in B.$$

In order to show that ρ is well-defined, we need to show that if $s < d_\delta(u, r^{-1}\mathcal{L}) < s \cosh \delta$, then the two alternative definitions of $\rho(u)$ agree. In this situation, $s/\cosh \delta < d_S(ru, \mathcal{L}) < s$, ru is contained in a flat piece of S, and the s-neighbourhood of ru is foliated. In fact, locally, $r \colon S_\delta \to S$ is just a projection from an equidistant surface at distance δ to a standard hyperbolic plane contained in hyperbolic 3-space. This decreases distances by a factor exactly equal to $1/\cosh \delta$. Hence

$$\rho(u) = \frac{1}{2s \cosh \delta} \int_{-s}^{s} t\mathrm{d}t = 0.$$

In other words, $\rho(u) = r(u)$, as required.

The above abuse of notation can easily lead to incorrect proofs. Let us re-express the idea in a more rigorous notation. Let $a \colon \mathrm{dom}\, a \to A$ be a parametrization of A, proportional to path-length, and similarly for $b \colon \mathrm{dom}\, b \to B$. We no longer worry about which points correspond to the parameter 0. Note that A and B are embedded as closed subspaces of V_μ and U_μ, respectively. They cannot return near themselves, because S is simply connected, and each geodesic of $\mathcal{L}$ separates S. Arguing by contradiction, if a return were to occur, there would be a subinterval of A with each endpoint on the same

geodesic, and with interior disjoint from $\mathcal{L}$. The construction of Section II.2.4
(*Extending a lamination to a pair of orthogonal foliations*) shows that this is
impossible.

The definition of ρ is then given by

$$b^{-1}\rho a(t) = \frac{1}{2s}\int_{-s}^{s} b^{-1}ra(t+u)\mathrm{d}u$$

provided $d_S(ra(t), \mathcal{L}) < s$. We now check, first that the definition of ρ at a point
is independent of changing the definition of a by a translation, then that is
independent of changing the orientation of a, and finally that it is independent
of an affine change, including a possible reversal of orientation, in b.

When working out the properties of ρ, it will be convenient to use finite outer
approximations, because in that case explicit calculations can be carried out.
We actually use a standard sequence of finite outer approximations, as defined
in Section I.1.13.1 (*Special pairs*).

We fix $u \in S_\delta$, and define the outer approximations S_n relative to u, as
explained in Section I.1.13.1 (*Special pairs*). Then $ru \in S_n \cap S$ and $u \in S_{n,\delta} \cap S_\delta$.
Let A, A_n, B, B_n, be the leaves in $S, S_n, S_\delta, S_{n,\delta}$ through u and ru, and let
a, a_n, b, b_n, be parametrizations, regarded as functions with values in $\mathbb{H}^3$. We
will choose these so that a_n approximates a and b_n approximates b, and so that
the parameters are equal at u and ru.

By Lemma II.2.6.2 (*Continuous dependence (H2)*), the line field $\mathcal{L}_{n,\mu}$ converges uniformly on compact sets to $\mathcal{L}_\mu$, and the line field $\mathcal{F}_{n,\mu}$ converges to
$\mathcal{F}_\mu$. Locally, we regard these line fields as being unit vector fields. (Recall
that these are fields on the hyperbolic plane, not on the convex hull boundary.) By Theorem II.A.3.2 (*Approximation inequality*), the solution curves for
$\mathcal{F}_{n,\mu}$ will converge uniformly to those of $\mathcal{F}_\mu$. It follows that the parametrized
curve $h_n^{-1}b_n$ converges uniformly to $h^{-1}b$. Since h_n converges uniformly
to h (Proposition I.1.12.6 (*Covering maps from the hyperbolic plane*)), b_n
converges uniformly to b.

By Section II.2.11.1 (*Continuous dependence (H3)*) we also know that
$\tau(\mathcal{F}_n, V_{n,\mu})$ converges uniformly to $\tau(\mathcal{F}, V_\mu)$. Once again, we apply Theorem II.A.3.2 (*Approximation inequality*) and we see that a_n converges
uniformly to a.

II.2.14.1. Lemma: Convergence to rho.   *For any point $u \in S_\delta$, we may
choose a standard sequence of outer approximations so that $u \in S_{n,\delta}$ and then
$\rho_n(u)$ converges to $\rho(u)$.*

Proof. Let $0 < x \leq s$ and let $k = k_1(x)$ be chosen as in Lemma II.2.7.9 (*Comparison with intrinsic metric*). We first show that, over an interval of length x, a, a_n, b, b_n are $1/(1 - k_1 x)$-bilipschitz, where the distance in the range is the path distance in S_δ, $S_{n,\delta}$, S and S_n, respectively. To see this, note that if $|t - t'| = y \leq x$, then, choosing a as an example,

$$(1 - k_1 x)y \leq d_\delta(a(t), a(t')) \leq y$$

as we see by appealing to Lemma II.2.7.9 (*Comparison with intrinsic metric*).

Let $x_n = d_S(ra(t), bb_n^{-1} r_n a_n(t))$. Using the lipschitz property of b^{-1}, we see that

$$|1 - k_1 x_n||b^{-1} ra(t) - b_n^{-1} r_n a_n(t)| \leq d_S(ra(t), bb_n^{-1} r_n a_n(t))$$
$$= d'(ra(t), bb_n^{-1} r_n a_n(t))$$

where d' is the path metric for the closure of the complement of the convex hull. Let ε_1 be a small positive number. Since b_n is uniformly near b, as we showed above, we may replace b by b_n on the right hand side of the above inequality, and we find that, for n large, the expression is bounded by

$$d'(ra(t), r_n a_n(t)) + \varepsilon_1 \leq d'(ra(t), ra_n(t)) + d'(ra_n(t), r_n a_n(t)) + \varepsilon_1$$
$$\leq d(a(t), a_n(t)) + 2\varepsilon_1 \leq 3\varepsilon_1$$

since r is distance decreasing, r_n converges uniformly to r on compact sets and a_n converges uniformly to a. Hence x_n converges uniformly to 0 and $b_n^{-1} r_n a_n$ converges uniformly to $b^{-1} ra$.

It follows that $b_n^{-1} \rho_n(u)$ converges to $b^{-1} \rho(u)$. Applying b to both sides, we see that $\rho_n(u)$ converges to $\rho(u)$.

$\square$

The convergence in the preceding lemma is at one point u only. If the sequence S_n is given, it only works for a countable dense set of points, namely the special points.

In order to make our estimates, we need a lemma. Let S be a finitely bent simply connected surface which is a component of the boundary of the convex hull X of a closed subset of S^2. One way S can arise is from a standard sequence of outer approximations. Let $\mathcal{L}$ be the bending lamination and let $\mathcal{L}_\mu$ and $\mathcal{F}_\mu$ be the foliations constructed in Section II.2.4 (*Extending a lamination to a pair of orthogonal foliations*).

II.2.14.2. Proposition: Limited angle in 3 dimensions. *Let*

$$f(t) = t - \frac{4}{5}(e^t - 2\sin(t/2) - \cos(t/2)).$$

Let $t_0 > 0$ be the first point where the derivative of f vanishes. Let $0 < t < t_0$ and let $s = f(t)$. Let A be a closed arc in an $\mathcal{F}_\mu$-leaf of length $2s$. The sum of the bending angles met by A (including the endpoints of A) is atmost

$$\frac{(\pi + 2t)\cos^2 t}{\cos t - \sin^2 t}.$$

Proof. Let $a, b \in A$ be the extreme points of $A \cap \mathcal{L}$. We may assume that $A = [a, b]$. Let $P_0, \cdots, P_k$ be the successive flat pieces whose closures meet A. Then $P_{i-1} \cap P_i$ is a geodesic $\lambda_i (1 \le i \le k)$. We have $a \in \lambda_1$ and $b \in \lambda_k$. The geodesics $\lambda_1, \cdots, \lambda_k$ are those geodesics of $\mathcal{L}$ which meet A. Let H_i be the supporting half space such that $P_i \subset \partial H_i$. Without loss of generality, we may assume that our convex subset X is equal to $H_0 \cap \cdots \cap H_k$. Then P_0 and P_k have exactly one boundary geodesic and, for $1 \le i < k$, P_i has two boundary geodesics. Let α_i be the common orthogonal geodesic or the common asymptotic endpoint of the two boundary geodesics λ_i and λ_{i+1}. We foliate each P_i in its entirety by the two foliations $G(\alpha_i)$ and $E(\alpha_i)$. By abuse of notation, we denote by G the first foliation (in which each leaf is a geodesic) and by E the second foliation. On each flat piece, the classical lipschitz constant of the vector fields tangent to the foliations is bounded by 1 (see Lemma II.2.6.1 (*Lipschitz (H2)*)). The foliations we have constructed agree with $\mathcal{L}_\mu$ and $\mathcal{F}_\mu$ on a neighbourhood of the interior of A. Inside P_0 and P_k, the leaf of E containing A does not necessarily agree with the leaf of $\mathcal{F}_\mu$ containing A.

Let α be the solution curve to the vector field E, with $\alpha(0)$ equal to the midpoint of A. Let P be the plane through $\alpha(0)$, which is orthogonal to E at $\alpha(0)$. Let B be the closed arc in $P \cap \partial X$ bounded by λ_1 and λ_k. (It has to be proved that $P \cap \partial X$ does meet λ_1 and λ_k.) We parametrize B proportionally to arclength by β, where $\beta(0) = \alpha(0)$, and we choose the direction of β to match that of α and of E. The method of proof will be to apply Lemma II.A.6.2 (*Limited angle*) to β in P. There are two factors which have to be taken into account, if β is used to estimate the angles met by α. First, the plane P is not orthogonal to the foliation $\mathcal{L}$, and so we have to multiply by a correction factor from Lemma II.A.4.3 (*Upper bound for theta*). Second, we have to make sure that β crosses all the angles which are crossed by α. This entails estimating the distance from $\alpha(t)$ to $\beta(t)$, which we do by considering β as an approximate solution to the vector field E.

Let $G(t)$ be the unit vector of G at $\beta(t)$. Let $v(t)$ be the unit tangent vector at $\beta(t)$ which is orthogonal to P, chosen so that $G(0) = v(0)$. Then $v(t)$ is a parallel field, so $dv/dt = 0$.

From Lemma II.2.5.1 (*Covariant derivatives of G and E*), we find that on P_i

$$\frac{d}{dt}\langle G, v\rangle = \langle \beta', E\rangle \tanh d_i \langle E, v\rangle$$

where d_i is the signed distance from α_i. This means that

$$\frac{d}{dt}\langle G, v\rangle \geq -\sqrt{1 - \langle G, v\rangle^2}.$$

Since $\langle G, v\rangle$ is continuous, and is analytic except at a finite number of points, we deduce that

$$\langle G, v\rangle \geq \cos t$$

provided $0 \leq t < \pi/2$. The factor

$$\frac{\cos^2(t)}{\cos t - \sin^2 t}$$

in the statement of the lemma we are now proving, is obtained immediately from Corollary II.A.4.3 (*Upper bound for theta*).

In the previous paragraph, a small problem was ignored. The theory of differential inequalities, as normally explained in textbooks, uses a lipschitz hypothesis on the coefficients which is not satisfied here. The initial condition was $\langle G, v\rangle = 1$, which is a problem since 0 is a non-lipschitz point for the square root function. One way round this is to define

$$u = \sqrt{1 - \langle G, v\rangle^2}$$

obtaining the differential inequality

$$\left|\frac{du}{dt}\right| \leq \sqrt{1 - u^2}$$

with initial condition $u = 0$. Now $\sqrt{1 - u^2}$ is lipschitz provided $|u| < 1$, and we deduce $|u(t)| \leq \sin t$, provided $0 \leq t \leq \pi/2$. Beyond $t = \pi/2$ solutions are not unique and we can only deduce $|u| \leq 1$.

Now let $\beta' = xG + yE$, where x and y are functions of t, which are well-defined provided $\beta(t)$ does not lie on a bending line. We have

$$\langle \beta', v\rangle = 0 = x\langle G, v\rangle + y\langle E, v\rangle.$$

Therefore $|y| \leq |x| \tan t$ and so $1 - x^2 = y^2 \leq x^2 \tan^2 t$. Therefore $|x| \geq \cos t$ and $|y| \leq \sin t$. We deduce that

$$\|\beta' - E\| \leq 2\sin(t/2).$$

Let d_S be the path distance in S and let $u(t) = d_S(\alpha(t), \beta(t))$. We wish to estimate $u(t)$. There is no loss of generality in assuming that $t > 0$. We apply Theorem II.A.3.2 (*Approximation inequality*) with lipschitz constant 1 and with $\varepsilon(s) = 2\sin(s/2)$. We see that

$$d_S(\alpha(t), \beta(t)) \leq 2 \int_0^2 e^{t-s} \sin(s/2)ds = \frac{4}{5}\left(e^t - 2\sin\frac{t}{2} - \cos\frac{t}{2}\right).$$

In P_k the foliation E consists of geodesics perpendicular to λ_k. Since A has length $2s$, where

$$s = t - \frac{4}{5}\left(e^t - 2\sin\frac{t}{2} - \cos\frac{t}{2}\right) \leq t$$

and we are dealing with half of A, we know that $\alpha(t)$ lies in P_k. In P_k α follows a geodesic path orthogonal to λ_k. Therefore $\alpha(t)$ is a distance at least $t - s$ from λ_k. But $d_S(\alpha(t), \beta(t)) \leq t - s$. Therefore $\beta(t)$ lies in P_k. Similarly, $\beta(-t)$ lies in P_0. Therefore the length of B is less than $2t$. By Lemma II.A.6.2 (*Limited angle*), the sum of the exterior angles of the polygon $P \cap X$, which are met by B, is bounded by $\pi + 2t$. The result now follows by multiplying $\pi + 2t$ by the factor

$$\frac{\cos^2(t)}{\cos t - \sin^2(t)}$$

found above.

$\square$

We now carry out a careful investigation of ρ, with the objective of finding its bilipschitz constants. The method is to show that it is path-bilipschitz. This means that we only need to investigate how far $\rho(z_1)$ is from $\rho(z_2)$ when z_1 and z_2 are infinitesimally near to each other. We will need both an upper and a lower bound in order to prove the lipschitz property for both ρ and ρ^{-1}.

First we deal with the trivial case, when rz_1 is not contained in the open s-neighbourhood of $\mathcal{L}$. In that case the distance in S_δ from z_1 to $r^{-1}\mathcal{L}$ is greater than or equal to $s \cosh \delta$. (Recall that $\mu = 2s$ and we have foliated a μ-neighbourhood of $\mathcal{L}$.) Then z_2, being extremely near to z_1, is at a distance greater than s from $r^{-1}\mathcal{L}$. It follows that $\rho z_i = rz_i$ for $i = 1, 2$. The path-lipschitz constant for ρ in this region is $1/\cosh \delta$ and the path-lipschitz constant for ρ^{-1} is $\cosh \delta$.

We now suppose that z_1 is contained in the open $s \cosh \delta$-neighbourhood of $\mathcal{L}$. Since z_2 is close, rz_2 is in the same neighbourhood. This means that the path of length $2s$, lying inside a leaf of the foliation $\mathcal{F}_{\delta,\mu}$, over which we integrate in order to define $\rho(z_1)$ and $\rho(z_2)$, exists. Let z_0 lie on the intersection of the $\mathcal{F}_{\delta,\mu}$-curve through z_1 and the $\mathcal{L}_{\delta,\mu}$-curve through z_2. We take foliation coordinates ϕ_δ centred at z_0 in S_δ and foliation coordinates ϕ centred at rz_0 in S. We have $z_1 = \phi_\delta(u, 0)$ and $z_2 = \phi_\delta(0, v)$, where u and v are small. By reversing the signs of the vector fields if necessary, we see that there is no loss of generality in taking $u \geq 0$ and $v \geq 0$. We write $r^2 = u^2 + v^2$.

Let a_1 be the $\mathcal{F}_{\delta,\mu}$-curve parametrized by arclength with $a_1(0) = z_0$. Let a_2 be the $\mathcal{F}_{\delta,\mu}$-curve parametrized by arclength with $a_2(0) = z_2$. Let b_1 be the $\mathcal{F}$-curve parametrized by arclength, with

$$b_1(0) = ra_1(0) = rz_0 = \phi(0,0).$$

Let b_2 be the $\mathcal{F}$-curve parametrized by arclength with

$$b_2(0) = ra_2(0) = r(z_2) = \phi(0, v/\cosh \delta).$$

Let $\tau(s)$ be an upper bound for the sum of the bending angles encountered by an arc A in an $\mathcal{F}_\mu$-leaf of length s. A value for $\tau(2s)$ is computed in Proposition II.2.14.2 (*Limited angle in 3 dimensions*).

II.2.14.3. Lemma: Estimate for r.   *If* $-s \leq t_1 \leq t_2 \leq s$ *then*

$$(t_2 - t_1)/\cosh \delta \geq b_1^{-1} ra_1(t_2) - b_1^{-1} ra_1(t_1)$$

$$\geq \sup\{0, t_2 - t_1 - \sinh \delta . \tau(t_2 - t_1)\}/\cosh \delta.$$

Proof.   The result for a general equidistant surface follows by taking a standard sequence of outer approximations. So we may assume our surface is finitely bent. We then apply Lemma II.2.14.2 (*Limited angle in 3 dimensions*). Note that $b_1^{-1} ra_1$ is constant along the inverse image of a bending line. The interval along which the function is constant has length at most $\tau(t_2 - t_1)\sinh \delta$. On the inverse image of a flat piece the function has slope $\cosh \delta$. The value of the function has the same sign as t.

$\square$

We define

$$I_1 = \frac{1}{2s} \int_{-s+u}^{s+u} b_1^{-1} r\, a_1(t)\, dt$$

and

$$I_2 = \frac{1}{2s} \int_{-s}^{s} b_2^{-1} r \, a_2(t) dt.$$

Then

$$\rho(z_1) = \phi(I_1, 0).$$

We have $\rho(z_2) = \phi(x, v/\cosh \delta)$, for some x. However, x is probably not equal to I_2, because ϕ does not in general preserve arclength, except along the axes.

II.2.14.4. Lemma: Estimate for integral. *Let*

$$v(s) = \sinh (\delta) \tau(s)(2s - \sinh (\delta) \, \tau(s))$$

if $\sinh (\delta)\tau(s) \le s$, and let $v(s) = s^2$ if $\sinh(\delta)\tau(s) \ge s$. Then

$$|I_1| \le \frac{v(s)}{4 s \cosh \delta} + O(u; s).$$

Proof. First we take $u = 0$. We get a maximum for I_1 by taking $b_1^{-1} r a_1$ to equal $t/\cosh \delta$ on $[0, s]$, to 0 on $[-\tau(s)\sinh \delta, 0]$ and to $(t + \tau(s)\sinh \delta)/\cosh \delta$ on $[-s, -\tau(s)\sinh \delta]$. We get $I_1 \le v(s)/4s \cosh \delta$. The result claimed now follows, even when $u \ne 0$, because the necessary correction can be estimated using the inequality $|b_1^{-1} r a_1(t)| \le |t|$.

$$\square$$

We now wish to estimate $I_1 - I_2$ from above and below. To do this we write

$$\psi_\delta(t) = \text{length } \phi|[0, t] \times v \text{ if } t \ge 0$$

and

$$\psi_\delta(t) = -\text{length } \phi|[t, 0] \times v \text{ if } t \le 0.$$

We have

$$\phi_\delta(t, v) = a_2(\psi_\delta t).$$

From Lemma II.2.8.3 (*Estimates for psi*), in the form presented in Lemma II.2.13.1 (*General equidistant*), we know that

$$|\psi'_\delta(t) - 1| \le k e^{k|t|} v + O(v^2; k, s)$$

and

$$|\psi_\delta(t) - t| \le (e^{k|t|} - 1)v + O(v^2; k, s)$$

Similarly we define ψ so that

$$\phi\left(t, \frac{v}{\cosh \delta}\right) = b_2(\psi(t)).$$

Then ψ satisfies the inequalities

$$(\cosh \delta)|\psi'(t) - 1| \le k e^{k|t|} v + O(v^2; k, s)$$

and

$$(\cosh \delta)|\psi(t) - t| \le (e^{k|t|} - 1)v + O(v^2; k, s).$$

Now

$$\phi(b_1^{-1} r a_1(t), 0) = r a_1(t) = r \phi_\delta(t, 0).$$

Hence

$$\phi(b_1^{-1} r a_1(t), \frac{v}{\cosh \delta}) = r \phi_\delta(t, v).$$

This can be rewritten as

$$b_2 \psi b_1^{-1} r a_1(t) = r a_2 \psi_\delta(t).$$

Hence

$$\psi b_1^{-1} r a_1(t) = b_2^{-1} r a_2 \psi_\delta(t).$$

We write

$$J_1 = \cosh \delta \int_{-s}^{s} b_1^{-1} r a_1(t)(1 - \psi_\delta'(t)) dt$$

$$J_2 = \cosh \delta \int_{-s}^{s} (b_1^{-1} r a_1(t) - b_2^{-1} r a_2 \psi_\delta(t)) \psi_\delta'(t) dt$$

$$J_3 = \cosh \delta \int_{\psi_\delta(s)}^{s} b_2^{-1} r a_2(t) dt - \cosh \delta \int_{\psi_\delta(-s)}^{-s} b_2^{-1} r a_2(t) dt$$

$$J_4 = \cosh \delta \int_{s}^{s+u} b_1^{-1} r a_1(t) dt - \int_{-s}^{-s+u} b_1^{-1} r a_1(t) dt.$$

The change of variable formula shows that

$$2s \cosh \delta (I_1 - I_2) = J_1 + J_2 + J_3 + J_4.$$

We have

$$|J_1| \le 2v \int_0^s |t| k e^{kt} dt + O(v^2; k, s)$$

$$\le \frac{2v}{k}(ks e^{ks} - e^{ks} - 1) + O(v^2; k, s).$$

In order to estimate J_2, we set $t_1 = b_1^{-1} ra_1(t)$. Then

$$b_1^{-1} ra_1(t) - b_2^{-1} ra_2 \psi_\delta(t) = t_1 - \psi(t_1) \le (e^{k|t|} - 1)\frac{v}{\cosh \delta} + \frac{O(v^2; k, s)}{\cosh \delta}.$$

We then have

$$|J_2| \le \cosh \delta \cdot \int_{-s}^{s} |t_1 - \psi(t_1)|\, (1 + O(v; k, s))dt.$$

Since $\cosh \delta |t_1| \le |t|$, we have

$$|J_2| \le 2v \int_0^s (e^{k|t|} - 1)dt + O(v^2; k, s)$$

$$= \frac{2v}{k}(e^{ks} - 1 - ks) + O(v^2; k, s).$$

Since $|b_2^{-1} ra_2(t)| \le t/\cosh \delta$, we have

$$|J_3| \le 2sv(e^{ks} - 1) + O(v^2; k, s).$$

We have

$$J_4 = \cosh \delta \cdot \int_0^u (b_1^{-1} ra_1(s + t) - b_1^{-1} ra_1(-s + t))dt.$$

It follows from Lemma II.2.14.3 (*Estimate for r*) that

$$u(2s - \tau(2s)\sinh(\delta)) \le J_4 \le 2us.$$

It follows that

$$I_1 - I_2 \le \frac{2(e^{ks} - 1)v + u}{\cosh \delta} + O(r^2; k, s)$$

and

$$I_1 - I_2 \ge \frac{1}{\cosh \delta}\left(\left(1 - \frac{\sinh(\delta)\tau(2s)}{2s}\right)u - 2(e^{ks} - 1)v\right) - O(r^2; k, s)$$

where $r^2 = u^2 + v^2$.

To get from $\rho(z_1)$ to $\rho(z_2)$, we travel along the $\mathcal{L}_\mu$-curve from $\phi(I_1, 0)$ to $\phi(I_1, v)$ and along the $\mathcal{F}_\mu$-curve from $\phi(I_1, v)$ to $\rho(z_2)$. The first curve has length w, where, by Lemma II.2.8.2 (*Length lipschitz*),

$$v \exp(-kI_1) - O(r^2; k, s) \le w \le v \exp(kI_1) + O(r^2; k, s).$$

The second curve has length $|\psi I_1 - I_2|$.

To estimate the squares of the appropriate lipschitz constants, we need to estimate the expression

$$\frac{|\psi I_1 - I_2|^2 + w^2}{u^2 + v^2}.$$

We write $u = r \cos\theta$ and $v = r \sin\theta$. Since u and v are being assumed positive, we can restrict to $0 \leq \theta \leq \pi/2$. We see that the square of the path lipschitz constant for ρ is bounded above by the maximum value of

$$\frac{(|I_1 - I_2| + |\psi I_1 - I_1|)^2 + w^2}{r^2}.$$

We write $I = (v(s)/4\, s \cosh\delta)$, where k is the classical lipschitz constant of the vector field and where v is defined in Lemma II.2.14.4 (*Estimate for integral*). Letting r tend to zero, we see that the square of the lipschitz constant of ρ is bounded above by the maximum value of

$$\frac{((2e^{ks} + e^{kl} - 3) \sin\theta + \cos\theta)^2 + e^{2kl} \sin^2\theta}{\cosh^2\delta}$$

taking the maximum as θ varies.

To obtain an upper bound for the lipschitz constant of ρ^{-1}, we have to take the reciprocal of the lim inf, as θ varies and as r tends to zero, of the expression

$$\frac{|\psi I_1 - I_2|^2 + e^{-2kl} v^2}{r^2}.$$

To this end, we define

$$\alpha(\theta, s, \delta) = \left(1 - \frac{\sinh(\delta)\tau(2s)}{2s}\right) \cos\theta - (2e^{ks} + e^{kl} - 3) \sin\theta.$$

The square of the reciprocal of the path lipschitz constant for ρ^{-1} is bounded below by minimizing over θ the value of

$$\frac{\inf(\alpha(\theta, s, \delta), 0)^2 + e^{-2kl} \sin^2\theta}{\cosh^2\delta}.$$

The homeomorphism claimed in the statement of Sullivan's theorem is the composite ρσ, where $\rho: S_\delta \to S$ is the map we have just been discussing and $\sigma: \Omega(S) \to S_\delta$ is the map discussed in Section II.2.3 (*From infinity to the epsilon surface*). The lipschitz constant of the composition of two maps is bounded by the product of their lipschitz constants. This gives lipschitz constants for $\rho\sigma$ and of $\sigma^{-1}\rho^{-1}$. Note that the limitations, which have to be imposed on s to make the proofs of the various lemmas work, have been explicitly computed in the lemmas concerned. Any particular choice of δ and s, within these limitations, will provide us with an example of a homeomorphism fulfilling the promised condition of Sullivan's theorem. By varying δ and s, in such a way that $\tau(2s)\sinh(\delta) < 2s$ and with s small enough so that the various lemmas work, we can find an upper bound for Sullivan's constant.

All the terms can be computed, and so we can find the maximum by a computer program which runs over a grid of all possible choices. To get a rigorous proof that the number we eventually determine is truly a possible value for Sullivan's theorem, we use the procedure just sketched to find a possible value for δ and s. The expressions above can then be maximized or minimized as functions of θ, using calculus.

II.2.15. Numerical results

The proof presented in the preceding pages of this article lead to actual numerical estimates, which we now present. (Actually, the search for definite numbers led to the proof which we have presented. But, following convention, we have hidden the fact that the discovery of the pure mathematical proofs was greatly aided by the use of computer calculations while trying out different routes to a proof.)

The maps given by our method occur by composing the nearest point projection to the surface at distance δ from the convex hull boundary with a map obtained by averaging over an interval of length $2s$. The best bilipschitz map ρ_{bl} occurs with $\delta = 0.0393$ and with $s = 0.131$. The lipschitz constant of ρ_{bl} is bounded above by 5.33 and the lipschitz constant of ρ_{bl}^{-1} is bounded above by 66.3. The quasi-conformal constant of ρ_{bl} is bounded above by 88.2.

The best quasiconformal map ρ_{qc} occurs with $\delta = 0.0325$ and with $s = 0.11$. The quasiconformal constant of ρ_{qc} is bounded above by 82.8. The lipschitz constant of ρ_{qc} is bounded above by 4.76 and the lipschitz constant of ρ_{qc}^{-1} is bounded above by 69.6.

II.2.16. Counterexample

We now show that Sullivan's theorem cannot be true for non-simply connected surfaces. We present examples such that any homeomorphism from the convex hull to the surface at infinity must have a large lipschitz constant. In the family of examples, the smallest possible lipschitz constants are unbounded. It would be nice to have better counterexamples than the one presented here, in order to delimit the situation more clearly in the non-simply connected case. It seems that one can produce a single example such that there is no lipschitz map in the correct homotopy class from the convex hull boundary to the surface at infinity, but we have not written down the details. We leave open various questions concerning possible maps from the surface at infinity to the convex hull boundary (i.e. the reverse direction from that of our counterexamples).

II.2.16.1. Theorem: Annulus.

(1) *Let $r > 1$ and let A_r be the annulus in the bounded by concentric circles of radius 1 and r, with centre at 0. Let λ_∞ be the length of the shortest closed geodesic in A_r with respect to the unique complete hyperbolic metric on A_r which is conformal with the given complex structure. Then*

$$\lambda_\infty = 2\pi^2 / \log r.$$

(2) *We regard the complex plane as the boundary of $\mathbb{H}^3$ via the upper half space model. Let C_r be the boundary of the convex hull in upper half space of the complement of A_r. Then C_r is an annulus with a complete hyperbolic metric induced from $\mathbb{H}^3$ (see Section 1.1.12 (The boundary is a complete hyperbolic manifold)). Let λ_{CH} be the length of the shortest closed geodesic in C_r. Then*

$$\lambda_{CH} = \frac{2\pi\sqrt{r}}{r-1}.$$

(3) *Any proper map of degree one from C_r to A_r has lipschitz constant at least*

$$\frac{\pi(r-1)}{\log r \sqrt{r}}$$

and this tends to infinity with r.

Proof. It is easy to check that the map

$$z \mapsto \exp\left(\frac{-i}{\pi} \log r \log z\right)$$

sends the upper half plane to A_r. In fact this map is the universal cover of A_r. The generating covering translation is the hyperbolic transformation of the upper half plane

$$z \mapsto \exp\left(\frac{2\pi^2}{\log r}\right).$$

This proves (1).

To prove (2), note that the convex hull boundary consists of the top of circular tunnel, whose base is the annulus A_r. The shortest geodesic will be invariant under rotations around the z-axis. We need only find the euclidean height at which the circle lies. The minimal length will be attained by the circle lying on C_r which is as near the z-axis (with respect to the hyperbolic metric) as possible. We compute that this circle lies at a euclidean distance $\sqrt{r}$ from 0, and then that its euclidean height is

$$h = \sqrt{r}\,\frac{r-1}{r+1}.$$

The euclidean radius of the circle is $2r/(r+1)$. The hyperbolic length of the circle is its euclidean length divided by its euclidean height. The remainder of the theorem now follows.

▯

Chapter II.3
Measured Pleated Surfaces

II.3.1. Introduction

In this chapter we will work with geodesic laminations in the hyperbolic plane. We recall that this is a set of mutually disjoint lines whose union is a closed subset of the hyperbolic plane. Given a lamination, the plane is the disjoint union of the lines of the lamination and the *flat pieces*, which are the complementary components. The term *facet* refers to either a line of the lamination or a flat piece. Note that a flat piece does not contain its boundary edges.

Given a geodesic lamination in the hyperbolic plane, with a complex valued transverse measure, we will define a map of the hyperbolic plane into hyperbolic 3-space, called a *quakebend map*. When the measure is real, this map will be an earthquake, and when the measure is purely imaginary, it will be a bending map. Both these concepts are due to Thurston.

We will work in the upper half 3-space model, and we will fix P to be the vertical half plane based on the real axis. We fix l^* to be the vertical half axis above the origin. We orient l^* vertically upwards, and we orient P in the usual way.

II.3.2. Finite quakebends

We start by defining a certain matrix in $SL(2, \mathbb{C})$, which will be important in what follows. Let $\zeta \in \mathbb{C}$. We define

$$E(l^*, \zeta) = \begin{bmatrix} e^{\zeta} & 0 \\ 0 & e^{-\zeta} \end{bmatrix}.$$

In general, given an oriented line l in $\mathbb{H}^3$ and a complex number 2ζ, we use the notation

$$E(l, \zeta) = AE(l^*, \zeta)A^{-1}$$

211

where A is a Möbius transformation mapping l^* onto l, preserving orientations. Let $\zeta = \log k + i\theta$, where $k > 0$ and $\theta \in \mathbb{R}$. Then $E(l, \zeta)$ is a transformation with axis l, which does not depend on the particular choice of A, rotating an angle 2θ about l and translating a distance $2\log k$ along l.

Having set up this notation, we proceed to consider a finite geodesic lamination $\mathcal{L}$ in P, consisting of mutually disjoint lines $l_0, l_1, \ldots, l_k$, such that l_j separates l_{j+1} from l_{j-1}. Orient l_0 so that the other lines lie to its right, and then orient the others to point in the same direction as l_0. Suppose also that a complex number $2\zeta_j$ is assigned to each line l_j. We will find the following notation convenient:

$$\zeta_j = \log k_j + i\theta_j, k_j, \theta_j \in \mathbb{R}, \quad k_j > 0.$$

Denote the $k+2$ components of $P \backslash \cup l_j$ from the left to the right by Δ_{-1}, $\Delta_0, \ldots, \Delta_k$.

II.3.2.1.

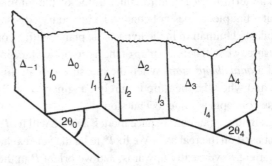

Figure II.3.2.1 *A more general measured pleated surface has self-intersections. The picture shows only bending – the shearing k_j along the line l_j is more difficult to draw.*

We are now ready to construct our "pleated surface" S step by step. Start with the extreme right line l_k. The transformation $E(l_k, \zeta_k)$ maps l_k to itself and rotates Δ_k to a half plane in $\mathbb{H}^3$, inclined to P along l_k at angle $2\theta_k$.

To the configuration $\Delta_{k-1} \cup E(l_k, \zeta_k)(\Delta_k)$ apply $E(l_{k-1}, \zeta_{k-1})$. The sector $E(l_{k-1}, \zeta_{k-1})(\Delta_{k-1})$ is inclined at angle $2\theta_{k-1}$ to P along l_{k-1}, while the line l_k is sent to $E(l_{k-1}, \zeta_{k-1})(l_k)$. Continue this process. We end up with the lines in $\mathbb{H}^3$,

$$l_0, E(l_0, \zeta_0)(l_1), \ldots, E(l_0, \zeta_0) \cdots E(l_{k-1}, \zeta_{k-1})(l_k).$$

Each successive pair of lines bounds a sector of the plane, isometric to one of the Δ_j, through those lines. At the jth line the two adjacent sectors are inclined at angle $2\theta_j$ $(0 \le j \le k)$.

Along with the pleated surface S there is a natural surjection $h: P \to S$. Namely,

$$h(x) = E(l_0, \zeta_0) \cdots E(l_j, \zeta_j)(x) \quad x \in \Delta_j, 0 \le j \le k$$

and $h(x) = x$ for $x \in \Delta_{-1}$. Thus on each piece Δj, h is an isometry to a flat piece of S. The sector on the right of l_j slides a distance $\log k_j^2$ relative to the sector on its left. On l_j itself, h is defined so that it slides exactly halfway as compared to the sectors on either side, that is, a distance $\log k_j$.

There are two special cases that deserve notice:

(1) For all j, $k_j = 1$, so that ζ_j is pure imaginary. This is the case that the construction consists of pure rotation about the lines. The map $h: P \to S$ preserves the length of paths.
(2) For all j, $\theta_j = 0$, so that ζ_j is pure real. Then $S = P$, but the lines $\{l_j\}$ have been moved in a particular manner. If in addition all the $k_j > 1$, the map h is called a left earthquake.

At each line $h(l_j)$ of S, it is natural to speak of the *medial support plane*. This is the plane through $h(l_j)$ that forms an angle θ_j with each of the two flat pieces on either side. Given a pair of lines $h(l_j), h(l_r), j < r$, the transformation:

$$E(l_j, \zeta_j/2)E(l_{j+1}, \zeta_{j+1}) \cdots E(l_{r-1}, \zeta_{r-1})E(l_r, \zeta_r/2)$$

maps the medial support plane at $h(l_j)$ onto that of $h(l_r)$. Later such mappings will form the basis for what we will call *quakebend cocycles*. Note the special role of the extreme lines.

II.3.3. Norms

We will need not only to compose, but also to add and subtract the matrices in $SL(2, \mathbb{C})$ associated with isometries. For this reason we introduce a norm defined on all 2×2 complex matrices. In the vector space $\mathbb{C}^2$ we will use the norm:

$$\|(z_1, z_2)\| = \max\{|z_1|, |z_2|\}.$$

As a linear transformation of $\mathbb{C}^2$, the matrix

$$T = \begin{bmatrix} a & b \\ c & d \end{bmatrix} \quad a, b, c, d \in \mathbb{C}$$

then has operator norm $\|T\|$, which satisfies

$$\|T\| = \max\{|a| + |b|, |c| + |d|\}.$$

In this way, in terms of an associated normalized matrix, we will define the norm of a Möbius transformation T.

II.3.3.1. Lemma: Norms and compactness. *The following conditions on a set X of matrices in $SL(2, \mathbb{C})$ are equivalent. Let $O = (0, 0, 1) \in l^*$ in upper half 3-space:*

(1) *The closure of X is compact.*
(2) *There is a constant M such that if $T \in X$ then $\|T\| \leq M$.*
(3) *There is a constant M such that if $T \in X$ then $\|T\| \leq M$ and $\|T^{-1}\| \leq M$.*
(4) *There is a hyperbolic ball B about O such that if $T \in X$ then $T(O) \in B$.*

II.3.3.2. Lemma: Geometry and norm. *Given a compact subset K of the upper half plane, there exists a constant M with the following property. Let $x_0 \in K$, v_0 be a unit tangent vector at x_0, and suppose that $A x_0 \in K$, for some $A \in PSL(2, \mathbb{R})$. Let v_1 be the result of parallel translating v_0 from x_0 to $A x_0$ along a geodesic. Then*

$$\frac{\|A - I\|}{M} \leq d(x_0, A x_0) + \|v_1 - A x_1\| \leq M \|A - I\|$$

for suitable choice of matrix representation of A.

This lemma is proved by writing A as a rotation about x_0 followed by a translation.

II.3.4. Products of rotations about geodesics

II.3.4.1. Lemma: Single-term comparison. *Let K_1 be a compact set of elements in $SL(2, \mathbb{C})$ and let K_2 be a compact subset of $\mathbb{C} \setminus 0$. Then there is a constant C_1, depending only on K_1 and on K_2, such that*

$$\|AEA^{-1} - BEB^{-1}\| \leq C_1 |\zeta| \|A - B\|$$

where $A,\ B \in K_1, \zeta \in K_2$ and

$$E = \begin{bmatrix} e^\zeta & 0 \\ 0 & e^{-\zeta} \end{bmatrix}.$$

Proof.   Clearly, $AEA^{-1} - BEB^{-1}$ is a real analytic function of the variables ζ, A, and B. This function vanishes when $\zeta = 0$ or when $A = B$. The result follows.

$\square$

Let $(\mathcal{L}, \mu) = \{l_j, 2\zeta_j\}$ be a system of parallel mutually disjoint and consistently oriented and indexed lines as discussed above. (Later on $(\mathcal{L}, \mu)$ will be a general geodesic lamination with a complex transverse measure.)

II.3.4.2. Lemma: Compactness of Bs.   *Given a compact set K in the hyperbolic plane P and given a number $M > 0$, there is another compact set K_1 in $\mathbb{H}^3$ with the following property. If $(\mathcal{L}, \mu)$ is any system, all of whose lines meet K, and if*

$$\sum |\operatorname{Re} \zeta_j| \leq M$$

then the image of $O = (0, 0, 1) \in l^*$ under

$$E(l_0, \zeta_0/2)E(l_1, \zeta_1) \cdots E(l_{k-1}, \zeta_{k-1})E(l_k, \zeta_k/2)$$

lies in K_1.

II.3.4.3.

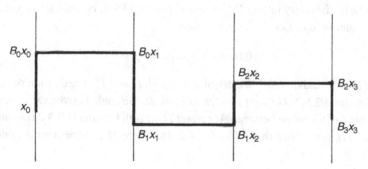

Figure II.3.4.3 *The broken geodesic which results from the quakebend construction.*

Proof.   Let K be contained in a disc around the origin of radius R. Let A be a geodesic arc from l_0 to l_k, meeting the lines of $\mathcal{L}$ successively in the points $x_0, \ldots, x_k$. The length of A is at most $2R$. We apply

$$B_i = E(l_0, \zeta_0/2) \cdots E(l_i, \zeta_i)$$

to the open interval (x_i, x_{i+1}). Here $0 \leq i < k$. We obtain a piecewise geodesic, where the pieces have the form $B_i(x_i, x_{i+1})$ and $(B_{i-1}(x_i), B_i(x_i))$. We add a geodesic piece at the beginning, from x_0 to $B_0(x_0)$, and a piece at the end from $B_{k-1}(x_k)$ to $B_{k-1}E(l_k, \zeta_k/2)(x_k)$. The sum of the lengths of these intervals does not exceed $2R + M$. We take K_1 to be the disc about O of radius $3R + M$.

$\square$

II.3.4.4. Lemma: Bunch of geodesics.   *Given a compact subset K of P and a number $M > 0$, there exists a constant $C_2 > 0$ with the following property. Let $(\mathcal{L}, \mu)$ be a system of lines as above, all of which intersect K, and such that $\Sigma|\mathrm{Re}\zeta_i| \le M$. Let l be any line in $\mathcal{L}$, and let $\zeta = \Sigma\zeta_i$. Then*

$$\|E(l_k, \zeta_k) \cdots E(l_0, \zeta_0) - E(l, \zeta)\| \le C_2\delta\Sigma|\zeta_i|$$

where δ is the length of any arc in K which cuts all the lines of $\mathcal{L}$.

Proof.   Let O be our fixed basepoint in P. We may assume that K is a disc of radius R with centre O and that γ is a geodesic arc in K from l_0 to l_k, which has length δ. We write $x_i = l_i \cap \gamma$ and $x = l \cap \gamma$. Let δ_i be the distance from x_{i-1} to x_i. Then

$$\delta_1 + \cdots + \delta_k \le \delta.$$

Let A be the isometry taking O to x and l^* to l, and let A_i be the isometry taking O to x_i and l^* to l_i. Let $T = A^{-1}A_i$. Then

$$d(O, TO) = d(x, x_i) \le \delta.$$

Since $l = Al^*$ and $l_i = A_il^*$ are disjoint or equal, l^* and Tl^* are disjoint or equal. By Lemma II.2.5.2 (*Disjoint geodesics mark 2*), the angle between $A^{-1}\gamma$ and l^* is within δ of the angle between $A^{-1}\gamma$ and Tl^*. From Lemma II.3.3.2 (*Geometry and norm*), we deduce that $\|T - I\| \le 2M\delta$, where M is a constant depending only on R.

We have

$$\|E(l_i, \zeta_i) - E(l, \zeta_i)\| = \|A_iE(l^*, \zeta_i)A_i^{-1} - AE(l^*, \zeta_i)A^{-1}\|$$
$$\le \|A\| \|A^{-1}A_iE(l^*, \zeta_i)A_i^{-1}A - E(l^*, \zeta_i)\| \|A^{-1}\|$$
$$\le C_3|\zeta_i|\delta$$

where C_3 is a constant depending only on R. The existence of the constant C_3 is deduced as follows. We know that A varies over a compact subset of $PSL(2, \mathbb{R})$, since $AO = x \in K$. Hence the norm of A and of A^{-1} is bounded by a constant depending only on R. We put $C_3 = 2C_1M$, where C_1 comes from Lemma II.3.4.1 (*Single-term comparison*).

Now note that we may apply Lemma II.3.4.2 (*Compactness of Bs*) to each of the composites $E(l_0, \zeta_0) \cdots E(l_i, \zeta_I)$ and $E(l, \zeta_i) \cdots E(l, \zeta_k)$ and deduce that each of these isometries has norm less than some constant C_4 depending only on R. Changing $E(l_i, \zeta_i)$ to $E(l, \zeta_i)$ one term at a time, we find that:

$$\|E(l_0, \zeta_0) \cdots E(l_k, \zeta_k) - E(l, \zeta)\| \leq \sum_{i=0}^{k} C_4^2 C_3 \delta |\zeta_i|.$$

$\square$

II.3.5. The quakebend cocycle

II.3.5.1. Definition. Let $\mathcal{L}$ be a geodesic lamination in the hyperbolic plane. We will define the concept of a transverse measure μ on $\mathcal{L}$, and refer to the pair $(\mathcal{L}, \mu)$ as a *measured lamination*. A transverse measure is a (finite valued) complex Borel measure on each closed finite geodesic segment α in the hyperbolic plane, subject to certain conditions. If α is entirely contained in a geodesic or in a flat piece of $\mathcal{L}$, then the measure is zero. The measure of α itself depends only on the facets in which the endpoints lie.

One example is the bending measure from a simply connected convex hull boundary component, pulled back to the hyperbolic plane by the isometry h of Chapter II.1.

Given a closed geodesic segment α, which is transverse to $\mathcal{L}$, and a $\delta > 0$, we define a δ-partition X as follows:

(1) It is finite partition $\{x_0 < x_1, \ldots, < x_k\}$ of α, where x_0 and x_k are the endpoints of α.
(2) The length of each subinterval $[x_{i-1}, x_i]$ is less than δ.
(3) The measure $\mu\{x_i\}$ is equal to zero for $1 \leq i \leq k$. The measures of the endpoints of α may or may not be zero.

Let $\mathcal{L}_\alpha$ be the measured lamination, which is the sublamination of $\mathcal{L}$ consisting of all lines of $\mathcal{L}$ meeting the closed segment α. Corresponding to any δ-partition of α, we can define a δ-partition of $\mathcal{L}_\alpha$ as follows. For each closed subinterval $[x_{i-1}, x_i]$, we choose (if there is one) a line of $\mathcal{L}_\alpha$ meeting it. For the two extreme intervals, we will choose the lines of $\mathcal{L}_\alpha$ through the endpoints x_0 and x_k if these exist. As some of the intervals may be disjoint from $\mathcal{L}$ or the same line may be chosen for two adjacent intervals, we relabel the lines $\{l_0, \ldots, l_k\}$ (by abuse of notation).

We now concentrate the measure of α on the lines l_j as follows. We associate with each line l_j the complex number ζ_j, where $2\zeta_j$ is the sum of the measures of the intervals $[x_{i-1}, x_i]$ with which it is associated. (The number of such intervals is either one or two.) As the intermediate partition points have measure zero,

$$2 \sum \zeta_j = \mu(\alpha).$$

II.3.5.2. Definition.   A *cocycle* defined on $\mathbb{H}^2$ is a map

$$B: \mathbb{H}^2 \times \mathbb{H}^2 \to PSL(2, \mathbb{C})$$

with the following properties:

(1) If $x, y, z \in \mathbb{H}^2$ then $B(x, y)B(y, z) = B(x, z)$.
(2) If $x \in \mathbb{H}^2$, then $B(x, x) = \mathrm{id}$.

It follows that $B(x, y) = B(y, x)^{-1}$.

II.3.5.3. Construction of quakebend cocycle.   Given a measured lamination $(\mathcal{L}, \mu)$ we will show how to construct a naturally associated cocycle, which we call the *quakebend cocycle*. In the next section we will construct a corresponding quakebend map $h: \mathbb{H}^2 \to \mathbb{H}^3$, which is an isometry on each facet of $\mathcal{L}$.

The construction will be carried out relative to the plane P, which is the half plane lying above the real axis in the upper half 3-space model of $\mathbb{H}^3$. We fix two points $x, y \in P$ and show how to define $B(x, y)$. Let α denote the closed geodesic segment from x to y. If α is contained in a facet of $\mathcal{L}$, we define $B(x, y)$ to be the identity, so we assume that α cuts $\mathcal{L}$ transversely.

For each $\delta = 1/n$, we will construct an approximation $B_n(x, y) \in PSL(2, \mathbb{C})$. For each $\delta = 1/n$, we choose a δ-partition of $\mathcal{L}_\alpha$. For clarity of notation we omit the index n, leaving it to be understood. The interval α is oriented from x_0 to x_k. The lines $l_0, \ldots, l_k$ are oriented from right to left, as we proceed in a positive direction along α. We define

$$B_n(x, y) = E(l_0, \zeta_0) \cdots E(l_k, \zeta_k)$$

with the following possible modifications. If l_0 passes through x_0, we replace ζ_0 by $\zeta_0/2$. If l_k passes through ζ_k, we replace ζ_k by $\zeta_k/2$.

By Lemma II.3.4.4 (*Bunch of geodesics*), $B_n(x, y)$ will converge to a limit as n tends to infinity. The usual arguments of taking common refinements show that the limit is independent of the sequence of partitions. We omit the details (which the reader would have no difficulty in supplying); in any case, we will give a more general treatment with details, which also includes a discussion of how the quakebend cocycle varies with the lamination in Section II.3.11 (*Varying the lamination*).

Note that $B(x, y)$ depends only on the facets containing x and y. Also notice that $B(x, y)$ depends only on the support of μ. That is to say, any lines of $\mathcal{L}$, which have neighbourhoods where the measure vanishes, can be deleted without changing $B(x, y)$.

We now show that the above construction does in fact define a cocycle. The only condition which we need to check is that $B(x, y)B(y, z) = B(x, z)$. This is clear when the geodesic segment $[x, z]$ meets the facet containing y. (It is in proving the cocycle property in this case that it becomes important to divide the ζs by two at the endpoints.) More generally, we form the sublaminations $\mathcal{L}_1$, $\mathcal{L}_2$, and $\mathcal{L}_3$ of $\mathcal{L}$, consisting of all lines of $\mathcal{L}$ which meet the geodesic segments $[x, y]$, $[y, z]$ and $[z, x]$, respectively. There exist points t_1 on $[x, y]$, t_2 on $[y, z]$ and t_3 on $[z, x]$, so that, on the interval $[x, y]$, t_1 separates the points on $\mathcal{L}_1 \cap \mathcal{L}_2$ from those on $\mathcal{L}_1 \cap \mathcal{L}_3$, and correspondingly for the other intervals. Note that $B(x, t_1) = B(x, t_3)$, $B(y, t_1) = B(y, t_2)$ and $B(z, t_2) = B(z, t_3)$. Factorizing by inserting the t_i, we obtain the desired equality. This proves that B is indeed a cocycle.

II.3.6. The quakebend map

We now define a map $h: P \to \mathbb{H}^3$, associated to the quakebend cocycle B. The definition depends on the choice of a basepoint $O \in P$. We define

$$h(z) = B(O, z)(z)$$

where $B(O, z)$ is the Möbius transformation already defined. It is easy to see that h maps the medial support plane P at O to the medial support plane at z (see Section II.3.1 (*Introduction – Measured Pleated Surfaces*)).

The quakebend map has the following properties:

(1) $h(O) = O$.
(2) Let l be a facet of $\mathcal{L}$. Then $h \mid l$ is a geodesic embedding of l.
(3) h is discontinuous at a point z if and only if z lies on a geodesic l which is an atom for the measure $\operatorname{Re} \mu$.

II.3.7. Invariance

So far we have discussed the quakebend cocycle $B(x, y)$ in terms of a fixed oriented plane P in upper half 3-space. We now wish to discuss the effect of varying this data. Let $T \in PSL(2, \mathbb{R})$. The pair $(\mathcal{L}, \mu)$ is sent to the pair $(T\mathcal{L}, T_*\mu)$. Let B_T be the quakebend cocycle corresponding to the new pair. We claim that

$$B_T(Tx, Ty) = TB(x, y)T^{-1}.$$

This is proved first for finite laminations, and then for general measured laminations by using partitions and taking limits.

Now let P_1 be any oriented hyperbolic plane in upper half 3-space, and let $T \in PSL(2, \mathbb{C})$ be any Möbius transformation taking the standard plane P to P_1. Let $(\mathcal{L}_1, \mu_1)$ be a measured lamination on P_1, and let $(\mathcal{L}, \mu)$ be the pullback under T to P. We define

$$B_1(Tx, Ty) = TB(x, y)T^{-1}$$

where B is the quakebend cocycle for $(\mathcal{L}, \mu)$. The preceding equation shows that the left-hand side is well defined.

Let G be a group of isometries of $\mathbb{H}^2$ which preserve $(\mathcal{L}, \mu)$. If $\mathcal{L}$ has a flat piece, then G must be discrete. To see this note that the flat piece must be sent into itself by any isometry in G which is near to the identity. The flat piece has at least two distinct boundary geodesics and these are mapped into themselves by elements of G which are near to the identity. The endpoints of the two geodesics give at least three points in the boundary which are preserved.

II.3.7.1. Lemma: Quakebend homomorphism.   *Let G be a subgroup of $PSL(2, \mathbb{R})$ which preserves a measured lamination $(\mathcal{L}, \mu)$. Then*

(1) $B(Tx, Ty) = TB(x, y)T^{-1}$ *for $T \in G$.*
(2) *The map $\chi: G \to PSL(2, \mathbb{C})$, defined by $\chi(T) = B(O, TO) \circ T$, is a homomorphism.*
(3) *The quakebend map h satisfies: $h \circ T = \chi(T) \circ h$ for $T \in G$.*

Proof.   Property (1) was proved above. To check that χ is a homomorphism we compute

$$\chi(T_1)\chi(T_2) = B(O, T_1O) \circ T_1 \circ B(O, T_2O) \circ T_2$$
$$= B(O, T_1O)B(T_1O, T_1T_2O)T_1T_2$$
$$= B(O, T_1T_2O)T_1T_2$$
$$= \chi(T_1T_2).$$

Similarly we prove that the quakebend map is equivariant.

$\square$

II.3.8. Deformations

We begin our study of deformations of a fuchsian group G with the following lemma.

II.3.8.1. Lemma: Holomorphic. *Let $(\mathcal{L}, \mu)$ be invariant under the fuchsian group G. Then for any given $z, x \in \mathbb{H}^2$, the bending cocycle $B_t(z, x)$, formed with respect to $(\mathcal{L}, t\mu)$ is a holomorphic function of $t \in \mathbb{C}$.*

To say that a Möbius transformation is a holomorphic function of a parameter t means that the matrix entries are holomorphic.

Proof. We have shown that

$$B_t(z, x) = \lim_{n \to \infty} B_{n,t}(z, x)$$

where $B_{n,t}(z, x)$ is a finite product of the form

$$\Pi_{j=0}^{r} A_j E_j(t) A_j^{-1}$$

and

$$E_j(t) = \begin{pmatrix} e^{t\zeta_j} & 0 \\ 0 & e^{-t\zeta_j} \end{pmatrix}.$$

If the $F_0, \ldots, F_r$ are the leaves of $\mathcal{L}_n$ cutting $[x, z]$, then $\zeta_j = \mu_n(F_j)$ (except that one may have to divide by two in the case of F_0 and F_r). For a fixed value of n, $B_{n,t}(z, x)$ is clearly holomorphic in t. Furthermore the convergence to $B_t(z, x)$ is uniform if t varies over a compact subset of $\mathbb{C}$. This fact is a consequence of the inequality

$$\| B_{n,t}(z, x) - \mathrm{id} \| \leq \hat{\mu}(X([x, z]))C_7$$

which will be obtained in the proof of Theorem II.3.11.9 (*Quakebends*). The lemma follows.

$\square$

For $t \in \mathbb{C}$, let $\chi_t \colon G \to PSL(2, \mathbb{C})$ denote the corresponding quakebend homomorphism.

II.3.8.2. Lemma: Injective. *Except for at most a countably infinite set of points $\{t_k\}$ in $\mathbb{C}$, the homomorphism χ_t is injective.*

Proof. If $T \in G$ is not parabolic and not equal to the identity, then the holomorphic function of t, $\mathrm{tr}^2 \chi_t(T)$, does not take the value 4 at $t = 0$. Therefore, except perhaps for a discrete set of values of $t \in \mathbb{C}$, $\mathrm{tr}^2 \chi_t(T) \neq 4$. If T is parabolic, choose T_1 so that $T_1 T T_1^{-1} T^{-1}$ is hyperbolic (see Beardon, 1983, p. 185). Again $\mathrm{tr}^2 \chi_t(T_1 T T_1^{-1} T^{-1})$ takes the value 4 only at isolated points. Except perhaps at some of these, $\chi_t(T) \neq \mathrm{id}$. Since G is countable, the set of $t \in \mathbb{C}$, for which χ_t is not injective, is countable.

$\square$

We refer the reader to Riley (1982) for further information.

II.3.9. Derivatives

We now work out the derivatives with respect to t of certain quantities associated to the measure $t\mu$ by quakebend constructions. The loxodromic transformation with multiplier $k^2 \in \mathbb{C}\backslash\{0\}$, $|k| > 1$, with repulsive fixed point 0 and attractive fixed point ∞, has the matrix representation

$$\begin{pmatrix} k & 0 \\ 0 & k^{-1} \end{pmatrix}.$$

If instead its attractive fixed point is $\xi_2 \in \mathbb{C}$ and its repulsive fixed point is $\xi_1 \in \mathbb{C}$, a normalized matrix is

$$\frac{1}{k(\xi_1 - \xi_2)} \begin{pmatrix} \xi_1 - \xi_2 k^2 & \xi_1 \xi_2 (k^2 - 1) \\ 1 - k^2 & \xi_1 k^2 - \xi_2 \end{pmatrix}.$$

The matrix also has an obvious interpretation if either ξ_1 or ξ_2 is equal to ∞.

The multiplier $k^2 \in \mathbb{C}\backslash\{0\}$, with $|k| > 1$, and the trace τ of a loxodromic transformation T are related by the formulas

$$\tau = k + k^{-1}$$

and

$$2k = \tau + (\tau^2 - 4)^{\frac{1}{2}}.$$

Since matrices in $PSL(2, \mathbb{C})$ are determined only up to a change of sign, the signs here need to be discussed. Writing $k = |k| \exp(i\arg k)$, we will assume that $-\pi/2 < \arg k \leq \pi/2$. Then k lies outside the unit circle and on the right-hand side of the complex plane. Then $\tau^2 - 4$ is never zero, and we select the branch of $(\tau^2 - 4)^{\frac{1}{2}}$ which is positive when $k > 1$.

Suppose that $k = k(t)$ is a differentiable function of the complex parameter t. We find that

$$\frac{d}{dt}(\log k) = \frac{1}{\sqrt{\tau^2 - 4}}\frac{d\tau}{dt}.$$

The number $\lambda(T) = 2\log|k| > 0$ is the length of the closed geodesic determined by T in the quotient $\mathbb{H}^3/\langle T \rangle$, with respect to the hyperbolic metric on $\mathbb{H}^3$. This means that

$$\frac{d\lambda(T)}{dt} = 2\mathrm{Re}\left\{\frac{1}{\sqrt{\tau^2 - 4}}\frac{d\tau}{dt}\right\}.$$

Each side depends only on the conjugacy class of T.

The formula for the second derivative is

$$\frac{d^2 \log k}{dt^2} = (\tau^2 - 4)^{-\frac{1}{2}}\left\{\frac{d^2\tau}{dt^2} - \frac{\tau}{\tau^2 - 4}\left(\frac{d\tau}{dt}\right)^2\right\}.$$

We have been considering general differentiable dependence of the multiplier and trace on a parameter t. Now suppose that the variation is given in a very explicit way. We consider the deformation

$$T_t = \begin{pmatrix} e^{\zeta t} & 0 \\ 0 & e^{-\zeta t} \end{pmatrix}\begin{pmatrix} a & b \\ c & d \end{pmatrix}, \quad ad - bc = 1.$$

The trace is

$$\tau(t) = ae^{\zeta t} + de^{-\zeta t}.$$

We see that

$$\frac{d\lambda(T)}{dt}\bigg|_{t=0} = 2\mathrm{Re}\left\{\frac{\zeta(a - d)}{\sqrt{(a+d)^2 - 4}}\right\}.$$

In terms of the attractive ξ_2 and repulsive ξ_1 fixed points of T, we find that

$$\frac{a - d}{\sqrt{(a+d)^2 - 4}} = \pm\frac{\xi_2 + \xi_1}{\xi_2 - \xi_1}$$

so that

$$\frac{d\lambda(T)}{dt}\bigg|_{t=0} = \pm 2\mathrm{Re}\left\{\zeta\frac{\xi_2 + \xi_1}{\xi_2 - \xi_1}\right\}.$$

If moreover a, b, c, $d \in \mathbb{R}$ and if ξ_1 and ξ_2 lie on opposite sides of the y-axis, then

$$\frac{\xi_2 + \xi_1}{\xi_2 - \xi_1} = \cos\phi$$

where, in the upper half plane, ϕ is the angle from the axis of

$$T = T_0 = \begin{pmatrix} a & b \\ c & d \end{pmatrix}$$

oriented from ξ_1 to ξ_2, to the line oriented from 0 to ∞. Therefore

$$\left.\frac{d\lambda(T)}{dt}\right|_{t=0} = \pm 2 \cos \phi \operatorname{Re}\zeta.$$

In fact the sign must be positive. To see this, take $\xi_1 \in \mathbb{R}$ very near 0 and $\xi_2 \in \mathbb{R}$ very near ∞. Then ϕ is near zero and T is nearly equal to the diagonal matrix

$$\begin{pmatrix} k & 0 \\ 0 & k^{-1} \end{pmatrix}.$$

It follows that the positive sign is correct here, and is therefore correct by analytic continuation, whenever ϕ exists (i.e. whenever the axis of T crosses the y-axis).

We have shown that

$$\left.\frac{d\tau/dt}{\sqrt{\tau^2 - 4}}\right|_{t=0} = \zeta \cos \phi.$$

The formula for the second derivative of $\log k$ shows that at $t = 0$

$$\frac{d^2 \log k}{dt^2} = \frac{\zeta^2 \tau}{\sqrt{\tau^2 - 4}}(1 - \cos^2 \phi).$$

Hence

$$\frac{d^2 \lambda(t)}{dt^2} = \frac{2\tau \sin^2 \phi \operatorname{Re}(\zeta^2)}{\sqrt{\tau^2 - 4}}$$

where we take the positive square root.

We now extend our formulas to the case where a hyperbolic transformation T of $\mathbb{H}^2$ keeps invariant the measured geodesic lamination $(\mathcal{L}, \mu)$ and we deform T using the measures $t\mu$. As in Section 3.8 (*Deformations*), we have the deformation of bending cocycles $B_t(O, TO)$, which varies holomorphically with t by Lemma II.3.8.1 (*Holomorphic*). The deformed transformation is

$$\chi_t(T) = B_t(O, TO) \circ T$$

which varies holomorphically with t by Lemma II.3.8.1 (*Holomorphic*). Therefore

$$\frac{d}{dt}\operatorname{tr}(\chi_t(T)) = \lim_{n \to \infty} \frac{d}{dt}\operatorname{tr}\{B_t(O, TO) \circ T\}.$$

We want to compute the derivative at $t = 0$. We find

$$\frac{d}{dt}(\text{tr}B_{n,t})\Big|_{t=0} = \sum_{j=0}^{r} \frac{d}{dt}\text{tr}(A_j E_j(t) A_j^{-1} T)\Big|_{t=0}.$$

When $t = 0$,

$$\text{tr}B_{n,t} \circ T = \text{tr}T = \tau.$$

We deduce that

$$\frac{d}{dt}\lambda_n(T)\Big|_{t=0} = 2\text{Re}\sum_{j=0}^{r} \zeta_j \cos \phi_j.$$

Let γ denote a closed segment of the axis of T, from a point y to a point Ty. We can choose y so that it either does not lie in any geodesic of $\mathcal{L}$, or so that it lies in a geodesic of $\mathcal{L}$ which is not an atom for μ. The orientation of γ determines an orientation for the leaves of $\mathcal{L}$ which cross γ. We insist that if (u, v) is an oriented basis for the tangent space of $\mathbb{H}^2$ at the intersection x of a geodesic $l(x) \in \mathcal{L}$ with γ, and if u points in the positive direction along γ and v is tangent to $l(x)$, then v points in the positive direction along $l(x)$. Let $\phi(x)$ denote the angle measured counterclockwise from the positive direction of γ to the positive direction of $l(x)$. By Lemma II.1.8.2 (*Disjoint geodesics*) $\phi(x)$ is a continuous function of $x \in \mathcal{L} \cap \gamma$. Therefore the integral we are about to write down exists. Letting n tend to infinity, we obtain the following result.

II.3.9.1. Theorem: First variation. *Let* $(\mathcal{L}, \mu)$ *be a measured geodesic lamination in* $\mathbb{H}^2$, *which is invariant under a hyperbolic transformation T. Let γ denote a closed segment of the axis of T from a point y to a point Ty. Let $\lambda(t)$ be the length of the geodesic associated to $\chi_t(T)$, resulting from quakebending according to the measure $t\mu$. Then*

$$\frac{d\lambda(t)}{dt}\Big|_{t=0} = 2\text{Re}\int_{\gamma} \cos \phi(x)\, d\mu(x).$$

This result also includes the case, not discussed above, where the axis of T is in $\mathcal{L}$. In that case quakebending leaves T unaltered up to conjugacy, so the length is independent of t. Also the measure induced by μ on γ is zero (by definition).

The above formula is a generalization of a formula of Kerckhoff (1983), who proved the result in the case of left earthquakes (when μ is real and positive). When μ is pure imaginary, corresponding to pure bending, the derivative is zero.

In this case $\lambda(t)$ has a local maximum at $t = 0$. (This can be seen geometrically and is proved analytically in the next section.)

II.3.10. Second variation

There is also a nice formula for the second derivative of $\lambda(t)$ at $t = 0$. We carry out some preliminary computations.

In the upper half plane consider two disjoint geodesics l_1 and l_2, with real endpoints ξ_{11}, ξ_{12} and ξ_{21}, ξ_{22}, respectively. Let l be the geodesic from 0 to ∞. We assume that both l_1 and l_2 cross l. We orient l_i from ξ_{i1} to ξ_{i2}. Let ϕ_i denote the angle from l_i to l. Let d denote the hyperbolic distance along l from $l_1 \cap l$ to $l_2 \cap l$. We have

$$\sin \phi_i = \frac{2\sqrt{-\xi_{i1}\xi_{i2}}}{\xi_{i2} - \xi_{i1}}$$

$$\cos \phi_i = \frac{\xi_{i2} + \xi_{i1}}{\xi_{i2} - \xi_{i1}}$$

$$e^d = \left(\frac{\xi_{21}\xi_{22}}{\xi_{11}\xi_{12}} \right)^{1/2}.$$

Let A_i denote a transformation which preserves $\mathbb{H}^2$ and maps l to l_i. Then

$$A_i = \frac{1}{\sqrt{\xi_{i2} - \xi_{i1}}} \begin{pmatrix} \xi_{i2} & \xi_{i1} \\ 1 & 1 \end{pmatrix}.$$

Let T be a hyperbolic transformation with axis l, represented by

$$T = \begin{pmatrix} k & 0 \\ 0 & k^{-1} \end{pmatrix}$$

with $k > 1$.

We can then compute that

$$\mathrm{tr} \left\{ A_1 \begin{pmatrix} 1 & 0 \\ 0 & -1 \end{pmatrix} A_1^{-1} A_2 \begin{pmatrix} 1 & 0 \\ 0 & -1 \end{pmatrix} A_2^{-1} T \right\}$$

$$= \tau \cos \phi_1 \cos \phi_2 + 2\cosh \left(\frac{\lambda}{2} - d \right) \sin \phi_1 \sin \phi_2.$$

Here $\tau = k + k^{-1}$ is the trace of T and $\lambda = 2\log k$ is the length of the closed geodesic in $\mathbb{H}^2/\langle T \rangle$.

Note that both sides of our formula are unchanged if A_1, A_2, and T are replaced by SA_1, SA_2, and STS^{-1} for some $S \in PSL(2, \mathbb{R})$. On the right the entries are then determined in terms of the axis $S(l)$ of STS^{-1} and the two lines $S(l_1)$ and $S(l_2)$. The angles satisfy $0 < \phi_i < \pi$.

From the formula for $B_{n,t}$ as a product of matrices of the form $A_j E_j(t) S_j^{-1}$, we then have

$$\left. \frac{d^2 B_{n,t}}{dt^2} \right|_{t=0} = \sum_{i,j=0}^{r} \zeta_i \zeta_j A_j \begin{pmatrix} 1 & 0 \\ 0 & -1 \end{pmatrix} A_j^{-1} A_i \begin{pmatrix} 1 & 0 \\ 0 & -1 \end{pmatrix} A_i^{-1} T.$$

The calculation of the traces of the summands shows that

$$\left. \frac{d^2 \operatorname{tr} B_{n,t}}{dt^2} \right|_{t=0} = \tau \left(\sum_{j=0}^{r} \zeta_j^2 \right) + \tau \left(\sum_{i \neq j} \zeta_i \zeta_j \cos \phi_i \cos \phi_j \right)$$
$$+ 2 \left(\sum_{i \neq j} \zeta_i \zeta_j \cosh \left(\frac{\lambda}{2} - d_{ij} \right) \sin \phi_i \sin \phi_j \right).$$

Here ϕ_i denotes the angle from the positively oriented axis of T to the leaf l_i and d_{ij} denotes the distance along the axis of T between the intersections with l_i and l_j. As usual, $\tau = \operatorname{tr} T$ and λ is the length of the shortest closed geodesic in $\mathbb{H}^2/\langle T \rangle$.

Now substitute in the formula for the second derivative of $\log k$, which appears in Section II.3.9 (*Derivatives*). Some cancellation takes place and we obtain

$$\left. \frac{d^2 \log k_n}{dt^2} \right|_{t=0} = \frac{2}{\sqrt{\tau^2 - 4}} \sum_{i,j=0}^{r} \zeta_i \zeta_j \cosh \left(\frac{\lambda}{2} - d_{ij} \right) \sin \phi_i \sin \phi_j$$

where $d_{ii} = 0$. The formula for the length is

$$\frac{d^2 \lambda_n(T)}{dt^2} = 2 \operatorname{Re} \frac{d^2 \log k_n}{dt^2}.$$

Letting n tend to infinity, and using the fact that uniform convergence of holomorphic functions implies uniform convergence of their derivatives, we obtain the following result.

II.3.10.1. Theorem: Second variation.

$$\frac{d^2\lambda(t)}{dt^2}\bigg|_{t=0} =$$

$$\frac{4}{\sqrt{\tau^2 - 4}}\mathrm{Re}\int_\gamma\int_\gamma \cosh\left(\frac{\lambda}{2} - d(x,y)\right)\sin\phi(x)\sin\phi(y)\,d\mu(x)\,d\mu(y).$$

II.3.10.2. Remarks.

(1) The second derivative is negative for pure bending with positive bending angles. It is positive for pure right or pure left earthquakes. In both cases the double integral is automatically real.

(2) If G is a fuchsian group acting on $\mathbb{H}^2$ and $\mathcal{L}$ consists of the lifts to $\mathbb{H}^2$ of a simple loop α on $\mathbb{H}^2/G$, then the counting measure μ is the only invariant transverse measure on $\mathcal{L}$, up to constant multiples. The resulting earthquakes are the one parameter family of Dehn shears at α. In this case our formula is the same as a special case of a formula of Wolpert (1983). His formula is for the mixed second partial derivatives for Dehn shears about distinct simple loops α_1 and α_2.

II.3.11. Varying the lamination

So far we have defined the quakebend cocycle by a process analogous to the definition of Riemann integration. However it is important to understand how the quakebends depend on the measured lamination itself: we will show that, in a certain sense, the dependence is continuous. This will enable us, in particular, to prove a uniqueness theorem.

To this end we recall some general facts about finite complex valued Borel measures on a compact metric space X. In our discussion, X will be the set of all geodesics meeting some fixed compact subset of the hyperbolic plane. We have discussed the concept of a geodesic lamination with a transverse measure. In an obvious way, this gives rise to a measure in the usual sense on the space of all geodesics in the hyperbolic plane. Conversely, a complex valued Borel measure v on the space of all geodesics in the hyperbolic plane defines a geodesic lamination $\mathcal{L}$ and a transverse measure μ, provided that different points (i.e. geodesics) in the support of v are disjoint geodesics in the hyperbolic plane. The lamination $\mathcal{L}$ is just the support of v.

By the Riesz Representation Theorem, such measures are in one-to-one correspondence with the bounded linear functionals on the space of complex valued continuous functions on X. The norm $\|\mu\|$ of the measure μ is the norm of this linear functional. It is also the total variation of the measure.

A sequence of measures μ_i converges *weakly* to a measure μ if, given any continuous complex valued function $f\colon X \to \mathbb{C}$, $\int f d\mu_n$ converges to $\int f d\mu$. For example, if on the unit interval we take μ_n to be the sum of a mass $+1$ at 0 with a mass -1 at $1/n$, then μ_n converges weakly to 0.

Any measure can be written canonically as the sum of a positive part, a negative part, an imaginary positive part and an imaginary negative part:

$$\mu = (\operatorname{Re}\mu)^+ - (\operatorname{Re}\mu)^- + i(\operatorname{Im}\mu)^+ - i(\operatorname{Im}\mu)^-.$$

We will also use the associated non-negative measure:

$$\hat{\mu} = (\operatorname{Re}\mu)^+ + (\operatorname{Re}\mu)^- + (\operatorname{Im}\mu)^+ + (\operatorname{Im}\mu)^-.$$

The example just given shows that if μ_n converges weakly to μ, then the same does not necessarily hold for one of these parts.

II.3.11.1. Definition. We say that a sequence of measures μ_n converges *totally* to a measure μ, if each of the four parts of μ_n converges weakly to the corresponding part of μ.

The next result is not necessary for this chapter, so we omit the proof. We thank Klaus Schmidt for providing us with the proof. The result is easy for a real valued measure, but is non-trivial for a complex valued measure.

II.3.11.2. Lemma: Total convergence. *Let X be a compact metric space. The sequence of measures μ_n converges totally to μ if and only if μ_n converges weakly to μ and $\|\mu_n\|$ converges to $\|\mu\|$.*

The convergence properties of quakebend cocycles do not fit naturally into established definitions of convergence in measure theory, and we will need to introduce a new type of convergence plus an endpoint condition.

II.3.11.3. Definition. Fix distinct points x_1 and x_2 and a closed hyperbolic disc K, containing x_1 and x_2 in its interior. The sequence of measured laminations $(\mathcal{L}_n, \mu_n)$ is said to converge to $(\mathcal{L}, \mu)$ *weakly, relative to $x_1, x_2, and K$,* if the following conditions are satisfied:

(1) On the set of geodesics meeting K, μ_n converges weakly to μ.
(2) There are continuous positive functions ϕ_1 and ϕ_2, with compact support in the space of all geodesics in the hyperbolic plane, such that ϕ_i is equal

to 1 on a neighbourhood of the set of all geodesics through x_i, and such that $\phi_i \mu_n$ converges totally to $\phi_i \mu$ $(i = 1, 2)$.

A sequence $(\mathcal{L}_n, \mu_n)$ that converges weakly relative to some x_1, x_2, and K, is said to satisfy the *endpoint condition* if, in addition,

(3) $\mu_n(X_i)$ converges to $\mu(X_i)$, where X_i is the set of all geodesics through x_i $(i = 1, 2)$.

It follows from the Uniform Boundedness Principle that condition (1) implies:

(4) The numbers $\|\mu_n | K \|$ are bounded.

To illustrate the above concepts, we consider measures on the interval $[-1, 1]$. These are converted into examples of measured laminations by taking orthogonal geodesics to a closed geodesic segment of length two in the hyperbolic plane. Consider the measure μ_n on $[-1, 1]$, which has two atoms, one at $1/n$ with value a certain constant c and the other at $-1/n$ with value $-c$. This converges weakly to the zero measure, but it does not satisfy condition (2) above, relative to $x_1 = 0$ and $x_2 = 1$. Note also that the quakebend cocycle $B(0, 1)$, corresponding to the zero measure, is the identity, while the cocycle $B_n(0,1)$, corresponding to μ_n is approximately a hyperbolic transformation with translation length c.

Our next example satisfies (1) and (2) above, but not (3). Here μ has a single atom at 0, with weight 2, and μ_n has a single atom at $1/n$ with weight 2. In this case $B(0, 1)$ is a hyperbolic transformation with translation distance 1, while $B_n(0, 1)$ is approximately a hyperbolic transformation with translation distance 2.

Note that the construction used in Section 3.5 (*The quakebend cocycle*), in the case that the lamination is not finite, gives an example of convergence with the endpoint condition.

The next lemma follows from the definition of bending measure, together with Lemma II.1.14.3 (*Convergence of bending laminations*).

II.3.11.4. Lemma: Bending measure and weak convergence.   *In the case of bending measure, the measures given by a standard sequence of outer approximations converge weakly to the bending measure.*

The bending measure, originally discussed as a non-negative measure, is most naturally thought of as a non-negative purely imaginary measure. For example, bending through an angle 2π gives the identity, corresponding to the fact that $e^{2\pi i} = 1$.

II.3.11.5. Theorem: Continuous dependence on a finitely supported measure. *Let $(\mathcal{L}_n, \mu_n)$ be a sequence of finite measured laminations, converging weakly to a measured lamination $(\mathcal{L}, \mu)$ relative to certain points x and x' contained in the interior of some closed hyperbolic disc K. Then the corresponding sequence of quakebends $B_n(x, x')$ has a convergent subsequence. Given two convergent subsequences, with limits the two isometries B_1 and B_2, respectively:*

(1) *If the convergence satisfies the endpoint condition then $B_1 = B_2$.*
(2) *More generally, let l and l' be the facets of $\mathcal{L}$ containing x and x'. Let 2ζ and $2\zeta'$ be the μ-measures of l and l'. Then there exist complex numbers w and w', with*

$$|\mathrm{Re}\, w| \leq |\mathrm{Re}\, \zeta| \quad and \quad |\mathrm{Im}\, w| \leq |\mathrm{Im}\, \zeta|$$

and similarly for w' and ζ', such that

$$B_1 = E(l, w)B_2 E(l', w').$$

Proof. The statement is clearly true if $x = x'$, so we assume this is not the case. Since the norms of the measures are bounded, we can apply Lemma II.3.4.2 (*Compactness of Bs*), to deduce that the $B_n(x, x')$ vary in a compact set. It follows that there exist convergent subsequences. We need only investigate the relationship of different limits.

Let γ be the geodesic through x and x'. By the definition of weak convergence relative to x and x', there is no loss of generality in assuming that $\mathcal{L}$ is equal to the support of μ. We may also assume that the support of μ_n is $\mathcal{L}_n$ for each n. Nor is there any loss of generality in assuming that all the leaves of all the laminations involved meet K. Let M be the bound on the norms of the μ_n given by the hypothesis of weak convergence.

II.3.11.6. Lemma: Case 1. *Suppose that for each n there is a geodesic $l_n \in \mathcal{L}_n$, such that l_n converges to γ, where γ is the geodesic containing x and x'. Then weak convergence with the endpoint condition satisfied implies $B_1 = B_2 = \mathrm{id}$. If $\mu(\gamma) = 0$, then weak convergence implies the endpoint condition, so that, once again, $B_1 = B_2 = \mathrm{id}$. If $\mu(\gamma) \neq 0$, the situation is as described in Theorem II.3.11.5 (Continuous dependence on a finitely supported measure).*

Proof. Case 1. Let X be the set of all geodesics of P through x and let X' be the set of all geodesics of P through x'. Recall that we are assuming that the support of μ is equal to $\mathcal{L}$. If $\mu(X) > 0$, then there must be some geodesic α of $\mathcal{L}$ through x. Since $\mathcal{L}$ is a lamination, there can only be one such geodesic, and so $\mu(\alpha) > 0$. By weak convergence, $\mathcal{L}_n$ contains a geodesic α_n which is near to α. Since l_n passes near x and is disjoint from or equal to α_n, Lemma II.2.5.2 (*Disjoint geodesics mark 2*) shows that l_n is near α_n. Hence $\gamma = \alpha$ and so $\mu(\gamma) > 0$. Similarly if $\mu(X') > 0$, $\mu(\gamma) > 0$.

We now assume that $\mu(\gamma) = 0$. It follows from the preceding paragraph that $\mu(X) = \mu(X') = 0$. We can therefore find neighbourhoods $N(X)$ of X and $N(X')$ of X' in the space of all geodesics in P, such that $\hat{\mu}(N(X)) < \varepsilon$ and $\hat{\mu}(N(X')) < \varepsilon$. Total convergence in $N(X)$ and $N(X')$ now shows that, for large values of n, $\hat{\mu}_n(N(X)) + \hat{\mu}_n(N(X')) < 8\varepsilon$. This shows that weak convergence together with the hypothesis $\mu(\gamma) = 0$ implies the endpoint condition.

Let β be a geodesic arc of length δ, whose midpoint is x and which is perpendicular to γ. Then each l_n can be assumed to meet β, and to be almost perpendicular to β. Let $f_0, \ldots, f_k$ be the sequence of geodesics of $\mathcal{L}_n$ which meet $[x, x']$. Then

$$B_n(x, x') = E(f_0, \zeta_0) \cdots E(f_k, \zeta_k)$$

where $2\zeta_i = \mu_n(f_i)$ for $1 \le i \le k$ and either $2\zeta_i = \mu_n(f_i)$ or $4\zeta_i = \mu_n(f_i)$ for $i = 0$ and $i = k$, depending on whether f_0 contains x and f_k contains x'.

For n large, $f_0, \ldots, f_k$ all meet β near x. Let β_n be the closed subinterval of β with endpoints $\beta \cap f_0$ and $\beta \cap f_k$. Then β_n converges to x. Let $w_n = \Sigma_i \zeta_i$. It follows from Lemma II.3.4.4 (*Bunch of geodesics*) $B_n(x, x')$ can be approximated by $E(\gamma, w_n)$. Since the convergence is total on $N(X)$, it is easy to see that any limit point w of the w_n must satisfy the following conditions. First, if the real part of $\mu(\gamma)$ is positive (negative), then so is the real part of w. Similarly for the imaginary part. Second, the absolute value of the real (imaginary) part of w is bounded by the absolute value of the real (imaginary) part of $\mu(\gamma)$. B_1 and B_2 are each of the form $E(\gamma, w)$ for a suitable value of w. The lemma now follows in the case of weak convergence. We also see that if $\mu(\gamma) = 0$, then w_n converges to zero, and so $B_n(x, x')$ converges to the identity.

If $\mu(\gamma) \ne 0$, it still remains to check the case of weak convergence with the endpoint condition. Then $\mu(X) = \mu(X') = \mu(\gamma) \ne 0$. By the endpoint condition, there are geodesics $\alpha_n \in X \cap \mathcal{L}_n$ and $\alpha'_n \in X' \cap \mathcal{L}_n$, such that $\mu_n(\alpha_n)$ and $\mu_n(\alpha'_n)$ are near $\mu(\gamma)$. Since l_n is disjoint from or equal to α_n and l_n passes near x, we see that α_n is near l_n. Therefore α_n is near γ. Similarly α'_n is near γ.

Let $N(\gamma)$ be a small neighbourhood of γ in the space of geodesics of P. Let $0 \le \phi \le 1$ be a continuous non-negative function on the space of geodesics, such that $\phi = 1$ on a smaller neighbourhood N_0 of γ, and such that ϕ is zero outside $N(\gamma)$. Let $\varepsilon > 0$. By total convergence on $N(\gamma)$, we have for large values of n,

$$\hat{\mu}_n(N_0 \setminus \{\alpha_n\}) = \hat{\mu}_n(N_0) - \hat{\mu}_n(\alpha_n)$$

$$\leq \int \phi d\hat{\mu}_n - \hat{\mu}_n(\alpha_n)$$

$$\leq \int \phi d\hat{\mu} - \hat{\mu}(\gamma) + \varepsilon$$

$$\leq \hat{\mu}(N(\gamma) \setminus \{\gamma\}) + \varepsilon$$

$$\leq 2\varepsilon$$

provided $N(\gamma)$ is small enough. Since $\mu_n(\alpha'_n)$ is near $\mu(\gamma)$, we see that $\alpha'_n \notin N_0 \setminus \{\alpha_n\}$ for n large. Hence $\alpha'_n = \alpha_n$ for n large. It follows that $B_n(x, x')$ is equal to the identity for n large.

$$\boxed{\text{Case 1}}$$

Let Q be a subset of the hyperbolic plane P. We will refine the notation used in Case 1, by using the notation $X(Q)$ to denote the set of all geodesics of P which contain points of Q. If Q is compact then $X(Q)$ is compact, and if Q is open then $X(Q)$ is open.

II.3.11.7. Lemma: Case 2. *Let γ be the geodesic containing x and x' and suppose there is a neighbourhood $N(\gamma)$ of γ in the space of all geodesics in P, such that, for all large values of n, $\mathcal{L}_n \cap N(\gamma) = \emptyset$. Then Theorem II.3.11.5 (Continuous dependence on a finitely supported measure) is true. If no geodesic of $\mathcal{L}$ meets $[x, x']$ transversely, then weak convergence alone implies the endpoint condition, and furthermore $B_1 = B_2 = \text{id}$.*

Note that the hypotheses do not mean that there is a neighbourhood of γ in $\mathbb{H}^2$ which is disjoint from each $\mathcal{L}_n$, but rather that if a geodesic of $\mathcal{L}_n$ meets γ then either the intersection point is near infinity or the angle of intersection is not too small.

Proof. Case 2. We suppose that there are two convergent subsequences. We first tackle the case when there are neighbourhoods $N(X)$ of X and and $N(X')$ of X', on which all the measures μ_n and μ are identically zero. We choose (u, u'), an interval on γ slightly larger than $[x, x']$ and contained in the interior of K.

We choose equally speed points:

$$x = x_0 < x_1 < \cdots < x_{2N} = x'$$

with $X([x_0, x_1]) \subset N(X)$, $X([x_{2N-1}, x_{2N}]) \subset N(X')$. We choose continuous functions $\lambda_i \colon [u, u'] \to [0, 1]$, such that λ_0 is supported in (u, x_1), λ_{2N} is supported in (x_{2N-1}, u'), λ_i is supported in (x_{i-1}, x_{i+1}) for $1 \le i \le 2N - 1$, and $\sum \lambda_i = 1$ on $[x, x']$.

Let $f_0, \ldots, f_k$ be the successive geodesics of $\mathcal{L}_n$ which meet $[x, x']$. Since $f_i \notin N(\gamma)$, it must meet γ transversely and at an angle which is not too small. Then

$$B_n(x, x') = E(f_0, \zeta_0) \cdots E(f_k, \zeta_k)$$

where $2\zeta_i = \mu_n(f_i)$ for $0 \le i \le k$.

For $1 \le i < 2N$, we choose $g_{n,i}$ to be a geodesic of $\mathcal{L}_n$, which meets (x_{i-1}, x_{i+1}), in a point which is as near as possible to x_i. Of course, it is possible that for certain values of i no such geodesic exists. In that case, we define $g_{n,i} = \emptyset$. Note that the geodesics $g_{n,i}$ meet γ in order from x to x'. Although we may sometimes have $g_{n,i} = g_{n,i+1}$, the order is never actually reversed. Let $\mathcal{L}_n'$ be the lamination consisting of the lines $g_{n,i}$.

We are now going to define two measures μ_n' and μ_n'', by concentrating μ_n in two different ways.

We define μ_n' as follows:

$$\mu_n'(g_{n,2i+1}) = \mu_n(x_{2i}, x_{2i+2}), \quad 0 \le i < N$$

$$\mu_n'(g_{n,2i}) = \mu_n(X(x_{2i})), \quad 0 < i < N.$$

Note that if $\mu_n(X(x_{2i})) \ne 0$ then our choice of $g_{n,2i}$ ensures that it passes through x_{2i}.

We define μ_n'' as follows:

$$\mu_n''(g_{n,i}) = \int_{[x,x']} \lambda_i \mu_n, \quad 0 < i < 2N.$$

For $0 \le i < N$ let

$$a_i = \mu_n(x_{2i}, x_{2i+1}) - \int_{(x_{2i}, x_{2i+1})} \lambda_{2i+1} \mu_n$$

and

$$b_i = \mu_n(x_{2i+1}, x_{2i+2}) - \int_{(x_{2i+1}, x_{2i+2})} \lambda_{2i+1} \mu_n.$$

Then

$$\mu'_n(g_{n,2i+1}) = a_i + \mu''_n(g_{n,2i+1}) - b_i, \quad 0 \le i < N$$

$$\mu'n(g_{n,2i}) = -b_{i-1} - \mu''_n(g_{n,2i}) - a_i, \quad 0 < i < N.$$

Care must be taken, when examining the above formulas, to distinguish between open and closed intervals.

We now estimate the finite quakebend cocycles $B_n(x, x')$, $B'_n(x, x')$ and $B''_n(x,x')$, corresponding to the measured laminations $(\mathcal{L}_n, \mu_n)$, $(\mathcal{L}'_n, \mu'_n)$ and $(\mathcal{L}''_n, \mu''_n)$, respectively. Applying the Lemmas II.3.4.4 (*Bunch of geodesics*) and II.3.4.2 (*Compactness of Bs*), we find that

$$\| B_n(x, x') - B'_n(x, x') \| \le C_5 \| \mu_n \| d(x, x')/2N$$

where C_5 is a constant depending only on K and an upper bound for the $\{\|\mu_n\|\}$. By Lemmas II.3.4.2 (*Compactness of Bs*) and II.3.4.1 (*Single-term comparison*), we obtain

$$\|B''_n(x, x') - B'_n(x, x')\| \le C_6 \|\mu_n\| d(x, x')/2N$$

where C_6 is a constant depending only on K and an upper bound for the $\{\|\mu_n\|\}$, since $2\|\mu_n\|$ bounds $\Sigma |a_i| + |b_i|$.

Putting the above two estimates together, we obtain

$$\|B_n(x, x') - B''_n(x, x')\| \le C_7 \|\mu_n\|/2N.$$

Now we are ready to let n go to infinity. There is a subsequence such that, for each i, the lines $g_{n,i}$ converge to a line g_i, $0 < i < 2N$. By weak convergence the finite measures μ''_n converge to the following finite measure

$$\mu''_\infty(g_i) = \int_{(x_{i-1}, x_{i+1})} \lambda_i \mu.$$

The sequence $\{B_n(x, x')\}$ converges to a limit $B_1^*(x, x')$, while $\{B''_n(x, x')\}$ converges to the finite product associated with the measure μ''_∞ on the lines $\{g_i\}$.

Suppose we have two such subsequences, giving rise to limiting measures v_1 and v_2. These are both finitely supported measures, but they are not necessarily equal, because the underlying laminations may not be equal. Let the corresponding isometries, obtained from the quakebend cocycles between x and x' be B_1^* and B_2^*, and let corresponding finite cocycles be B''_1 and B''_2. Then we will see that

$$\|B_1^* - B_2^*\| \le C_8/2N.$$

To prove this we write each of B_1'' and B_2'' as a product of $2N - 1$ terms in the usual way. Corresponding terms in this product are now compared by Lemma II.3.4.1 (*Single term comparison*). There are two cases to consider (we recall that $\mathcal{L}$ is equal to the support of μ). If $\mathcal{L}$ meets the interval (x_{i-1}, x_{i+1}), then so will each $\mathcal{L}_n$ for large n, and the geodesics of $\mathcal{L}_n$ which meet this subinterval will be fairly near those of $\mathcal{L}$. Alternatively, $\mathcal{L}$ may be disjoint from (x_{i-1}, x_{i+1}). In that case weak convergence shows that the corresponding factors are in any case each equal to the identity.

Finally let N tend to infinity. We see that the limits B_1 and B_2 in the statement of the theorem must be equal. This completes the proof when the measures are zero near the endpoints.

To understand the general case, we will work only at the endpoint x, since consideration of the other endpoint is identical. So we continue to assume that the measures are all identically zero in some neighbourhood of X'. Choose a small neighbourhood $N(X)$ of X and let ϕ be a continuous function with $0 \le \phi \le 1$, which is equal to 1 on X and to 0 outside $N(X)$. We need to compare the isometry $B_{n,1}$, defined by the quakebend cocycle between x and x' corresponding to the measure μ_n, with the isometry $B_{n,2}$, defined by the quakebend cocycle between x and x' corresponding to the measure $(1 - \phi)\mu_n$. Each of these is a product of elementary terms, one for each geodesic of the finite lamination $\mathcal{L}_n$.

Let

$$2\zeta_n = \frac{\mu_n(X)}{2} + \int_{(x,x')} \phi\mu_n.$$

Lemma II.3.4.4 (*Bunch of geodesics*) enables us to deduce that $B_{n,1}$ is approximately equal to $E(l, \zeta_n)B_{n,2}$ for some geodesic l through x. If $\mathcal{L}$ contains a geodesic passing through x, then l is equal to this geodesic. The reader may now verify the conclusions of the lemma without difficulty.

> Case 2

Continuation, proof of Continuous dependence on a finitely supported measure: Now suppose that the sequence $(\mathcal{L}_n, \mu_n)$ satisfies the hypotheses of the theorem, but satisfies the hypotheses of neither Lemma II.3.11.6 (*Case 1*) nor of Lemma II.3.11.7 (*Case 2*). Then there is a subsequence $n(i)$ and a geodesic $l_{n(i)} \in \mathcal{L}_{n(i)}$, such that $l_{n(i)}$ converges to γ, and there is a subsequence $m(i)$ and a neighbourhood $N(\gamma)$ of γ in the space of all geodesics, such that $N(\gamma) \cap \mathcal{L}_{m(i)} = \emptyset$ for each i. In view of the lemmas we have proved, there is no loss of generality in assuming that B_1 is obtained from the first subsequence and B_2 from the second.

The existence of the second subsequence shows that $\mu(\gamma) = 0$. By Lemma II.3.11.6 (*Case 1*) we have $B_1 = \text{id}$. The existence of the first subsequence shows that no geodesic of $\mathcal{L}$ meets $[x, z]$ transversely. Lemma II.3.11.7 (*Case 2*) now shows that $B_2 = \text{id}$. This completes the proof of the theorem.

Continuous dependence on a finitely supported measure

II.3.11.8. Remark. All of the limits allowed by the statement of (2) of the above theorem actually arise.

We can now prove the uniqueness theorem for the quakebend cocycle. We recall that P is the standard hyperbolic plane in $\mathbb{H}^3$, lying above the real axis in the upper half space model.

II.3.11.9. Theorem: Quakebends. *There is a unique cocycle B, associated to any given geodesic lamination $\mathcal{L}$ in P, with a complex transverse measure μ:*

(1) *$B(x, x')$ does not change as x varies over the points of any one facet. Similarly for x'.*

(2) *Let x lie in a geodesic f of $\mathcal{L}$, and let x' tend to x from one side of f. We orient f by letting (u, v) be an oriented basis at x, letting u point to the side of f on which x' lies, and taking v to be tangent to f. Then $B(x, x')$ converges to the isometry $E(f, \mu(f)/4)$.*

(3) *If $(\mathcal{L}_n, \mu_n)$ converges weakly to $(\mathcal{L}, \mu)$, relative to x and x', with the endpoint condition satisfied, then $B_n(x, x')$, associated to $(\mathcal{L}_n, \mu_n)$, converges to $B(x, x')$, associated to $(\mathcal{L}, \mu)$.*

(4) *For any isometry T of $\mathbb{H}^2$ we have*

$$B(Tx, Tx') = TB(x, x')T^{-1}.$$

Proof. We have not quite proved that the quakebend cocycle does indeed satisfy all the conditions of the theorem. To prove (2) we note from Lemma II.3.4.4 (*Bunch of geodesics*) that for a finite lamination $\mathcal{L}$, we have

$$||B(x, x') - \text{id}|| \leq \hat{\mu}(X([x', x]))C_8$$

where C_8 depends only on x' and x. The same inequality then follows when $\mathcal{L}$ is not finite, by taking limits. We then fix $f \in \mathcal{L}$, and find a neighbourhood N of f in the space of geodesics, such that $N \setminus \{f\}$ has small $\hat{\mu}$-measure.

The cocycle property then shows that $B(x, x_n)$ is a Cauchy sequence of isometries, as x_n converges to x from one side. To see that the limit is $E(f, \mu(f)/4)$, we approximate by a finite sequence, converging weakly with the endpoint condition.

Assertion (3) follows by approximating $(\mathcal{L}_n, \mu_n)$, and then using these approximations to construct a sequence of finitely supported measured laminations which converge, with the endpoint condition satisfied, to $(\mathcal{L}, \mu)$. We then use Theorem II.3.11.5 (*Continuous dependence on a finitely supported measure*).

To prove uniqueness, note that, the cocycle condition together with condition (2), determine B for finite laminations. Then condition (3) determines B on laminations which are not finite.

Appendix

II.A.1. The lipschitz property

In our investigations, we will stick to metric spaces where the metric is equal to the infimum of the lengths of rectifiable paths connecting two points. These are called *path spaces* and the metrics are called *path metrics*. Nearly all the time we will be working with the most common type of path space, namely Riemannian manifolds. In that case the original definition of the length of a (smooth) path is given by integrating the norm of the tangent vector to the path, the metric is given by the infimum of the lengths of smooth paths, and a short proof is needed to show that this metric is a path metric.

We want to talk about lipschitz maps and vector fields. There are a number of slight but annoying variations in the possible definitions one might choose. These differences arise from the fact that a subset U, connected by rectifiable paths, of a path space M, has two distinct metrics on it. The metric d_M is measured using all paths in M, and the metric d_U is measured using all paths in U. Thus if $x, y \in U$, $d_M(x, y) \leq d_U(x, y)$, and in general we will not have equality.

II.A.1.1. Definition. Let $\phi: U \to V$ be a map between two metric spaces. We say that ϕ is K-path-lipschitz if for any rectifiable path γ in U, length($\phi \circ \gamma$) $\leq K$ length(γ). We say that ϕ is K-path-bilipschitz or a K-path-quasi-isometry if ϕ is a homeomorphism onto V and if both ϕ and ϕ^{-1} are K-path-lipschitz.

If U and V have metrics which are not necessarily path metrics, for example, if the metric is induced from a path metric on a larger space, then the usual notions of lipschitz and bilipschitz (equivalently, quasi-isometry) apply. If ϕ is K-lipschitz, then ϕ is also K-path-lipschitz. If the domain is a path space with its path metric, and if ϕ is K-path-lipschitz, then ϕ is also K-lipschitz. Thus there is no distinction between the two notions if the domain is a path space with its path metric.

II.A.2. Distances in the tangent bundle

We want to discuss the notion of *lipschitz* vector fields on subsets U of M, where U is connected by rectifiable paths. The notion of lipschitz vector fields will always be interpreted in the sense of path-lipschitz. Let $\pi: TM \to M$ be the projection of the tangent bundle. Let $X: U \to TM$ be a vector field over U. This means simply that $\pi \circ X = \mathrm{id}_U$. A path metric can be defined on U by using the infimum of lengths of paths. If a path metric is also defined on TM, then X is a map between path spaces and "lipschitz" has its usual sense as a map. Note that the metric we will use on U is given by the infimum of lengths of paths in U, and *not* paths in M.

II.A.2.1. Metric on tangent bundle.
We briefly remind readers of the construction of the standard Riemannian metric on the tangent bundle of any Riemannian manifold M. At each point $u \in TM$, we have the horizontal subspace defined by the Riemannian connection on M. This gives us a direct sum decomposition of $T_u(TM)$ as the direct sum of two orthogonal copies of $T_{\pi u}M$, namely the horizontal and vertical subspaces, and hence defines an inner product on the direct sum. This makes the manifold TM into a Riemannian manifold of dimension $2 \dim M$. Every Riemannian manifold becomes a metric space by taking the infimum of the lengths of paths over all paths joining two points, and this is the metric we use on TM.

It is easy to see that the metric can also be defined as follows. If u, $v \in TM$, let γ be a piecewise smooth path from πu to πv, whose derivative is never zero. Let $\tau: T_{\pi u}M \to T_{\pi v}M$ be parallel translation along γ. We define $\Theta(u, v) = ||\tau u - v||$. Then the square of the distance from u to v is the infimum over all paths γ from πu to πv of

$$d_\gamma(u, v)^2 = \Theta(u, v)^2 + \mathrm{length}(\gamma)^2. \tag{A}$$

To justify the above assertion, note that, over a small piece of γ, parallel translation gives us a canonical identification of any two tangent spaces, and hence the tangent bundle is metrically the product of an interval and $\mathbb{R}^{\dim M}$.

This enables us to define the path-lipschitz constant of a vector field X, by taking u and v to be the values of X at different points of M.

II.A.2.2. Classical lipschitz constant.
Classically, the lipschitz constant of a vector field is not the same constant as the constant we have just defined. The reason for this is that a vector field on $\mathbb{R}^n$ is classically regarded as a map from $\mathbb{R}^n$ to $\mathbb{R}^n$. Our definition regards it as a map from $\mathbb{R}^n$ into the tangent bundle of $\mathbb{R}^n$, and the tangent bundle is equal to $\mathbb{R}^n \times \mathbb{R}^n$ with the product metric.

It is possible to define an analogue for the classical lipschitz constant of a vector field, even when the manifold M has a non-trivial tangent bundle. It is important for us to do this, because we will prove that the classical estimate for an approximate solution to a differential equation remains true in the case of a Riemannian manifold, provided the lipschitz constant of the vector field, which appears in the classical estimate, is correctly defined. Moreover the classical estimate turns out to be easy to compute within the case of a general Riemannian manifold.

II.A.2.3. Definition.
Let X be a continuous vector field on M. Then K is the *classical lipschitz constant* of X if

$$K = \sup_\gamma \frac{||\tau_\gamma X(\gamma(0)) - X(\gamma(1))||}{\text{length}(\gamma)}$$

where $\gamma \colon [0,1] \to M$ varies over all C^1-paths, and τ_γ is parallel translation along γ.

For an arbitrary continuous vector field K may be infinite. If M is compact (or has some compact closure in a larger manifold where X is defined) and if X is C^1, then K is finite, as we will see in Section II.A.2.6 (*Covariant derivative*).

II.A.2.4. Lemma: Lipschitz constant.
Let M be a Riemannian manifold, and let X be a vector field on X:

(1) *Then X is lipschitz in the path sense if and only if the classical lipschitz constant is finite.*

(2) *If K is the classical lipschitz constant, then the path-lipschitz constant is equal to $\sqrt{K^2 + 1}$.*

(3) *We get the same supremum if we vary only over geodesic paths in M.*

Proof. Suppose we know that for any rectifiable path γ,

$$\text{length } X \circ \gamma \le k \text{ length } \gamma.$$

Then this will also be true if γ is C^1 and

$$k^2 (\text{length } \gamma)^2 \ge (\text{length } X \circ \gamma)^2 \ge ||\tau_\gamma X\gamma(0) - X\gamma(1)||^2 + (\text{length } \gamma)^2.$$

It follows that the classical lipschitz constant is bounded by $\sqrt{k^2 - 1}$. So if X is k-path lipschitz, then it is K classically lipschitz, and $K^2 \le k^2 - 1$.

If X is lipschitz with classical constant K, then K will be an upper bound for the same ratios if we restrict γ to be a geodesic.

II.A.2.5. Definition.   We recall that a subset K of a Riemannian manifold is said to be *convex*, if any two points of K can be joined by a unique geodesic arc in K.

If x and y are two points in the same convex neighbourhood in M, let α be the geodesic joining them. Let τ be parallel translation along α from the tangent space at x to the tangent space of y. Then

$$d_{TM}(X(x), X(y))^2 \leq d(x, y)^2 + ||\tau_\alpha X(x) - X(y)||^2.$$

We define

$$K_0 = \sup_\gamma \frac{||\tau_\gamma X(\gamma(0)) - X(\gamma(1))||}{\text{length}(\gamma)}$$

when γ ranges over geodesics. Suppose K_0 is finite. Let $\beta\colon I \to M$ be any rectifiable path in M, and let $0 = t_0 < \cdots < t_N = 1$ be any partition of the domain of β, chosen so that successive points are mapped into the same convex neighbourhood in M. Then

$$\sum d_{TM}(X\beta(t_i), X\beta(t_{i+1})) \leq \sum \{d(\beta(t_i), \beta(t_{i+1}))^2$$
$$+ ||\tau_i X\beta(t_i) - X\beta(t_{i+1})||^2\}^{1/2}$$

(where τ_i is the parallel translation along a short geodesic)

$$\leq \sqrt{K_0^2 + 1} \sum d(\beta(t_i), \beta(t_{i+1}))$$
$$\leq \sqrt{K_0^2 + 1}\ \text{length}\,(\beta).$$

It follows that $X \circ \beta$ is rectifiable and that

$$\text{length}\,(X \circ \beta) \leq \sqrt{K_0^2 + 1}\ \text{length}\,(\beta).$$

Therefore X is path-lipschitz and we have

$$k^2 \leq K_0^2 + 1 \leq K^2 + 1 \leq k^2.$$

This completes the proof of the lemma.

$\square$

II.A.2.6. Covariant derivative.   If X is a C^1-vector field, the classical lipschitz constant of X is the supremum of $||\nabla_v X||$, as v varies over all unit vectors. This follows from the previous lemma, because the covariant derivative along a geodesic can be regarded as the ordinary directional derivative. We can then argue in the standard way to deduce that a lipschitz constant is equal to the norm of the derivative.

The advantage of the path-lipschitz constant over the classical constant is that a vector field is treated just like any other map. In particular, the lipschitz constant of the composite of two lipschitz maps is the product of the lipschitz constants of the maps, and we will have occasion to use this when one of the two maps is a vector field. In addition, it makes sense to talk of the path-lipschitz constant of a vector field defined on a subset which is connected by rectifiable paths, but which is not a smooth submanifold. We will have occasion to use this too. The advantage of the classical constant is that, in practice, it is easier to compute with, basically because we know how to manipulate covariant derivatives. The disadvantage of using both concepts in the same paper (which we do) is that it becomes necessary to constantly distinguish the two.

The following lemma is obvious.

II.A.2.7. Lemma: Lipschitz sum. *Let X and Y be vector fields, which have classical lipschitz constants K_1 and K_2, respectively. Then their sum has classical lipschitz constant $(K_1 + K_2)$ and minus the field, $-X$, has classical lipschitz constant equal to K_1.*

II.A.3. Approximate solutions to differential equations

The next result is standard in the classical theory of differential equations. However, it seems not to be well known in the generality presented here. We would like to thank C.T.C. Wall for suggesting that a straight-forward approach to this question might be successful, instead of the much more roundabout method we were using previously.

II.A.3.1. Definition. Let Z be a vector field on a manifold M. Let $a < 0 < b$. Let $\varepsilon: (a, b) \to (0, \infty)$ be a continuous function. An ε-*approximate* solution of Z is a differentiable map $\beta: (a, b) \to M$, such that for each $t \in (a, b)$,

$$\|\beta'(t) - Z(\beta(t))\| \leq \varepsilon(t).$$

II.A.3.2. Theorem: Approximation inequality. *Let M be a complete or a convex (see Definition II.A.2.5 (Convex subset)) Riemannian manifold with a lipschitz vector field Z. Let the classical lipschitz constant of Z be $K > 0$. Let γ: $(a, b) \to M$ be a solution curve for Z and let $\varepsilon: (a, b) \to (0, \infty)$ be a continuous* *function and* $\beta\colon (a, b) \to M$ *be an* ε-*approximate solution to* Z. *Suppose that* $d(\gamma(0), \beta(0)) \le \delta$. *Then*

$$d(\gamma(t), \beta(t)) \le \delta e^{K|t|} + \left| \int_0^t \varepsilon(s) e^{K|t-s|} ds \right|.$$

Proof. There are two cases, depending on whether t is positive or negative. The two inequalities corresponding to the two cases are equivalent to the following two inequalities. Let $L(t) = d(\gamma(t), \beta(t))$ and suppose $a \le t' \le t \le b$. Then

$$L(t') e^{-K(t-t')} - \int_{t'}^t \varepsilon(s) e^{-K(t-s)} ds \le L(t) \le L(t') e^{K(t-t')} + \int_{t'}^t \varepsilon(s) e^{K(t-s)} ds.$$

We prove this first when M is convex. Suppose that $\beta(t) \ne \gamma(t)$ for some value of $t \in (a, b)$. Let $\alpha_t \colon [0, L(t)] \to M$ be the geodesic from $\beta(t)$ to $\gamma(t)$, parametrized by arclength. Let

$$u_t = \frac{\partial}{\partial s} (\alpha_t(s)) \bigg|_s = 0 \quad \text{and} \quad v_t = \frac{\partial}{\partial s} (\alpha_t(s)) \bigg|_{s=L(t)}.$$

Then $\|u_t\| = \|v_t\| = 1$. By the First Variation Formula for arclength, we have

$$\frac{\mathrm{d}}{\mathrm{d}t} L(t) = \langle v_t, \gamma'(t) \rangle - \langle u_t, \beta'(t) \rangle.$$

Let τ_t be parallel translation along α_t from $\gamma(t)$ to $\beta(t)$. Then

$$\langle v_t, \gamma'(t) \rangle = \langle u_t, \tau_t \gamma'(t) \rangle.$$

Now

$$\|\tau_t \gamma'(t) - Z(\beta(t))\| \le L(t) K \quad \text{and} \quad \|Z(\beta(t)) - \beta'(t)\| \le \varepsilon(t).$$

It follows that

$$\left| \frac{\mathrm{d}}{\mathrm{d}t} L(t) \right| \le \left| \langle u_t, \tau_t \gamma'(t) - \beta'(t) \rangle \right| \le L(t) K + \varepsilon(t).$$

This gives two inequalities for the derivative. Multiplying by the integrating factors e^{Kt} or e^{-Kt} and integrating, we obtain the two inequalities desired, provided that $\beta(s)$ is distinct from $\gamma(s)$ throughout the open interval (t', t).

Continuing with the hypothesis that M is convex, we now suppose for the moment that $L(0) > 0$. We choose t_1 as large as possible, such that the right-hand inequality holds when $t' = 0$ and $0 \le t < t_1$. We will assume that $t_1 < b$ and prove a contradiction. By continuity, the inequality then holds on the closed interval $[0, t_1]$. Equality must hold at t_1. By continuity, $L(t) > 0$ for t near t_1. But then we must have

$$L(t) \le L(t_1)e^{K(t-t_1)} + \int_{t_1}^{t} \varepsilon(s)e^{K(t-s)}ds = \delta e^{Kt} + \int_{t_1}^{t} \varepsilon(s)e^{K(t-s)}ds$$

for $t > t_1$ and for t near to t_1. This is a contradiction, which proves the right-hand inequality when $L(0) > 0$, M is convex, $t' = 0$ and $t > 0$.

To prove the left-hand inequality when $L(0) > 0$, M is convex, $t' = 0$ and $t > 0$, again choose t_1 maximal so that the left-hand inequality is satisfied. Once again we suppose that $t_1 < b$ and prove a contradiction. The left-hand inequality becomes an equality at t_1. If $L(t_1) > 0$, we argue as for the right-hand inequality. If $L(t_1) = 0$, we see that the left-hand side of the inequality is zero. But the expression is strictly decreasing in t and lengths are always positive. Therefore the inequality is true for all $t > t_1$. This contradiction proves the left-hand inequality when $L(0) > 0$, M is convex, $t' = 0$ and $t > 0$.

To prove the two inequalities when $L(0) > 0$, M is convex, $t' < 0$ and $t = 0$, we simply reverse the direction of time. Note that $-Z$ also has classical lipschitz constant equal to K.

The next step is to prove the two inequalities for $t' < t$, when M is convex and $L(0)$ may be zero. If $L(t)$ is identically zero, the inequalities are obvious. If there is some point where $L(t) > 0$, then we prove the two inequalities by reparametrizing so that we recover the hypothesis $L(0) > 0$.

To prove the two inequalities if M is complete but not convex, we choose $t_1 > 0$ maximal so that the right-hand inequality holds. We have a geodesic path of length $L(t_1)$ from $\beta(t_1)$ to $\gamma(t_1)$. We subdivide the geodesic with points

$$\beta(t_1) = x_0, x_1, \ldots, x_N = \gamma(t_1)$$

chosen in such a way that the shortest geodesic arc from x_i to x_{i+1} is contained well within a convex open subset of M. Let $\gamma_1, \ldots, \gamma_N$ be solution curves for Z, with $\gamma(t_i) = x_i$. Then, for $t > t_1$ and for t near enough to t_1,

$$L(t) \le \sum_{i=1}^{N} d(\gamma_{i-1}(t), \gamma_i(t)) + d(\gamma_N(t), \beta(t))$$

$$\le \sum_{i=1}^{N} d(x_{i-1}, x_i)e^{K(t-t_1)} + \int_{t_1}^{t} \varepsilon(s)e^{K(t-s)}ds$$

$$\leq L(t_1)e^{K(t-t_1)} + \int_{t_1}^{t} \varepsilon(s)e^{K(t-s)}ds$$

$$= \delta e^{Kt} + \int_{t_1}^{t} \varepsilon(s)e^{K(t-s)}ds.$$

This contradiction proves the right-hand inequality for $t \geq 0$. The left-hand inequality is deduced by reversing the direction of time.

$\square$

Note that the theorem above is false if M is neither complete nor convex. A counterexample is constructed by taking the unit vertical field in the euclidean plane, from which the closed positive y-axis is removed. The classical lipschitz constant of this vector field is zero. In particular, it also has classical lipschitz constant K for each $K > 0$. The inequality is clearly not satisfied (recall that we are using the path metric, which is larger than the euclidean metric).

Let S_ε be the ε-distant surface from a finitely bent convex hull boundary (see Section II.2.2 (*The epsilon distant surface*)). We will need the above results also for vector fields on S_ε which are lipschitz with respect to the intrinsic metric.

II.A.3.3. Corollary: Approximation inequality – piecewise version.
The results of Theorem II.A.3.2 (Approximation inequality) *hold also on* S_ε.

Proof.   The proof consists of going through the proof of the theorem and noting that every step goes through. We use the fact that the distance function is differentiable, proved in Proposition II.2.2.1 (*Geometry of a piecewise smooth manifold*). (The reason for stating the above result separately is that, in stating Theorem II.A.3.2 (*Approximation inequality*), we were following the convention in differential geometry that Riemannian manifolds have C^∞ metrics.)

$\square$

II.A.4. Angle measurement

Suppose that P_1 and P_2 are oriented planes in $\mathbb{R}^3$ intersecting in the line l, and that P_1 is not equal to P_2 with the opposite orientation. (They are allowed to be equal with the same orientation.) Let P be one of the two oriented planes orthogonal to l at a point $p \in l$, and denote by θ, $-\pi < \theta < \pi$, the dihedral angle between P_1 and P_2, measured in a clockwise direction on P, from the normal to P_1 to the normal to P_2. Take a second oriented plane P^* that crosses

l transversely at *p*. We will ultimately be interested only in the situation that *P**
is close to *P*. That is, that the unit normal vector *n* to *P* at *p* is close to the unit
normal vector *n** to *P** at *p*. Of course, *n* lies in *l*.

Consider the oriented lines $P_1 \cap P^*$, $P_2 \cap P^*$, which cross at *p*. Set

$$\phi_1 = \text{angle between } n \text{ and } P_1 \cap P^*, 0 < \phi_1 < \pi$$

$$\phi_2 = \text{angle between } n \text{ and } P_2 \cap P^*, 0 < \phi_2 < \pi$$

$$\theta^* = \text{angle in } P^* \text{ between } P_1 \cap P^* \text{ and } P_2 \cap P^*.$$

Note that, since all the planes are oriented, each of these angles makes sense
(given two unoriented lines which intersect, their angle of intersection is either
θ or $\pi - \theta$). We will find a formula relating θ^* to θ.

II.A.4.1.

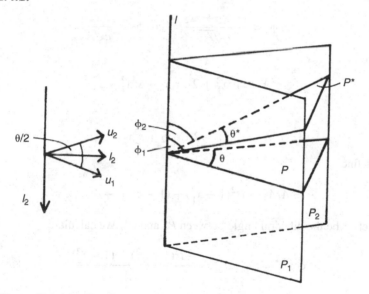

Figure II.A.4.1 *The diagram on the left is in the plane of P. It shows u_1 on
$P \cap P_1$ and u_2 on $P \cap P_2$. On the right we show horizontal plane P and the nearly
horizontal plane P*, in which the angles θ and θ^* respectively are measured.*

We fix the origin for $\mathbb{R}^3$ space at the intersection of the three planes P_1, P_2,
and *P*. We take *P* as the plane $x_3 = 0$, and $n = (0, 0, 1)$. Let n_i be the normal
unit vector to P_i $(i = 1, 2)$. Let u_i be a unit vector in P_i, orthogonal to *n*. We
choose an orthonormal basis e_1, e_2, e_3 with $e_3 = n$, and so that

$$u_1 = e_1 \cos \frac{\theta}{2} + e_2 \sin \frac{\theta}{2} \quad \text{and} \quad u_2 = e_1 \cos \frac{\theta}{2} - e_2 \sin \frac{\theta}{2}.$$

Let

$$n^* = x_1 e_1 + x_2 e_2 + x_3 e_3$$

with $x_1^2 + x_2^2 + x_3^3 = 1$ and with x_3 near the identity. We will write $s = \sin(\theta/2)$, $c = \cos(\theta/2)$, and $t = \tan(\theta/2)$.

Let u_i^* be the unit normal vector in P_i, which is near to u_i and orthogonal to n^*. Then

$$u_i^* = \cos\phi_i u_i + \sin\phi_i e_3.$$

We deduce that

$$\sin\phi_1 = \frac{-cx_1 + sx_2}{c\sqrt{A_1}}, \quad \cos\phi_1 = \frac{x_3}{c\sqrt{A_1}}$$

and

$$\sin\phi_2 = \frac{-cx_1 + sx_2}{c\sqrt{A_2}}, \quad \cos\phi_2 = \frac{x_3}{c\sqrt{A_2}}$$

where

$$c^2 A_1 = c^2 x_1^2 + 2scx_1 x_2 + s^2 x_2^2 + x_3^2$$

and

$$c^2 A_2 = c^2 x_1^2 - 2scx_1 x_2 + s^2 x_2^2 + x_3^2.$$

We define

$$A = A_1 A_2 = (t^2(1 - x_1^2) + (1 - x_2^2))^2 - 4t^2 x_1^2 x_2^2.$$

Let θ^* be the dihedral angle between P_1^* and P_2^*. We calculate

$$\cos\theta^* = \langle u_1^*, u_2^* \rangle = \frac{-t^2(1 - x_1^2) + (1 - x_2^2)}{\sqrt{A}}.$$

Let $t^* = \tan(\theta^*/2)$. Then

$$\frac{(t^*)^2}{t^2} = \frac{1 - \cos\theta^*}{t^2(1 + \cos\theta^*)} = \frac{4(1 - x_1^2 - x_2^2)}{(\sqrt{A} - t^2(1 - x_1^2) + (1 - x_2^2))^2}.$$

If $x_1^2 < 1/2$ and $x_2^2 < 1/2$, then $1 - x_1^2 > x_1^2$ and $1 - x_2^2 > x_2^2$. Hence $A > 0$, because

$$A > (t^2(1 - x_1^2) - (1 - x_2^2))^2$$

unless $t = 0$. If $t = 0$, then $A = (1 - x_2^2)^2 > 1/4$.

The condition $x_1^2 < \frac{1}{2}$ and $x_2^2 < \frac{1}{2}$ also ensures that both the numerator and the denominator in the expression for $(t^*)^2/t^2$ are non-zero. It follows that t^* and t^*/t are analytic functions of t, x_1, and x_2, which can be extended to be analytic at the value $t = 0$, provided P_1 and P_2 are not equal planes with opposite orientations and that $x_i^2 < \frac{1}{2}$ for $i = 1, 2$. We also see that t^*/t is positive, by analytic continuation from $x_1 = x_2 = 0$. We write

$$f(t, x_1, x_2) = t^*/t$$

and f is an analytic function in the region where $x_1^2 < 1/2$ and $x_2^2 < 1/2$.

We have

$$\frac{\theta^*}{\theta} = \frac{\arctan(f(t, x_1, x_2) \cdot t)}{\arctan t}.$$

This can be estimated using the Mean Value theorem, with f regarded as the variable. Since t is an analytic function of θ, we deduce that θ^*/θ is an analytic function $h(\theta, x_1, x_2)$, when $-\pi < \theta < \pi$, and $x_i^2 < 1/2$ for $i = 1, 2$. We also see that, given $\varepsilon > 0$ there is a $\delta > 0$ such that, if $x_1^2 + x_2^2 < \delta$, then

$$(1 - \varepsilon)\theta \leq \theta^* \leq (1 + \varepsilon)\theta.$$

> **Theta – star theta inequality**

We will prove a much more precise estimate of this kind.

II.A.4.2. Theorem: Theta – theta star estimate. *Set $Z = x_1^2 + x_2^2$. Then, when $Z < 1/2$,*

$$\left| \frac{\theta^*}{\theta} - 1 \right| \leq \frac{1}{1 - Z} - \frac{1}{\sqrt{1 - Z}}.$$

Proof. **Theta – theta star estimate:** We first estimate

$$\frac{\theta^*}{\theta} - 1 = \frac{\arctan(\lambda t)}{\arctan t} - 1$$

where $\lambda = t^*/t$ and t and t^* are as defined above. The derivative with respect to λ is

$$\frac{t}{(1 + \lambda^2 t^2) \arctan t} = \frac{1 + t^2}{1 + \lambda^2 t^2} \frac{t}{(1 + t^2) \arctan t}.$$

This quantity is positive, and we would like to find an upper bound for it. We may assume that $t > 0$. The first factor on the right is monotonic in t^2 and is therefore bounded by $\max\{1, 1/\lambda^2\}$. The second factor is monotonic decreasing and it tends to zero as t tends to infinity. Therefore its maximum, 1, is achieved at $t = 0$. It follows that the derivative of $(\theta^*/\theta - 1)$ is bounded by $\max\{1, 1/\lambda^2\}$. By the Mean Value theorem,

$$\left|\frac{\theta^*}{\theta} - 1\right| \le |\lambda - 1| \max\left\{1, \frac{1}{\lambda^2}\right\}.$$

This shows that our task is to estimate λ. We will at first work with the estimates already made for λ^2.

We refer to the calculations and notation above. To simplify the notation, we write $T = t^2$, $X_1 = x_1^2$, and $X_2 = x_2^2$. Then,

$$A = (T(1 - X_1) + (1 - X_2))^2 - 4TX_1X_2.$$

Consider the function

$$Y = \sqrt{A} - T(1 - X_1) + (1 - X_2)$$

whose square is the denominator of $f(t, x_1, x_2)$, as a function of T. We want to find a maximum and minimum value as T varies. If $X_1 = 0$, then $Y = 2(1 - X_2)$, and if $X_2 = 0$, then $Y = 2$, so Y is independent of T in these cases, which can therefore be ignored. So we now assume that $X_1 > 0$ and $X_2 > 0$.

To find the maximum and minimum, we differentiate, obtaining

$$\frac{dY}{dT} = \frac{1}{2\sqrt{A}}\frac{dA}{dT} - (1 - X_1).$$

We compute that this can only equal zero if $X_1 + X_2 = 1$; but we are assuming that $X_1 < 1/2$ and $X_2 < 1/2$, so the derivative is non-zero. We also find that this derivative is negative if $T = 0$. Therefore Y has a maximum at $T = 0$ and tends to its minimum as T tends to infinity.

The maximum value of Y is $2(1 - X_2)$. The minimum value is

$$\frac{2(1 - X_1 - X_2 - X_1X_2)}{1 - X_1}.$$

The first statement is immediate. The second follows by expanding in powers of $1/T$.

The previous paragraph shows that a maximum value for

$$\lambda^2 = \frac{(t^*)^2}{t^2} = \frac{4(1 - X_1 - X_2)}{Y^2}$$

is

$$\frac{(1 - X_1 - X_2)(1 - X_1)^2}{(1 - X_1 - X_2 - X_1 X_2)^2}$$

and a minimum value is

$$\frac{1 - X_1 - X_2}{(1 - X_2)^2}.$$

We have to estimate how far these values are from 1.

We have $Z = X_1 + X_2$. We decrease the minimum value for λ^2, replacing it with $1 - Z$.

The maximum value of λ^2, written in terms of Z, is

$$\frac{(1 - Z)(1 - X_1)^2}{(1 - Z - X_1(Z - X_1))^2}.$$

By taking the logarithmic derivative with respect to X_1, we see that this function is monotonic decreasing as a function of X_1, provided that $Z < 1/2$, which is our assumption. So the maximum value is increased by taking $X_1 = 0$, when the expression becomes $1/(1 - Z)$.

We are looking for estimates for $\theta^*/\theta - 1$. Suppose first that $\lambda \geq 1$. We have

$$\lambda \leq \frac{1}{\sqrt{1 - Z}}$$

and $1/\lambda^2 \leq 1$. If $\lambda < 1$, we need a lower bound. We have

$$\lambda \geq \sqrt{1 - Z} \quad \text{and} \quad 1/\lambda^2 \leq \frac{1}{1 - Z}.$$

It follows that

$$\left| \frac{\theta^*}{\theta} - 1 \right| \leq \max \left\{ \frac{1}{\sqrt{1 - Z}} - 1, \frac{1 - \sqrt{1 - Z}}{1 - Z} \right\}.$$

The first of these functions is smaller than the second for $0 < Z < 1$.

> **Theta – theta star estimate**

II.A.4.3. Corollary: Upper bound for theta. *If $x_1^2 + x_2^2 < \sin^2 t$, where $t < \pi/4$, then*

$$\theta \leq \frac{\cos^2 t \, \theta^*}{\cos t - \sin^2 t}.$$

This follows by substitution in the inequalities we have already found.

II.A.5. Distances to geodesics

II.A.5.1. Lemma: Nearest geodesics. *Let l_1, l_2, and l_3 be disjoint geodesics in $\mathbb{H}^2$, arranged so that there is a region C whose boundary in $\mathbb{H}^2$ consists of l_1, l_2, and l_3. Let $z \in C$ be at a distance less than u from l_1 and less than v from l_2. Suppose that $\sinh u \cdot \sinh v < 1$. Then the distance of z from l_3 is greater than*

$$\operatorname{arcsinh}\left(\frac{1 - \sinh u \cdot \sinh v}{\sinh u + \sinh v}\right).$$

Proof. Imagine l_1 as moving rigidly in $\mathbb{H}^2$, together with its u-neighbourhood and imagine l_2 moving with its v-neighbourhood. Let K be the intersection of these two neighbourhoods, which we may suppose is non-empty. We move the three lines so as to decrease the distance of l_3 to K. First we move l_3 towards K until its endpoints are equal to endpoints of l_1 and l_2. Then we hold l_3 and l_2 fixed, and "rotate" l_1 about its common fixed point with l_3 until its other endpoint becomes a common endpoint with l_2. Clearly, this changes K and decreases its distance to l_3. Arguing in this way, we see that we may as well assume that l_1, l_2, and l_3 are sides of an ideal triangle.

We examine this triangle in the upper half-plane model. Let l_1 be a vertical line through the origin, and let l_2 be the line $x = 1$. Then l_3 is a semicircle, centred at $(\frac{1}{2}, 0)$. We recall that, if d is the distance of a point (x, y) from l_1, then $\sinh d = x/y$.

The curve of points at distance u from l_1 is represented by a euclidean straight line l_4 through $(0, 0)$ at slope $1/\sinh u$ and the curve of points at distance v from l_2 is represented by a euclidean straight line l_5 through $(1, 0)$ at slope $1/\sinh v$. The lines l_4 and l_5 meet at a point A whose coordinates are

$$\left(\frac{\sinh u}{\sinh u + \sinh v}, \frac{1}{\sinh u + \sinh v}\right).$$

The distance of A to l_3 is most easily calculated by inverting the figure in the circle with centre the origin and radius 1. This inversion preserves l_1 as a set, and interchanges l_2 and l_3. In this way, we compute that the distance of A to l_3 is equal to the expression in the statement of the lemma.

The geodesic l_3 has the equation $\sinh u \cdot \sinh v = 1$. The hypothesis implies that A is inside the ideal triangle. A is on the boundary of the open set K, and is nearer to l_3 than any point of K. The easiest way to see this is to note that, along the portions of l_4 and l_5 which bound K, the distance from l_3 increases as one moves away from A. This is a consequence of the formula we have already found for A, applied to such a moving point. Hence any point of K is at a distance from l_3, which is greater than the expression in the statement of the lemma.

$\square$

II.A.5.2.

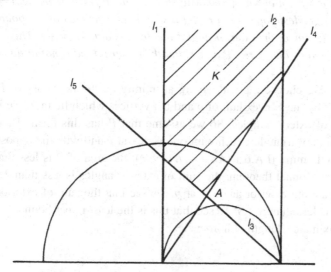

Figure II.A.5.2 *The lines l_1, l_2, and l_3 are the sides of an ideal triangle. Inversion in the circle drawn, with centre at the foot of l_1, interchanges l_3 and l_2 showing an ideal triangle and the circle of inversion.*

II.A.6. Maximum angle

In this section we prove an estimate which is the key to why the averaging process in the proof of Sullivan's theorem works.

II.A.6.1. Lemma: Area of triangle. *If A is a hyperbolic triangle with one side of length s, then area A/s < 1.*

Proof. Note first that we may assume the other two sides have infinite length, because this increases the area. We need only prove the result when s is very small. We denote the area by dA and the length by ds. Then dA/d$s = \sin\theta$, where θ is the angle between one of the infinite sides and the very short side. We see this by using the upper half plane model with the infinite sides vertical, and carrying out an integration with respect to the y-coordinate to find dA.

$\square$

The proof of the following result, which is more elegant than our original proof, was suggested by Francis Bonahon.

II.A.6.2. Lemma: Limited angle. *Let P be a non-compact convex polygon with a finite number of geodesic sides in the hyperbolic plane. Let A be a piecewise geodesic open arc of length s, which forms part of the boundary of P. Then the sum of the exterior angles met by A is at most $\pi + s$. This is a least upper bound as one varies P and A, but the upper bound is not attained.*

Proof. We choose a point p lying at infinity on the boundary of P. If we replace P by the convex hull of p and the vertices which lie in A, we increase the sum of exterior angles. So we assume that P has this form. Then P is a finite union of triangles, each with one vertex at p and with the opposite edge on A. By Lemma II.A.6.1 (*Area of triangle*), the area of P is less than s. By the Gauss–Bonnet theorem, the sum of exterior angles is less than $2\pi + s$. If we subtract the exterior angle π at p, we see that the sum of exterior angles along A is less than $\pi + s$. To see that this is the least upper bound, we take A to approximate a length of horocycle. ◻

References

Beardon, A.F. (1983), *The Geometry of Discrete Groups*, Springer, New York.
Canary, R.D., Epstein, D.B.A. and Green, P. Cambridge University Press. This volume.
Douady, A. (1979), "L'espace de Teichmüller," *Astérisque*, **66–67**, 127–137, Société Mathématique de France.
Epstein, D.B.A. (1984), "Transversely hyperbolic 1-dimensional foliations," *Astérisque*, **116**, 53–69. Société Mathématique de France.
Hewitt, E. and Stromberg, K. (1965), *Real and Abstract Analysis*, Springer-Verlag, Berlin.
Kerckhoff, S.P. (1983), "The Nielsen realization problem," *Ann. Math.*, **117**, 235–265.
Riley, R. (1985), "Holomorphically parametrized families of subgroups of SL(2,$\mathbb{C}$)," *Mathematika* **32**, 248–264.
Sullivan, D.P. (1981), *Travaux de Thurston sur les groupes quasi-fuchsiens et les variétés hyperboliques de dimension 3 fibrés sur S^1*, Séminaire Bourbaki 554, Lecture Notes 842, Springer Verlag.
Thurston, W.P. (1979), *The Geometry and Topology of 3-Manifolds*, Princeton University Mathematics Department. Part of this material – plus additional material – will be published in book form by Princeton University Press.
Wolpert, S. (1983), "On the symplectic geometry of deformations of a hyperbolic surface," *Ann. Math.*, **117**, 207–234.

Addendum 2005

In the 18 years since our original paper was written, there have been major advances in the understanding of Sullivan's theorem, in terms of new proofs of the theorem itself and of the universal constant K that appears in it. There is better understanding the nearest point retraction which plays a major role in our analysis. Also two new measures of bending have proved useful, complementing what we dealt with in Section II.1. Aside from these new developments, there is need to set straight the parameterization used in Proposition III.1.3.5 for earthquakes. The addition of some important formulas related to those already displayed in Section II.1 seems appropriate as well.

We want to thank Vlad Markovic for his helpful comments.

1. Sullivan's theorem

We begin by setting some new notation. Let $\Omega \subset \mathbb{C}$ be a simply connected region, not the whole plane. Here we regard $\mathbb{C} \subset \mathbb{S}^2$ as the boundary of $\mathbb{H}^3$. The hyperbolic convex hull of the closed set $\mathbb{S}^2 \backslash \Omega$ has a single relative boundary component in $\mathbb{H}^3$ and this faces Ω. We denote it by $\text{Dome}(\Omega)$.

The statement of Sullivan's theorem is as follows.

Theorem 1.0.1. (Sullivan, 1981, Section 2). *There exists a universal constant $K > 1$ with the following property. For any simply connected region $\Omega \subset \mathbb{C}$, $\Omega \neq \mathbb{C}$, there exists a K quasi-conformal mapping Φ: $\text{Dome}(\Omega) \to \Omega$ such that:*

- *Φ extends continuously to the common boundary $\partial \text{Dome}(\Omega) = \partial \Omega$ and is the identity there.*
- *If Ω is invariant under a group of Möbius transformations Γ, then Φ is equivariant under Γ.*

255

There are two new proofs both of which are relatively short and avoid the need for interpolation between $S = \mathrm{Dome}(\Omega)$ and its ε-distance surface S_ε. One was proposed by Bishop (2002). We will show below how to make his idea into a complete proof. The other, presented by Epstein *et al.* (2003), is perhaps more natural and arises from the method of angle scaling.

1.1. Bishop's approach. *Step 1.* In Section II.2.2 we introduced the surface S_ε at distance ε from S and using Lemma I.1.3.6 showed that it is C^1. Then we introduced the "nearest point map" $\sigma\colon \Omega \to S_\varepsilon$ and showed in Theorem II.2.3.1 that it is L_ε-bilipschitz and K_ε-quasi-conformal, for explicit constants L_ε, K_ε, providing $\varepsilon < \log 2$.

Let $r_\varepsilon\colon S_\varepsilon \to S$ be the nearest point map; in this case it is a continuous retraction which is not in general one-to-one. However by the triangle inequality it does satisfy

$$d_\varepsilon(x,y) - 2\varepsilon \le d_S(r_\varepsilon(x), r_\varepsilon(y)) \le d_\varepsilon(x,y) + 2\varepsilon.$$

Here $d_\varepsilon(\cdot, \cdot)$ is the induced hyperbolic path metric on S_ε.

The composition $r = r_\varepsilon \circ \sigma\colon \Omega \to S$ is the nearest point retraction $\Omega \to S$. It satisfies

$$\frac{1}{L_\varepsilon} d_\Omega(x,y) - 2\varepsilon \le d_S(r(x), r(y)) \le L_\varepsilon d_\Omega(x, y) + 2\varepsilon$$

where $d_\Omega(\cdot,\cdot)$ denotes the hyperbolic metric on Ω. According to Theorem II.2.3.1, r is at most 4-Lipschitz. New results to be discussed in Section 2 below show that any nearest point retraction r is actually at most 2-Lipschitz and that 2 is the smallest universal constant.

By Proposition I.1.12.6 there is an isometry $\iota\colon S \to \mathbb{D}$. Choose a Riemann map $\Phi\colon \mathbb{D} \to \Omega$. Both ι and Φ are isometries in the respective hyperbolic metrics. Therefore, denoting the hyperbolic metric in $\mathbb{D}$ by $d(\cdot, \cdot)$, we end up with the inequality for $h = \iota \circ r \circ \Phi$:

$$\frac{1}{L_\varepsilon} d(x,y) - 2\varepsilon \le d(h(x), h(y)) \le L_\varepsilon d(x,y) + 2\varepsilon \quad \forall x, y \in \mathbb{D}.$$

The continuous mapping $h\colon \mathbb{D} \to \mathbb{D}$, which is in general not a homeomorphism, is therefore a *quasi-isometric* mapping of $\mathbb{D}$.

Step 2: Digression on quasi-isometric mappings. A mapping $h\colon \mathbb{D} \to \mathbb{D}$, which is in general not a homeomorphism or even continuous, is called *quasi-isometric* (in the hyperbolic metric) if there are constants $L \geq 1$ and $a \geq 0$ such that, in the hyperbolic metric $d(\cdot, \cdot)$ on $\mathbb{D}$,

$$\frac{1}{L}d(x,y) - a \leq d(h(x), h(y)) \leq Ld(x,y) + a \quad \forall x, y \in \mathbb{D}.$$

While for $a > 0$ the inequality gives no information when $d(x, y)$ is small, it is restrictive when x, y are far apart. In the hyperbolic metric it forces on h the following remarkable properties which we shall state without proof:

- If γ is a geodesic ray ending at $\zeta \in \partial\mathbb{D}$, then $h(\gamma)$ has a well-defined endpoint on $\partial\mathbb{D}$ which is denoted by $h(\zeta)$.
- There exists $M = M(L, a)$ such that if γ' is a geodesic ray ending at $h(\zeta)$ then $d(h(z), \gamma') < M$, for all $z \in \gamma$, close enough to ζ.
- The boundary mapping h: $\partial\mathbb{D} \to \partial\mathbb{D}$ is a homeomorphism, and is quasi-symmetric.
- If $h(\zeta) = \zeta$ for all $\zeta \in \partial\mathbb{D}$, there exists $B = B(L, a)$ such that $d(z, h(z)) < B$, for all $z \in \mathbb{D}$.

These results were established by Efrimovich and Tihomirova (1964) and De-Spiller (1970); later they formed the foundation for Mostow's proof of his rigidity theorem. They are true in any hyperbolic space $\mathbb{H}^n$ (when $n > 2$ the conclusion in the third item should read "quasi-conformal" rather than "quasi-symmetric") and in more general spaces as well.

Step 3: Completion of the proof. We ended step 1 with the quasi-isometry $h = \iota \circ r \circ \Phi$ of $\mathbb{D}$. We now know it has well-determined boundary values and in fact $h(\zeta)$, $\zeta \in \partial\mathbb{D}$, is quasi-symmetric. Now we bring in the Douady–Earle extension D: $\mathbb{D} \to \mathbb{D}$ of the boundary values (Douady and Earle, 1986) which has the advantage of equivariance. The map D is both bilipschitz in the hyperbolic metric, and it is quasi-conformal; the Lipschitz and quasi-conformal constants are functions of ε.

Consider the mapping

$$F = \iota^{-1} \circ D \circ \Phi^{-1} \colon \Omega \to S = \text{Dome}(\Omega).$$

It is both quasi-conformal and bilipschitz in the hyperbolic metrics. But why does it extend continuously to be the identity on $\partial\Omega$? For example, $\partial\Omega$ may not be a Jordan curve or even locally connected.

The fact that F does so extend is the deepest element of the proof. That this is so is established in Theorem 7.5 (Epstein and Markovic, 2005).

Finally suppose there is a group Γ of Möbius transformations acting on Ω and hence on S as well. The maps Φ, ι carry Γ over to groups G_1, G_2 acting on $\mathbb{D}$. The map $\iota \circ \Phi$ defined on $\partial\mathbb{D}$ conjugates G_1 to G_2. The Douady–Earle extension also has this property. Therefore $F\colon \Omega \to S$ will commute with the elements of Γ.

1.2. The $K = 2$ conjecture.   We proved in Section II.2.15 that for all cases including the equivariant, $K < 82.8$, and improved this in his work (Epstein *et al.*, 2003) to $K < 13.88$. When equivariance is not required, Bishop (2002) proved that $K \leq 7.82$. But what is the smallest universal constant K? We will see in the examples below that it cannot be less than 2. It has been conjectured to be 2.

We will break this conjecture into two separate ones:

(1) Without any requirement of equivariance, the smallest universal constant satisfies $K = 2$.
(2) Consider only regions Ω which are invariant under a discrete group Γ of Möbius transformations for which Ω/Γ has finite area. The smallest universal constant K_{eq} for this class satisfies $K_{\mathrm{eq}} = 2$.

The latter is a conjecture of Thurston (refer to Hyperbolic structures on 3-manifolds II, available from http://arXiv.org.)

Actually *neither conjecture is true*.

It is shown by Epstein *et al.* (2003) or (2004) that $K_{\mathrm{eq}} > 2$. There is an open set of quasi-fuchsian groups for which the equivariant constant is necessarily greater than 2.

The answer to the general conjecture is also negative. Here the counterexample is a certain logarithmic spiral (Epstein and Markovic, 2005). It requires that $K > 2.1$.

In contrast, there are many examples of specific cases where $K \leq 2$. We will exhibit some simple ones below. For general euclidean convex regions, at the end of a complicated, lengthy analysis one discovers that $K < 2$ (Epstein *et al.*, 2005). The domes of such regions are automatically C^1-surfaces.

1.3. A class of examples.   Assume that the domain Ω is the wedge of angle $\pi < \alpha \leq 2\pi$. Its complement is the acute wedge of angle $0 \leq 2\pi - \alpha < \pi$ and in the case $\alpha = 0$, Ω is the complement of a line segment to ∞.

The convex hull boundary S consists of two hyperbolic half planes joined along the bending line. We will find a 2-quasi-conformal mapping of it onto Ω. To this end, represent S as the orthogonal projection of the convex hull boundary into Ω. The projection consists of the two quarter planes bounded by the edges of Ω and the half lines ℓ_1, ℓ_2 orthogonal to the edges of Ω at its vertex. There is the additional data that a point of euclidean distance s along ℓ_1 must be identified with the point of distance s along ℓ_2.

We have to find a quasi-conformal mapping of the two components of the projection onto the full Ω that is the identity on the boundary of the wedge, and which identifies the two half lines ℓ_1, ℓ_2.

Set $\beta = \alpha/\pi$ so that $1 < \beta \le 2$. The conformal map

$$f: z \mapsto z^{\frac{2\pi}{\alpha}}$$

sends Ω onto the complement of the slit $\mathbb{R}_+ = [0, \infty]$ and it sends the rays ℓ_1, ℓ_2 to the rays ℓ'_1, ℓ'_2 at angles $\pm(\pi^2/\alpha)$. Follow this by the mapping

$$h: z = r e^{i\theta} \mapsto r e^{i\beta\theta} = z^{\frac{1+\beta}{2}} \bar{z}^{\frac{1-\beta}{2}}$$

applied to the union of the two regions bounded by ℓ'_1, ℓ'_2. It sends the union of the two regions onto the complement of $\mathbb{R}_+$, identifying symmetric points on the rays ℓ'_1, ℓ'_2. It is quasi-conformal with maximal dilatation

$$K = \frac{\alpha}{\pi} \le 2.$$

The composition $g(z) = f^{-1} \circ h \circ f(z)$ sends the union of the two regions bounded by ℓ_1, ℓ_2 onto the complement Ω of the wedge, and it identifies the two points at distance t along ℓ_1 and ℓ_2 for any t so that it can be interpreted as a map from the convex hull boundary. It is the identity on the boundary of the wedge. It is quasi-conformal with maximal dilatation $K = \alpha/\pi \le 2$.

We will next examine the boundary values on the upper half plane model of $\mathbb{H}^2$.

A euclidean rotation of $\alpha - \pi/2$ will rotate one region to be adjacent to the other, identifying corresponding points on ℓ_1 and ℓ_2. This is in effect the isometry Ψ of the convex hull boundary to $\mathbb{H}^2$. We interpret the composition

$$h^* = f^* \circ f^{-1} \circ h \circ f \circ \Psi^{-1}(z) = z^{\frac{\pi+\alpha}{2}} \bar{z}^{\frac{\pi-\alpha}{2}}, \quad f^*(z) = z^{\frac{\pi}{\alpha}}$$

as a $K = \alpha/\pi$ quasi-conformal mapping of the upper half plane onto itself. The boundary values are

$$h^*(x) = x^{\frac{\alpha}{\pi}} \quad x \ge 0, \qquad h^*(x) = -|x|^{\frac{\alpha}{\pi}} \quad x \le 0.$$

It is known that h^* is the unique extremal quasi-conformal mapping for its boundary values Strebel (1976). Therefore the universal K in Theorem 1.0.1 cannot be less than 2. The limiting case of a half-infinite line ($\alpha = 2\pi$) requires that $K = 2$ exactly.

It is illuminating to compute the Lipschitz constant for the nearest point retraction r in this case. Suppose Ω is the complement of the horizontal line $\mathbb{R}_+ = [0, \infty)$. Set $w = |w| e^{i\theta}$ for $w \in \Omega$. The Lipschitz constant for r at a point $w \in \Omega$ with $\pi/2 \leq \theta \leq 3\pi/2$ is $2 \sin(\theta/2)$. At a point $w \in \Omega$ with $0 \leq \theta \leq \pi/2$ the Lipschitz constant is $1/\cos(\theta/2)$. Thus the largest constant 2 is assumed only for points along the negative real axis.

The inside of a wedge. To complete the picture we also consider the acute wedge Ω of angle $2\beta = 2\pi - \alpha$, $\pi < \alpha \leq 2\pi$, so that $0 < \beta < \pi/2$. The convex hull boundary S is half of a cone surface. The family of semicircles of length $\pi r \sin \beta$, $0 < r < \infty$ sweep out S.

The conformal map $w = \log z$ sends Ω onto the horizontal strip $\{w \in \mathbb{C} : -\beta < \operatorname{Im} w < \beta\}$. It also is a conformal map of Dome (S) onto the horizontal strip $\{w \in \mathbb{C} : -\pi(\sin\beta)/2 < \operatorname{Im} w < \pi(\sin\beta)/2\}$.

The affine mapping $u \mapsto u$, $v \mapsto 2\beta/\pi \sin \beta v$, where $w = u + iv$, is a K-quasi-conformal map which is in effect a map which maps S onto Ω pointwise fixing $\partial\Omega$. Here $K = \pi \sin \beta/2\alpha$, with $1 < K < \frac{\pi}{2} \sim 1.57$ for $0 < \beta < \frac{\pi}{2}$. If the mapping between the horizontal strips is pulled to the upper half plane, the induced boundary map on $\mathbb{R}$ is $x \mapsto x|x|^c$, where $c = (\pi \sin \beta)/2\beta - 1$. The map in the upper half plane is $z \mapsto z|z|^c$. It too is extremal for its boundary values.

The limiting case of the interior of the wedge can be regarded to be an infinite strip $\Omega = \{z : -a < \operatorname{Im} z < a\}$, $a > 0$. The dome S is a half cylinder of radius a. By similar considerations we find that there is a K-quasi-conformal map $S \to \Omega$ that fixes $\partial\Omega$. In this case, $K = \pi/2$.

Another interesting choice is the hyperbolic orthogonal projection $S \to \Omega$. Parameterize each semicircle cross-section by θ, $0 < \theta < \pi$. The maximal dilatation of the orthogonal projection is then $K(\theta) = 1 + \sin \theta \leq 2$.

2. The nearest point retraction $r: \Omega \to \mathbf{Dome}(\Omega)$

Theorem II.2.3.1 tells us that the Lipschitz constant associated with r does not exceed 4. A number of recent papers have further explored the nature of the constant. In fact the smallest universal constant L for nearest point retractions is 2. The proof is given in Theorem 3.1 (Epstein *et al.*, 2004). That $L = 2$ is best possible is seen from the example above for the limiting case $\alpha = 2\pi$ of a half-infinite line.

Vlad Markovic made the following provocative conjecture. Suppose that Ω is invariant under a group Γ for which Ω/Γ has finite hyperbolic area. Let L denote the lipschitz constant for the nearest point retraction $r: \Omega \to \operatorname{Dome}(\Omega)$.

The conjecture is that the lipschitz constant of *every* lipschitz homeomorphism f in the homotopy class of r is at least L. If this conjecture holds, these examples would provide an explicit, non-elementary, class of lipschitz maps that minimize the lipschitz constant in a homotopy class of lipschitz homeomorphisms (*cf.* "Minimal stretch maps between hyperbolic surfaces", available from http://arXiv.org.).

The inverse r^{-1} is of course not a mapping in general. But in the case that there are no isolated leaves in the bending lamination, r is a homeomorphism. Yet, again in the group case, r^{-1} is not lipschitz (and not quasi-conformal either). The proof is presented by Epstein et al. (2003). It is a consequence of the following fact. In the group case (still with no isolated bending lines), r^{-1} takes the bending lines, which cover a set of area zero on the dome, to a closed set of circular arcs, which has positive area in Ω.

On the other hand, in the recent paper (Bridgeman, 2003), improving the result of the earlier work by Bridgeman and Canary (2005), it is shown that in the group case, r has a *homotopy* inverse which is 4.4502-lipschitz.

Compare this with what we found in Section II.2.15 for our constructed quasi-conformal map $\Phi: Dome(\Omega) \to \Omega$, which is automatically equivariant if there is a group preserving Ω. Our proof also establishes that Φ is bilipschitz. Namely, the Lipschitz constant for Φ does not exceed 69.6 while the lipschitz constant for Φ^{-1} does not exceed 4.76. Here it is Φ^{-1} that is in the homotopy class of r.

3. Measures of bending

There are two notions that have been introduced to describe the amount of bending. Bridgeman (1998) introduced the notion of the *average* or *long range* bending of a given dome. Namely define

$$B(\tau; \beta) = \frac{1}{\text{Len}(\tau)} \int_\tau d\beta = \frac{i(\tau, \beta)}{\text{Len}(\tau)}$$

where τ is a finite length transverse segment in the dome. Here β denotes the (positive) bending measure on $Dome(\Omega)$ as defined in Section II.1.11, and $i(\tau, \beta)$ is the β-measure of τ. For an infinite length transversal τ, $B(\tau; \beta)$ is taken as the supremum over all finite length subintervals.

It follows from Lemma II.A.6.2 and Theorem II.A.4.2 that there is some finite, universal upper bound to average bending.

The best universal upper bound known for B is given by Bridgeman (2003), improving the earlier studies by Bridgeman and Canary (2003). Namely, for all domes,

$$B(\tau; \beta) \leq 3.4502.$$

In the references, the given upper bound is stated under the assumption that a group acts on the dome. However according to Bridgeman (personal communication), the assumption is not necessary; the upper bound holds for all domes, whether or not a group is acting.

It was suggested by Bridgeman (1998) that 1 may be an upper bound for average bending. However this cannot be the case: The long range bending for the dome determined by a certain logarithmic spiral is approximately 1.54 (Epstein and Markovic, 2005).

In contrast to average bending, the *roundness* of Ω, or *short range bending* of Dome(Ω), is defined (Epstein *et al.*, 2004) as the norm

$$||\beta|| = \sup_\tau \int_\tau d\beta = \sup_\tau i(\tau, \beta).$$

Here $\tau \subset \text{Dome}(\Omega)$ is a transverse segment of *unit* hyperbolic length. The attribution "roundness" is justified by the fact that the norm is zero if and only if $\Omega \subset \mathbb{S}^2$ is a round disk.

There exists a universal constant C, $\pi + 1 \leq C \leq 4.8731$, such that for every dome,

$$||\beta|| \leq C.$$

The upper bound 2π on C is established in Theorem 1.3 (Bridgeman, 2003). The lower bound $\pi + 1$ is given by a simple example in Figure 8 (Epstein *et al.*, 2003).

The norm $||\mu||$ can be equally well defined for measured laminations with arbitrary signed Borel measures – or general pleated surfaces $\Psi_\mu : \mathbb{H}^2 \to \mathbb{H}^3$, embedded or not. As such it has an important property. It is shown in Theorem 4.2 (Epstein *et al.*, 2004) that there is a universal constant $0 < C < 0.97$ with the following property. If $||\mu|| < C$ then Ψ_μ is necessarily a bilipschitz embedding which extends to $\partial \mathbb{H}^2$ and maps it to a quasi-circle. In particular if μ is a positive bending measure then its Ψ_μ-image is a dome over a quasi-disc Ω. In this case Ω satisfies the $K = 2$ conjecture (1) of Section 1.2 above and, if an appropriate group is acting, conjecture (2) as well.

4. Amended formulas

We start by rescaling the formula in theorem under Section II.3.9.1. To explain why this is necessary consider the case that μ and t are real. We had used the parameter t to represent the right earthquake along μ of distance $2t$. So as to be in agreement with the usual notation, we must replace our "t" by a new parameter which in terms of the original is $t/2$. With this reparameterization the formula in the case of a real measure μ becomes

$$\frac{\mathrm{dLen}(\lambda(t))}{\mathrm{d}t}\Big|_{t=0} = \int_{\gamma} \cos \phi(x)\, \mathrm{d}\mu(x). \tag{1}$$

Here λ is the measured lamination with one leaf ℓ which is the axis of a hyperbolic Möbius transformation T and measure taken as the unit atomic measure. The length $\mathrm{Len}(\lambda)$ is the translation length of T along ℓ and $\mathrm{Len}(\lambda(t))$ is its change after a quakebend $t\mu$. The angle $\phi(x)$ of intersection of ℓ and μ at x is measured counter-clockwise from the positive direction of ℓ to the positive directions of the leaf of μ containing x.

Note that the length of λ can be expressed by the integral of the product measure over the quotient surface:

$$\mathrm{Len}(\lambda) = \int\int_{\mathbb{H}^2/\langle T \rangle} \mathrm{d}\lambda \times \mathrm{d}\rho$$

since $\mathrm{d}\lambda$ is the counting measure across ℓ. Here $\mathrm{d}\rho$ is the hyperbolic metric.

Instead take λ to be the lift to $\mathbb{H}^2$ of an arbitrary positive, bounded measured lamination on a finite area quotient surface $S = \mathbb{H}^2/G$ for a fuchsian group G. "Bounded" means that $\int_{\tau} \mathrm{d}\lambda$ is uniformly bounded for all transverse segments τ of unit length. In particular no leaf can end at a cusp on S. For simplicity, in the remainder of this section all measured laminations will be assumed to be *positive and bounded*.

Extending the idea of length to the general case, the *length* of the measured lamination (λ, ρ) is defined by the analogous integral

$$\mathrm{Len}(\lambda) = \int\int_{S} \mathrm{d}\lambda \times \mathrm{d}\rho.$$

When μ is another positive, bounded measured lamination transverse to λ, Equation (1) becomes

$$\frac{\mathrm{dLen}(\lambda(t))}{\mathrm{d}t}\Big|_{t=0} = \int\int_{S} \cos \phi\, \mathrm{d}\lambda \times \mathrm{d}\mu. \tag{2}$$

The integral is effected by covering $\lambda \cup \mu$ by small, non-overlapping quadrilaterals $\{Q\}$ with the following properties: The "horizontal" edges are contained in leaves of λ, the "vertical" edges are contained in leaves of μ, and no leaf of $\lambda \cap Q$ or $\mu \cap Q$ crosses the same edge of Q twice. Then define the product measure on Q in terms of the generalized geometric intersection numbers as

$$d\lambda \times d\mu(Q) = i(h, \lambda) \cdot i(v, \mu).$$

Here h denotes a "horizontal" edge and v, a "vertical" edge of Q. The angle ϕ is defined at each point of intersection of a leaf of λ and μ as earlier.

For further details about the generalized length and product measures see Kerckhoff (1986).

Note that if we interchange λ and μ then the new angle of intersection ϕ_1 satisfies $\phi_1 = \pi - \phi$. The formulas then tell us that

$$\left. \frac{d\text{Len}(\mu(t))}{dt} \right|_{t=0} = - \left. \frac{d\text{Len}(\lambda(t))}{dt} \right|_{t=0}.$$

The above formulas result in an important estimate valid when the angle of intersection satisfies $|\phi(\zeta)| < \delta$ for all points of intersection ζ. Namely,

$$(1 - \delta)i(\mu, \lambda) < \left. \frac{d\lambda(t)}{dt} \right|_{t=0} < \left(1 - \frac{\delta}{2}\right) i(\mu, \lambda). \qquad (3)$$

Here $i(\mu, \lambda)$ denotes the (generalized) geometric intersection number.

From Equation (3) we deduce

$$\left. \frac{d \log \text{Len}(\lambda(t))}{dt} \right|_{t=0} < \left(1 - \frac{\delta}{2}\right) \frac{i(\lambda, \mu)}{\text{Len}(\lambda)}. \qquad (4)$$

Suppose λ is transverse to both μ_1 and μ_2. Let $\text{Len}(\lambda_i(t))$ denote the change of $\text{Len}(\lambda)$ by right earthquaking for distance t along μ_i, $i = 1, 2$. For the ratio we find

$$\left(\frac{1 - \delta}{1 - \frac{\delta}{2}}\right) \frac{i(\lambda, \mu_1)}{i(\lambda, \mu_2)} < \left. \frac{\frac{d \log \text{Len}(\lambda_1)}{dt}}{\frac{d \log \text{Len}(\lambda_2)}{dt}} \right|_{t=0} < \left(\frac{1 - \frac{\delta}{2}}{1 - \delta}\right) \frac{i(\lambda, \mu_1)}{i(\lambda, \mu_2)}. \qquad (5)$$

These formulas are required for the proof of Theorem 5.1 in Thurston (1986a).

Theorem 4.0.1. *For a given measured lamination λ on a closed surface S, regard $\log \text{Len}(\lambda)$ as a function on the Teichmüller space $\text{Im}(S)$. Given a point*

$\zeta \in \text{Im}(S)$, *suppose* λ, λ_1 *are measured laminations on the surface* S_ζ *corresponding to* ζ *which have the following property. For every tangent vector* X *to* $\text{Im}(S)$ *at* ζ,

$$\frac{d \log (\text{Len}(\lambda))dt}{dX}\bigg|_\zeta = \frac{d \log (\text{Len}(\lambda_1))}{dX}\bigg|_\zeta \quad \forall X.$$

Then $\lambda_1 = c\lambda$ *for some constant* $c > 0$.

That is, if at a point ζ all the "directional derivatives" are equal, then the laminations differ by a constant multiple.

References

Bishop, C.J. (2002), "Quasiconformal Lipschitz maps, Sullivan's convex hull theorem, and Brennan's conjecture", *Arkiv für Math.*, **40**, 1–26.

Bishop, C.J. "An explicit constant for Sullivan's convex hull theorem," In *Proceedings of 2002 Ahlfors-Bers Colloquium*, Contemporary Math, Amer. Math. Soc.

Bridgeman, M. (1998), "Average bending of convex pleated planes in hyperbolic 3-space," *Inv. Math.*, **132**, 381–391.

Bridgeman, M. (2003), "Bounds on the average bending of the convex hull of a Kleinian group," *Mich. J. Math.*, **51**, 363–378.

Bridgeman, M. and Canary, R.D. (2003), "From the boundary of the convex core to the conformal boundary," *Geom. Dedicata.*, **96**, 211–240.

Bridgeman, M. and Canary, R.D. "Bounding the bending of a hyperbolic 3-manifold," *Pacific J. Math*, **218**, 299–314 (2005).

De-Spiller, D.A. (1970), "Equimorphisms and quasiconformal mappings of the absolute," (transl.), Soviet Math. Dokl. **11**, 1324–1328.

Douady, A. and Earle, C. (1986), "Conformally natural extensions of homeomorphisms of the circle," *Acta Math.*, **157**, 23–48.

Efrimovich, V. and Tihomirova, E. (1964), "Equimorphisms of hyperbolic spaces," *Izv. Akad. Nauk. SSSR*, **28**, 1139–1144.

Epstein, D.B.A., Marden, A. and Markovic, V. (2003), "Complex earthquakes and deformations of the unit disk," preprint.

Epstein, D.B.A., Marden, A. and Markovic, V. (2003a), "Complex angle scaling," in *Kleinian Groups and Hyperbolic 3-Manifolds*, ed., C. Series, LMS Lecture Notes, Cambridge University Press.

Epstein, D.B.A., Marden, A. and Markovic, V. (2004), "Quasiconformal homeomorphisms and the convex hull boundary," *Ann. Math.*, **159**, 305–336.

Epstein, D.B.A., Marden, A. and Markovic, V. (2005), "Convex regions in the plane and their domes, "*Proc. Lon. Math. Soc*, to appear.

Epstein, D.B.A., and Markovic, V. (2005), "The logarithmic spiral: a counterexample to the $K = 2$ conjecture," *Ann. Math.*, **165**, 925–957.

Kerckhoff, S. (1986), "Earthquakes are analytic," *Comm. Math. Helv.*, **60**, 3–48.

Strebel, K. (1976), "On the existence of extremal Teichmüller mappings," *Jour. d'Anal. Math.* **30**, 441–447.

Sullivan, D.P. (1981), "Travaux de Thurston sur les groupes quasi-fuchsiens et les variétés hyperboliques de dimension 3 fibrés sur $\mathbb{S}^2$," In *Springer Lecture Notes*, Vol. **842**, Springer-Verlag.

Thurston, W.P. (1986) "Hyperbolic structures on 3-manifolds II," **GT981039** from *http://arXiv.org*.

Thurston, W.P. (1986a) "Minimal stretch maps between hyperbolic surfaces," **GT9801039** from *http://arXiv.org*.

PART III

Earthquakes in 2-dimensional Hyperbolic Geometry

W.P. Thurston, Department of Mathematics, Cornell University,
Ithaca, NY 14853, USA

Chapter III.1
Earthquakes in 2-dimensional Hyperbolic Geometry

III.1.1. Introduction

A hyperbolic structure on a surface is related to any other by a left earthquake (to be defined presently). I proved this theorem in the case of closed surfaces several years ago, although I did not publish it at that time.

Stephen Kerckhoff made use of the earthquake theorem in his proof of the celebrated Nielsen Realization Conjecture (Kerckhoff, 1983), and he presented a proof in the appendix to his article.

The original proof in some ways is quite nice, but it has shortcomings. It is not elementary, in that it makes use of the understanding and classification of measured laminations on a surface as well as the classification of hyperbolic structures on the surface. It also uses some basic but non-elementary topology of $\mathbb{R}^n$. Given this background, the proof is fairly simple, but developed from the ground-up it is complicated and indirect.

In this chapter, I will give a more elementary and more constructive proof of the earthquake theorem. The new proof is inspired by the construction and analysis of the convex hull of a set in space. The new proof also has the advantage that it works in a very general context where the old proof would run into probably unsurmountable difficulties involving infinite-dimensional Teichmüller spaces.

The first part of this chapter will deal with the hyperbolic plane, rather than with a general hyperbolic surface. This will save a lot of fussing over extra definitions and cases. The general theory will be derived easily from this, in view of the fact that the universal cover of any complete hyperbolic surface is the hyperbolic plane.

The standard definition for a hyperbolic structure on a manifold is an equivalence class of hyperbolic metrics, where two metrics are equivalent when there is a diffeomorphism isotopic to the identity which acts as an isometry from one metric to the other. However, this definition is not appropriate for studying hyperbolic structures on manifolds via their universal covers, since by this definition there is only one possible structure on the universal cover.

The solution to this difficulty is to use a smaller equivalence relation which keeps track of more information, that is, information which describes the behaviour of the metric at infinity.

III.1.1.1. Definition.   A *relative hyperbolic structure* on $(\mathbb{H}^2, S_\infty^1)$ is a complete hyperbolic metric on the hyperbolic plane, up to the relation that two metrics are equivalent if there is an isometry ϕ between them which:

(a) is isotopic to the identity, and
(b) which extends continuously to be the identity at infinity.

(The condition on isotopy is redundant here, but it will be needed for the more general case.)

There are relative hyperbolic structures of varying quality. Let *adjective* be a quality which describe maps of the circle to itself: for instance, continuous, Lipschitz, quasi-symmetric, etc. We suppose that an *adjective* map composed on the left and right with any element of $PSL(2, \mathbb{R})$ is still *adjective*. Then a relative hyperbolic structure is *adjective* at infinity if there is an isometry to the standard hyperbolic structure which extends to continuously to a map which is *adjective* at infinity.

It is well-known that a homeomorphism between two closed hyperbolic surfaces lifts to a map between their universal covers which extends uniquely to a continuous map on $\mathbb{H}^2 \cup S_\infty^1$. Therefore, a hyperbolic structure on a surface determines a continuous relative hyperbolic structure on $(\mathbb{H}^2, S_\infty^1)$.

One nice thing about studying continuous relative hyperbolic structures on $(\mathbb{H}^2, S_\infty^1)$ is that there is an easy classification of them. For any continuous relative hyperbolic structure h, there is a homeomorphism f_h of the disk which takes h isometrically to the standard hyperbolic structure on the hyperbolic plane. Clearly, $f_h|S_\infty^1$ is determined by the relative hyperbolic structure h up to postcomposition (i.e., composition on the left) by homeomorphisms of the circle which come from isometries of the hyperbolic plane, that is, elements of $PSL(2, \mathbb{R})$.

III.1.1.2. Proposition: Homeomorphisms classify hyperbolic structures.   *There is a one-to-one correspondence between the set of right cosets:*

$$PSL(2, \mathbb{R}) \backslash \text{homeomorphisms}(\mathbb{H}^2)$$

and the set of continuous relative hyperbolic structures on $(\mathbb{H}^2, S_\infty^1)$.

Proof. Homeomorphisms classify hyperbolic structures: The only remark needed to complete the proof is that every homeomorphism of the circle extends to a homeomorphism of the disk. Any such homeomorphism pulls back a continuous relative hyperbolic structure. Every relative hyperbolic structure arises in this way. Two relative hyperbolic structures are equivalent iff they are in the same right coset of $PSL(2, \mathbb{R})$.

> **homeomorphisms classify hyperbolic structures**

III.1.2. What are hyperbolic earthquakes?

A left earthquake is intuitively a change in hyperbolic structure which is obtained by shearing toward the left along a certain set of geodesics, the *faults*; the union of these geodesics is called the *fault zone.*

The simplest example of a left earthquake on the hyperbolic plane is obtained by cutting along a geodesic g and gluing the two half planes A and B back together again by attaching the boundary of B to the boundary of A by an isometry which moves to the left a distance d along the boundary of A, as viewed from A. (In other words, the displacement is in the positive sense with respect to the orientation of $g = \partial A$ induced from A.) Such an earthquake is an *elementary* left earthquake. The description is exactly the same if the roles of A and B are interchanged: the attaching map is replaced by its inverse, and the induced orientation of g is reversed, so the attaching map still shifts to the left by a distance d.

For this definition to make sense, it is necessary that our hyperbolic plane be equipped with an orientation. This orientation will be preserved throughout; whenever we speak of isometries of the hyperbolic plane, they will be understood to preserve the orientation.

The cutting and regluing procedure, strictly speaking, does not define a new hyperbolic structure on the hyperbolic plane, since the map – called an *earthquake map* – between the hyperbolic plane and the new hyperbolic manifold is discontinuous. This difficulty can be resolved by approximating the earthquake map by continuous maps. However, since we are working with the hyperbolic plane, an easier way to think of it is in terms of the circle at infinity, where the limit of the earthquake map is continuous. To identify the two hyperbolic planes which differ by a given elementary earthquake, we can use any homeomorphism between the two hyperbolic planes which extends the continuous map at infinity. In any event, the earthquake determines a relative hyperbolic structure on $(\mathbb{H}^2, S^1_\infty)$.

In a similar way, we can construct left earthquakes whose faults are finite sets of disjoint geodesics. We could also construct left earthquakes whose faults

III.1.2.1.

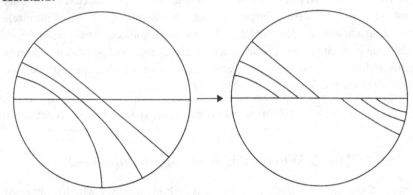

Figure III.1.2.1   *The effect of an elementary left earthquake.*

are countable sets of geodesics, although if the countable set has accumulation points we would have to worry about considerations of summability for the set of displacements. What we really need, however, is a definition for left earthquakes whose faults consist of an arbitrary closed set of disjoint geodesics.

III.1.2.2. Definition.   A *geodesic lamination* λ on a Riemannian surface is a closed subset L (the locus of λ) of the surface together with a foliation of the set by geodesics. In other words, for any point p in the locus of λ, there is a neighbourhood U of p on the surface such that $L \cap U$ is given the structure of a product $X \times (0, 1)$ where each $x \times (0, 1)$ is a piece of a geodesic; the geodesics in the product structures assigned to different neighbourhoods are required to match.

The geodesics in these product structures can be extended uniquely and indefinitely in both directions, assuming the Riemannian metric is complete, although the extension will repeat periodically, if the geodesic is closed.

These geodesics are called the *leaves* of λ. The locus of a lamination is the disjoint union of its leaves.

The components of the complement of a lamination are its *gaps*. Sometimes it is useful to refer to the metric completions of the gaps, with respect to their Riemannian metrics (so that they become complete Riemannian surfaces with geodesic boundary). We will call these the *complete gaps*.

The gaps together with the leaves of a lamination, are its *strata*.

III.1.2.3. Definition.   If λ is a geodesic lamination on a hyperbolic plane, S, a λ-*left earthquake map E* is a (possibly discontinuous) injective and surjective map E from the hyperbolic plane to the hyperbolic plane which is an isometry on each stratum of λ. The map E must satisfy the condition that for any two strata $A \neq B$ of λ, the *comparison isometry*,

$$\mathrm{cmp}(A, B) = (E|A)^{-1} \circ (E|B) \colon \mathbb{H}^2 \to \mathbb{H}^2$$

is a hyperbolic transformation whose axis weakly separates A and B and which translates to the left, as viewed from A. Here, $(E|A)$ and $(E|B)$ refer to the isometries of the entire hyperbolic plane which agree with the restrictions of E to the given strata. A line l weakly separates two sets A and B if any path connecting a point $a \,\varepsilon\, A$ to a point $b \,\varepsilon\, B$ intersects l. Shifting to the left is defined as before: the direction of translation along l must agree with the orientation induced from the component of A in $\mathbb{H}^2 - l$. One exceptional case is allowed: if one of the two strata is a line contained in the closure of the other, then the comparison isometry is permitted to be trivial.

A *left earthquake* on the hyperbolic plane is a structure consisting of two copies of the hyperbolic plane (the source and target for the earthquake), laminations λ_s and λ_t on the source and target hyperbolic planes, and a λ_s-earthquake map between them which sends the strata of λ_s to the strata of λ_t. A *right earthquake* is defined similarly. The inverse of a left earthquake is a right earthquake. It is obtained by interchanging source and target hyperbolic planes, source and target laminations, and using the inverse of the left earthquake map.

Why not allow earthquakes to shift both to the right and to the left? There is no particular reason, except that the theory becomes very different. A comparison can be made between the properties of earthquakes and the properties of functions $\mathbb{R} \to \mathbb{R}$. Left and right earthquakes are like not necessarily continuous monotone functions (either non-increasing or non-decreasing). A general earthquake would be like a completely general function. The theory of monotone functions is neat and useful, and closely related to the theory of measures on the line. The theory of general functions (without additional conditions and structure) is terrible.

This definition of an earthquake is a bit different from the definitions that have been used previously in the case of hyperbolic surfaces of finite area, where less care was needed because geodesic laminations on such surfaces automatically have measure 0. To motivate and justify the present definition, we prove

III.1.2.4. Proposition: Simple left earthquakes are dense. *If λ is a finite lamination, a map $E \colon \mathbb{H}^2 \to \mathbb{H}^2$ which is an isometry on each stratum of λ is a left earthquake map iff the comparison maps for adjacent strata are hyperbolic transformations which translate the separating leaf of λ to the left.*

Left earthquake maps with finite laminations are dense in the set of all left earthquake maps, in the topology of uniform convergence on compact sets.

This says that left earthquakes could have been alternatively defined as the closure of the set of left earthquakes which have finite laminations. Such earthquakes we will call *simple* left earthquakes.

Proof. Simple left earthquakes are dense: First we need to verify that if s and t are hyperbolic transformations with disjoint axes such that they translate in the "same" direction, then the axis of $s \circ t$ separates the axes of s from t. (To translate in the "same" direction means that the region between the axes of s and t induces an orientation on the axes so that which agrees with one of the translation directions and disagrees with the other.) One way to see this is by looking at the circle at infinity: in the interval cut off by the axis of s (or by the axis of t), both s and t are moving points in directions which agree. This forces $s \circ t$ to have a fixed point in each of the two intervals between the axes of s and t, since the composition maps one of these intervals into itself, and the inverse of the composition maps the other of these intervals into itself.

By induction on the number of strata separating two strata of the finite lamination λ of the proposition, it follows that the comparison isometry of an arbitrary pair of strata shifts to the left if all the comparison isometries for adjacent strata do.

The proof of the density of simple left earthquakes among all left earthquakes is quite easy.

First, observe that the image of a compact set by a left earthquake map has a bounded diameter, hence a compact closure. The proof of this is in analogy to the theorem that the image of compact interval by a monotone function is bounded and is left to the reader.

Given any left λ-earthquake E, then for any compact subset K of the hyperbolic plane, there is a finite subset of strata which intersect K such that the union of the images of this finite set of strata in the graph of $E \subset \mathbb{H}^2 \times \mathbb{H}^2$ is ε-dense in the graph of E restricted to K.

Define a new left earthquake F by the condition that the isometries obtained by restricting F to its strata are exactly the isometries obtained by restricting E to the chosen finite set of strata. The lines separating adjacent strata of F are axes of certain comparison isometries for E. These are disjoint, as any two are separated by at least one of the strata, so they form a lamination which is respected by F. The simple earthquake map F approximates E in the set K.

> **simple left earthquakes are dense**

III.1.2.5.

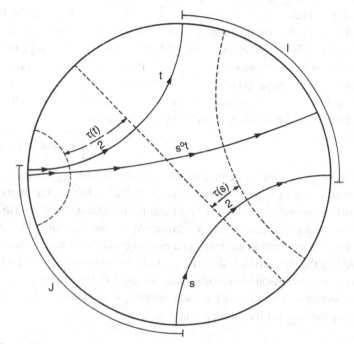

Figure III.1.2.5 *The composition of two hyperbolic isometries which translate in the "same" direction is a hyperbolic isometry whose axis separates the axes of the original two isometries which also translates in the same direction. The interval I is mapped into itself by the composition, while the interval J is mapped over itself. The fixed points of the composition within these two intervals are the endpoints of its axis. Any translation is the product of two reflections.*

Even though earthquake maps are not (in general) continuous, they have an extension which is continuous at infinity:

III.1.2.6. Proposition: Quake at infinity. *If E is a λ-earthquake map from $\mathbb{H}^2$ to $\mathbb{H}^2$, then there is a unique map*

$$E_\infty: S^1_\infty \to S^1_\infty$$

such that E together with E_∞ form a map which is continuous at each point $x \in S^1_\infty$.

Proof. Quake at infinity: First we will make sure that E_∞ is well-defined.

If the closures of A and of B at infinity intersect, then $E|A$ and $E|B$ agree in this intersection (which consists of one or two points), because the axis of the comparison isometry cmp(A, B) must hit these points at infinity.

If a point $x \in S^1_\infty$ is not in the closure of any stratum of λ, then it has a neighbourhood basis (in D^2) consisting of neighbourhoods bounded by leaves l_i of λ. The diameters of the image leaves $E(l_i)$ must go to zero, for otherwise a limit point in $\mathbb{H}^2$ of their images would have to be on the image of some stratum which would violate the fact that E preserves separation properties. This determines a unique point which is defined to be $E_\infty(x)$.

This defines an injective and surjective map of the circle to itself which preserves the circular order. Any such map is continuous.

> **quake at infinity**

Just as in the case of an elementary earthquake, an arbitrary left earthquake determines a relative hyperbolic structure on $(\mathbb{H}^2, S^1_\infty)$. In fact, the earthquake map up to the equivalence relation of post-composition (composition on the left) with isometries of $\mathbb{H}^2$ determines a relative hyperbolic structure. This equivalence relation on earthquake maps is a nice one: the equivalence classes are described by the lamination associated with a left earthquake, together with the collection of comparison isometries. Later, we shall see that the set of comparison isometries is determined by a much simpler, well-understood structure: a transverse measure for the lamination.

III.1.3. Associating earthquakes to maps of the circle

Here is the main theorem:

III.1.3.1. Theorem: Geology is transitive. *Any two continuous relative hyperbolic structures on $(\mathbb{H}^2, S^1_\infty)$ differ by a left earthquake. The earthquake is unique, except that there is a range of choices for the earthquake map on any leaf where it has a discontinuity. The choices are maps which all have the same image, but differ by translations ranging between the limiting values for the two sides of the leaf.*

Equivalently, every continuous orientation-preserving map of S^1_∞ to itself arises as the limiting value E_∞ of a left earthquake map E, unique up to the same set of ambiguities.

III.1.3.2. Corollary: Rightward geology. *Any two continuous relative hyperbolic structures on $(\mathbb{H}^2, S^1_\infty)$ differ by a right earthquake, unique except on leaves where the earthquake has a discontinuity.*

The corollary immediately follows by reversing orientation. It is rather curious that the effect of a left earthquake can also be obtained by a right earthquake. The reader may enjoy puzzling out the right earthquake map which corresponds to an elementary left earthquake. In the next section, after we have proven the earthquake theorem, we will show how this example and similar ones can be worked.

Proof. Geology is transitive: Let $f: S^1 \to S^1$ be any orientation preserving homeomorphism. We will prove the theorem by looking at the left coset of the group of isometries of hyperbolic space in the group of homeomorphisms of the circle which passes through f, that is, $PSL\,(2, \mathbb{R}) \cdot f$. This coset is a 3-manifold, homeomorphic to $PSL(2, \mathbb{R})$ which we name C. Choose a reference point $x_o \in \mathbb{H}^2$. Then $PSL(2, \mathbb{R})$ has a fibration over $\mathbb{H}^2$, with projection map $\gamma \mapsto \gamma(x_o)$. The fibres are circles. This fibration carries over to a fibration of $p: C \to \mathbb{H}^2$ via the homeomorphism $(\gamma \leftrightarrow \gamma \circ f)$.

Some elements of C have fixed points as they act on the circle, others have none. If a homeomorphism h of the circle has at least one fixed point, then there is a unique lift of the h to a homeomorphism $\tilde{h}$ of the universal cover of the circle (namely $\mathbb{R}$) which also has fixed points. We will call h an *extreme left* homeomorphism if it has at least one fixed point, and if $\tilde{h}$ satisfies $\tilde{h}(x) \geq x$ for all $x \in \mathbb{R}$. Less formally, this says that h moves points counterclockwise on S^1, except for those points that it fixes. Let $XL \subset C$ be all extreme left homeomorphisms in C.

III.1.3.3. Lemma: XL is a plane. *The fibration p maps XL homeomorphically to the hyperbolic plane.*

Proof. XL is a plane: Let $F \subset C$ be any fibre of p, and let $f_o \in F$ be any element of the fibre. The other elements of F are obtained by post-composing f_o with a isometry which fixes $p(F)$. Using appropriate coordinates we can take the isometries which fix $p(F)$ to act as rotations of the circle.

The proof is easiest to complete in the universal cover. Let $\tilde{f}_o$ be any lift of f_o to $\mathbb{R}$. The set of all lifts of all elements of F are the homeomorphisms $\tilde{f}_o + T$, for constants T. Obviously, exactly one of these has the property that the periodic function $\tilde{f}_o(x) + T$ has its minimum value 0.

We have demonstrated that $p|XL$ gives a one-to-one correspondence between XL and $\mathbb{H}^2$. Construction of a continuous inverse reduces to a choice of T depending continuously on f_o. This follows from the fact that the infimum of a function depends continuously on the function, using the uniform topology on functions.

> **XL is a plane**

With any element $g \in XL$ is associated its fixed point set $\text{fix}(g) \subset S^1$. Let $H(g) \subset D^2 = \mathbb{H}^2 \cup S^1_\infty$ be the convex hull of $\text{fix}(g)$. The convex hull is taken in the hyperbolic sense; this agrees with the euclidean convex hull if we use the projective model for $\mathbb{H}^2$, so that the compactified hyperbolic plane is the ordinary disk in the euclidean plane, and hyperbolic straight lines are euclidean straight lines.

The sets $H(g)$ often consist of a single point on the S^1_∞. In fact, for any $g \in XL$ and any $x \in \text{fix}(g)$, if g is post-composed with a parabolic transformation which fixes x and moves all other points counterclockwise, the resulting homeomorphism is also in XL and has x as its sole fixed point.

Other possibilities are that $\text{fix}(g)$ consists of two points, so that $H(g)$ is a compactified hyperbolic line, or that $\text{fix}(g)$ is a finite set of points, so that $H(g)$ is a compactified ideal polygon, or that $\text{fix}(g)$ is infinite.

III.1.3.4. Lemma: Convex hulls do not cross. *The sets $H(g)$ do not cross each other. Formally, if g_1 and g_2 are elements of XL such that the $H(g_i)$ are distinct sets which are not points, then $H(g_2)$ is contained in the closure of a single component of $D^2 - H(g_1)$.*

Proof. Convex hulls do not cross: Consider any two elements g_1 and g_2 of *XL*. Suppose that l_i [$i = 1, 2$] is a line connecting two fixed points of g_i. Suppose the two lines intersect other than at their endpoints. The (hyperbolic) angle of intersection of their image under an element $g \in C$ does not depend on the choice of g.

It follows from the fact that g_1 is an extreme left homeomorphism that a certain one of the two angles formed by l_1 and l_2 cannot decrease. Since g_2 is also an extreme left homeomorphism, the complementary angle cannot decrease. This means that all four endpoints of the two lines must be fixed by g_1 and g_2. Since an element of $PSL(2, \mathbb{R})$ which fixes three or more points is the identity, it follows that $g_1 = g_2$.

The proposition readily follows by applying this reasoning to lines joining various pairs of points in $H(g_i) \cap S^1_\infty$.

> **convex hulls do not cross**

As you probably suspect by now, the sets $H(f)$ for $f \in XL$ will turn out to be the closed strata for an earthquake map. The remaining difficulty in constructing an earthquake map is to show that these strata cover all of D^2. The reason they cover the whole disk is related to Brouwer's theorem that any map of the disk to itself which is the identity on the boundary is surjective. The correspondence

$f \mapsto H(f)$ which maps XL to subsets of the disk is, in fact, much like a continuous map from the disk to the disk. To picture this similarity, consider the set $G \subset \mathbb{H}^2 \times \mathbb{H}^2$ which describes the graph of this correspondence. In other words, $G = \{(p(f), x) | x \in H(f)\}$. It should seem plausible that G is topologically a disk, and that it can be approximated by a disk which is the graph of a continuous function. What we will prove is related but slightly different.

Here is a continuity property for H.

III.1.3.5. Proposition: Continuity of H. *For any $f \in XL$ and any neighbourhood U of $H(f) \subset D^2$, there is a neighbourhood V of f in XL such that*

$$g \in V \Rightarrow H(g) \subset U.$$

Intuitively, this says that as f moves a little bit, what $H(f)$ can do is move a little bit and decrease to a subset of its former size.

Proof. Continuity of H: We can assume that the neighbourhood U is convex. In that case, it suffices to take V small enough so that the fixed points of homeomorphisms $g \in V$ are contained in $U \cap S^1_\infty$. This can easily be accomplished in view of the fact that there is a lower bound to the distance that f moves points in the complement of $U \cap S^1_\infty$.

> continuity of H

The set XL is homeomorphic to an open disk; it has a compactification $\overline{XL}$ to a closed disk, copied from the compactification of the hyperbolic plane. Extend the map H to the circle at infinity for XL (which we also call S^1_∞) by the rule that

$$H(x) = \{x\} \quad [x \in S^1_\infty].$$

III.1.3.6. Proposition: Continuity of H bar. *The preceding statement remains true for the extension of H to $\overline{XL}$.*

Proof. Continuity of H bar: What we need to show is that for any $x \in S^1_\infty$ and for any neighbourhood U of x in S^1, there is a neighbourhood V of x in XL consisting of homeomorphisms whose fixed point sets are all contained in U.

Any $\gamma \in PSL(2, \mathbb{R})$ can be written in the form $\gamma = \tau \circ \rho$, where τ is the hyperbolic element whose axis passes through x_o and takes x_o to $\gamma(x_o)$, and ρ is an elliptic element, fixing x_o. Let us take x_o to be the centre of the disk.

If $\gamma(x_o)$ is near $x \in S^1_\infty$, then τ translates a great distance. If $\gamma \circ f_o$ has any fixed point y which is not close to x, then $\tau^{-1}(y)$ is close to the point diametrically opposite hx. Consider what $\gamma \circ f_o$ does to points in a neighbourhood of y. A point y_1 close to y in the clockwise direction is first sent to a point clockwise from $f_o(y)$ (how far is estimated by constants for uniform continuity of f_o). Then ρ sends y and y_1 to a pair of points spaced exactly the same amount; since $\rho \circ f_o(y) = \tau^{-1}(y)$ is near the point antipodal to x, y_1 is also near this antipodal point. Now τ expands the interval between the two points by a tremendous factor. If τ is chosen to overcome whatever shrinking f_o caused on the interval, then y_1 moves clockwise, so $\gamma \circ f_o$ was not an extreme left homeomorphism.

The only possibility is for an extreme left homeomorphism near x is that its fixed points are near x.

continuity of H bar

To prove that every point in $\mathbb{H}^2$ is in some $H(f)$, we will replace H with a continuous map by an averaging technique. We can do the averaging using the euclidean metric on the disk, considered as the projective model of $\mathbb{H}^2$. This disk is identified with both the compactified hyperbolic plane and $\overline{XL}$. There is an obvious measure defined on the set $H(f)$: either 2-dimensional, 1-dimensional or 0-dimensional Lebesgue measure. Let $h(f)$ be the centre of mass of $H(f)$, taken with respect to this measure.

Choose a bump function β around the origin with small support, that is, a positive continuous function with integral 1 having support contained in an ε-neighbourhood of the origin. The convolution $\beta * h$ of β with h is a new function from XL the plane. ($\beta * h$ is the average of the values of h over a small neighbourhood in XL, with weighting governed by β.) $\beta * h$ is continuous, since the difference of its values at nearby points can be expressed as the integral of the bounded function h times a small function (the difference of copies of β with origin shifted to the two nearby points).

For any $f \in XL, \beta * h(x)$ is contained in the convex hull of the union of the $H(g)$ where g ranges over the ε-neighbourhood of f. By proposition under Section III.3.6 (*Continuity of H bar*), this is contained in a small neighbourhood of $H(f)$.

The map $\beta * h$ is close to the identity near S^1_∞. Therefore (by consideration of degree), it is surjective onto all except possibly a small neighbourhood of S^1_∞ whose size depends on ε. For any $x \in \mathbb{H}^2$, let f be an accumulation point of elements f_i such that $\beta_i * h(f_i) = x$, as the diameter of the support of β_i goes to zero. It follows that $x \in H(f)$.

The union of lines of $H(XL)$ with boundary components of 2-dimensional regions of $H(XL)$ forms a lamination λ.

To complete the proof of existence of an earthquake, we must construct a λ-earthquake map which agrees with $f_\circ$ at infinity. For each stratum A, choose an $f_A \in XL$ such that $H(f) \supset A$. There is only one choice possible except for some cases when A is a line. Define E on the stratum A to be γ_A, where γ_A is determined by the equation

$$f_\circ = \gamma_A \circ f_A.$$

It is clear that E is a one-to-one correspondence from $\mathbb{H}^2$ to itself, and that it agrees with $f_\circ$ at infinity. What remains to be shown is that any comparison isometry

$$\text{cmp}(A, B) = \gamma_A^{-1} \circ \gamma_B = f_A \circ f_B^{-1}$$

is hyperbolic, that its axis weakly separates A and B, and that it translates to the left as viewed from A.

Let I and J be the two intervals of S^1_∞ which separate A from B, named so that I is to the left of J as viewed from A. (We allow the possibility that one or both of them is a degenerate interval, that is, a point.) The two homeomorphisms f_A and f_B^{-1} both map I into itself. The cmp(A, B) maps I into itself, so by the Brouwer fixed point theorem it must have a fixed point somewhere in that interval. Similarly, cmp$(B, A) = (\text{cmp}(A, B))^{-1}$ must have a fixed point in J.

Therefore, cmp(A, B) is hyperbolic, and its axis separates A from B. The direction of translation along the axis is from the interval J to the interval I, since I the isometry maps I inside itself. It translates to the left as viewed from A, so the map E that we have constructed is indeed a left earthquake map.

The uniqueness part of the main theorem is easy. Suppose that E' is any earthquake having the same limiting values on S^1_∞ as the earthquake map E we have constructed. If A is any stratum of E', then the composition h of E' with an isometry which makes h the identity on A acts as an extreme left homeomorphism on S^1_∞, with two or more fixed points. Therefore A is a stratum of E, and the two left earthquakes agree on A unless it is a leaf where they are discontinuous.

> **geology is transitive**

III.1.4. Examples

Given a map of the circle to itself, the main theorem gives a reasonably constructive procedure to find a left earthquake with the given limiting values. (How constructive the procedure really is depends on the form in which the map of the circle is given, and whether certain inequalities involving its values can be answered constructively. Technically, the theorem is not constructive except under very restrictive hypotheses. The natural stratification of the boundary of the convex hull of a curve in $\mathbb{R}^3$ into 0-dimensional, 1-dimensional, and 2-dimensional flat pieces is a quite similar problem, and it is not technically constructive, either. For most practical purposes, both procedures are reasonably constructive.)

To illustrate, consider an elementary left earthquake map L which shifts by a distance d along a fault l, and let us find the corresponding right earthquake. Let p be a point on l; choose coordinates so that p is at the center of the disk. There is an extreme left homeomorphism h in the coset of L which acts as a hyperbolic transformation which moves a distance of $d/2$ on the two sides of l, in opposite directions. If h is composed with a suitable clockwise rotation, an extreme right homeomorphism g will be obtained. The fixed points of g cannot be at the ends of l, so g is differentiable at its fixed points. The qualitative picture immediately shows that the derivative of an extreme left homeomorphism (when it exists) can only be 1.

III.1.4.1.

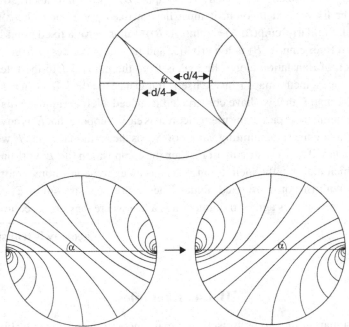

Figure III.1.4.1 *The faults of the right earthquake which corresponds to an elementary left earthquake consists of lines making a constant angle α to the fault of the left earthquake. The angle α is such that after the endpoints are transformed by the left earthquake, the lines which now join them meet the left earthquake fault with the complementary angle π−α.*

A little reflection shows that the stratum containing p is a leaf. It can be constructed by drawing the two lines perpendicular to l at a distance of $d/4$ on either side of p; the leaf is one of the diagonal lines m connecting their endpoints on S^1_∞. The right earthquake map R corresponding to L has as its lamination the set of all lines meeting l at the same angle as m. R acts as the identity on l, and takes each leaf of its lamination to a leaf meeting l in the complementary angle at the same point.

There are other maps $f_{\alpha,\beta}$ which can be described in a similar way, acting as the identity on l and sending the foliation by lines which meet l with angle α to the foliation by lines which meet l with angle β. These are not earthquake maps except when α and β are complementary angles. The intuitive reason is that the spacing between the leaves of the foliation is not preserved except when α and β are complementary. The part of the definition that is violated is that the axes of the comparison isometries cannot meet the strata being compared.

A similar procedure works to find the right earthquake corresponding to any simple left earthquake. The leaves of the lamination for the right earthquake fall into a finite number of groups. The leaves in any group make a constant angle to a line which is the axis of a comparison isometry between two of the strata for the left earthquake. Whether or not there is a group of leaves for a given comparison isometry, and if so, how big it is, depends on the amount of shearing along the faults of the left earthquake and the spacing of the leaves.

III.1.5. Earthquakes on hyperbolic surfaces

As indicated in the introductory section, the earthquake theorem carries over without difficulty to general complete hyperbolic surfaces, even surfaces of infinite genus or infinitely many ends. To make this generalization, we have to extend the definitions.

III.1.5.1. Definition. A *left earthquake map* between two surfaces is a one-to-one, surjective map between the surfaces, together with a lifting to a map between the universal covers which is a left earthquake. A right earthquake map is defined similarly.

If there is a closed leaf in the earthquake lamination on the surface, then the map between surfaces does not determine the lifting: there are an infinite number of different liftings, differing by twists along the geodesic, which would give different earthquakes, even though the physical map is the same. It is because of these examples, which it is useful to distinguish (so that an earthquake map determines a well-defined homotopy class of continuous maps) that the lifting is included as part of the data.

III.1.5.2. Definition.   If M^2 is a complete hyperbolic surface, the *Fuchsian enlargement* of M^2 is defined to be

$$\overline{M}^2 = (\mathbb{H}^2 \cup \Omega)/\pi_1(M^2)$$

where Ω is the domain of discontinuity of $\pi_1(M^2)$ as it acts on the circle at infinity. $\overline{M}^2$ is a surface with boundary.

Relative hyperbolic structures are defined on $(M^2, \partial\overline{M}^2)$ just as for $(\mathbb{H}^2, S^1_\infty)$. A relative hyperbolic structure is a complete hyperbolic metric on M^2, up to homeomorphisms which extend to homeomorphisms on $\overline{M}^2$ which are isotopic to the identity while remaining constant on ∂M^2. A relative hyperbolic structure M_1^2 is *adjective* if there is a representative metric in the equivalence class so that the identity map extends to a homeomorphism which is *adjective* from $\partial\overline{M}^2$ to $\partial\overline{M}^1$.

Note that cusps need not correspond in arbitrary relative hyperbolic structures. For example, if τ is a hyperbolic transformation and $M^2 = \mathbb{H}^2/\tau$ is the quotient cylinder, there are relative hyperbolic structures on $(M^2, \partial M^2)$ having either end a cusp. The same is true if τ is parabolic. For continuous relative hyperbolic structures, an end cannot change from having a cusp to having a boundary component of $\overline{M}^2$, or vice versa.

The significance of the Fuchsian enlargement of a complete hyperbolic surface comes about through its close relation to the compactification of the hyperbolic plane.

III.1.5.3. Proposition: Continuity at infinity.   *If* $h\colon M_1 \to M_2$ *is a homeomorphism between hyperbolic surfaces and* $\tilde{h}\colon \mathbb{H}^2 \to \mathbb{H}^2$ *is a lift of* h *to the universal covers, then* h *extends to a continuous map* $\overline{h}\colon \overline{M}_1 \to \overline{M}_2$ *iff* $\tilde{h}$ *extends to a continuous map* $\tilde{\overline{h}}\colon \overline{\mathbb{H}} \to \overline{\mathbb{H}}$.

Proof. Continuity at infinity: If $\tilde{h}$ extends continuously to $\tilde{\overline{h}}\colon \overline{\mathbb{H}}^2 \to \overline{\mathbb{H}}^2$, a continuous extension of h is directly constructed as $\tilde{\overline{h}}|\mathbb{H}^2 \cup \Omega$ modulo $\pi_1(M_1)$.

Conversely, suppose that $\overline{h}$ is a continuous extension of h. A neighbourhood basis of the circle at infinity for $\tilde{M}_2$ can be formed using two kinds of sets: half planes bounded by axes of hyperbolic elements, and half planes bounded by lines which have both endpoints in Ω. (Actually, the first kind of set suffices if $\Omega = \emptyset$, and the second kind of set suffices otherwise.) Since a monotone function can fail to be injective only on a countable number of intervals, all but a countable set of the arcs with endpoints in Ω have pre-images which converge at both ends, and these will suffice to form a neighbourhood basis.

For any point p on the circle at infinity for $\tilde{M}_1$, there is a unique point q on the circle at infinity for $\tilde{M}_2$ so that a neighbourhood basis for p maps to a family of sets cofinal in a neighbourhood basis for q. This means there is a unique continuous extension $\tilde{h}$.

> **continuity at infinity**

We remark that the analogous proposition in 3 dimensions is false, which may explain why this proof sounds a bit strained. In order for there to be a homeomorphism between universal covers of hyperbolic 3-manifolds, it is necessary that there be a homeomorphism between their Kleinian enlargements, but in addition it is necessary (at least) that ending laminations for any geometrically infinite ends be the same.

III.1.5.4. Corollary: Earthquakes on surfaces. *If M^2 is any complete hyperbolic surface, and if N^2 is any other continuous relative hyperbolic structure on $(M^2, \overline{\partial M^2})$, then there is a left earthquake between $(M^2, \overline{\partial M^2})$ and $(N^2, \overline{\partial N^2})$.*

Proof. Earthquakes on surfaces: Since there is a homeomorphism between the Fuchsian enlargements, by the preceding proposition there is a homeomorphism between the compactified universal covers. By Theorem III.1.3.1 (*Geology is transitive*) there is an earthquake between the universal covers. The earthquake can be taken to be equivariant with respect to the action of the fundamental group of M, if choices are made consistently at its discontinuities. This gives an earthquake map between the surfaces.

> **earthquakes on surfaces**

III.1.6. The measure and cause of earthquakes

Part of the data comprising an earthquake is the source lamination λ_s, but there is additional information, namely a transverse measure for λ_s, which is determined by the map. The transverse measure measures the amount of shearing along the faults. Mathematically, the shearing can be determined from the translation distances of comparison isometries. Given any transverse arc and any partition of it into subintervals, associated with each partition element is a comparison isometry between the strata containing its endpoints. The composition of these isometries is the comparison isometry for the strata of the two endpoints of the entire interval. To specify the transverse shearing measure, we require that the sum of the translation lengths of the comparison isometries for the partition should approximate the total measure of the arc.

We need to check that these approximations converge. The translation distance of a hyperbolic isometry is the log of its derivative at the repelling fixed point on the circle at infinity. If coordinates are chosen so that the axis of the isometry is a diameter of the circle, the norm of the derivative at a nearby point p on the circle differs by $O(T \cdot d^2)$, where d is the distance of p from the fixed point and T is the translation distance. More intrinsically, it follows that the translation distance of the composition of two hyperbolic transformations with disjoint nearby axes pointing the same direction differs from the sum of translation distances by $O(T \cdot d^2)$, where d now represents the distance between the axes and T is the smaller of the two translation distance of the composition.

Applying this iteratively for a comparison isometries of a partition, we conclude that the sum of translation distances for a partion differs from the translation distance T of the comparison isometry for the endpoints of a transverse arc by at most $O(T \cdot l^2)$, where l is the length of the arc. This implies the sums for partitions must converge, as the partition is refined.

III.1.6.1. Proposition: Metering earthquakes. *Associated with any earthquake is a transverse lamination on its source lamination, which approximates the translation distances of comparison isometries for nearby strata.*

For any two earthquakes with the same source lamination and the same shearing measure, there is an isometry between the targets which conjugates the two earthquake maps.

Proof. Metering earthquakes: We have already established assertion of the first paragraph, that there is a well-defined measure associated with an earthquake.

It remains to show that different earthquakes are distinguished by their measures.

As a special case, we need to make sure that a non-trivial earthquake has a non-trivial measure. Actually, this case is quite clear: for any transverse arc with total measure 0, there are arbitrarily fine partitions such that the total translation distance is arbitrarily close to 0. By the preceding discussion, it follows that the translation distance of the composition is arbitrarily close to 0. Another way to visualize this is to keep track of how far each of the comparison isometries for neighbouring points moves the centre of the disk. It never gets very far, so the comparison isometry is the identity.

Given any two earthquake maps which have the same source lamination and the same shearing measure, consider now the inverse of one earthquake map composed with the other. We do not immediately know that this is an earthquake. However, we can still consider the comparison isometries. For any two nearby strata, the comparison isometry is the composition of two hyperbolic isometries whose axes are geometrically restricted so that they nearly coincide, and whose translation distances are equal but moving in opposite directions. The composition is close to the identity, its distance from the identity is $O(T \cdot d)$. As the partition is refined, the total size of the comparison isometries goes to 0.

> **metering earthquakes**

The question of existence of an earthquake, given a lamination and a transverse invariant measure, is more delicate. There is always an earthquake of sorts, but without some restrictions on the measure, it can easily happen that the target surface is not complete. For instance, the three-punctured sphere can be divided into two ideal triangles. There is a sort of earthquake which shears to the left by a distance of 1 along each of the edges; the target surface is the interior of a pair of pants whose boundary components all have length 1. (This and related examples are discussed in [Thurston].)

III.1.6.2. Definition. An earthquake is *uniformly bounded* if for every constant $a \geq 0$ there is a constant C such that for any two strata A and B whose distance does not exceed a, the comparison isometry cmp(A, B).

A measured lamination is *uniformly bounded* if for every constant $a \geq 0$ there is a constant C such that any transverse arc of length less than or equal to a has total measure less than C.

To show that an earthquake or a measured lamination is uniformly bounded, it clearly suffices to produce a constant C for one value of a greater than 0.

III.1.6.3. Proposition: Making bounded earthquakes. *For any uniformly bounded measured lamination μ, there is a uniformly bounded earthquake E_μ having μ as shearing measure.*

III.1.7. Quasi-symmetries and quasi-isometries

A *quasi-isometry* of a metric space is a homeomorphism h for which there is a constant K such that for all x, y:

$$d(h(x), h(y)) \leq K \, d(x, y) \quad \text{and} \quad d(x, y) \leq K \, d(h(x), h(y)).$$

For a Riemannian manifold, this is the same as saying that h is bi-Lipschitz with constant K.

There is a certain sense in which uniformly bounded earthquakes can be approximated by quasi-isometries. This is not uniform approximation as maps: the uniform limit of continuous maps is always continuous. Instead, we consider again the graph of the earthquake map as a relation. At the discontinuities, where there is ambiguity in the earthquake map, adjoin all possible choices of the earthquake map to the graph, thus obtaining a canonical earthquake relation.

III.1.7.1. Proposition: Approximating by quasi-isometries. *The graph of a uniformly bounded earthquake can be uniformly approximated by a quasi-isometric diffeomorphism. The approximation can be done in a natural way.*

There are two methods which work well, both using smoothing. Method (a) is to integrate a vector field, flowing through the one-parameter family of earthquakes. Method (b) is to construct a foliation extending the lamination to a neighbourhood, together with a transverse foliation, and to use these as local coordinates for a leaf-preserving diffeomorphism. The second method is more difficult to do in a natural way, so that it does not depend on *ad hoc* choices, so we will describe the first.

Proof. Approximating by quasi-isometries: Sketch. Construct a 3-manifold fibred by hyperbolic 2-manifolds resulting from earthquakes with the same faults, but linearly increasing measure. Construct a transverse vector field by averaging the local isometries coming from infinitesimal earthquake maps between slices.

Now integrate.

> **Approximating by quasi-isometries**

There is a famous characterization by Ahlfors of which maps of the circle can extend to quasi-conformal maps of the disk. This ties in quite nicely with the theory of earthquakes.

III.1.7.2. Definition. A homeomorphism $f: S^1 \to S^1$ is *quasi-symmetric* if there are constants K and $\varepsilon > 0$ such that, for $0 < t < \varepsilon$ and $x \in S^1$

$$\frac{1}{K} \leq \frac{|f(x+t) - f(x)|}{|f(x-t) - f(x)|} \leq K.$$

III.1.7.3. Theorem: Extending quasi-symmetry. *A quasi-symmetric homeomorphism of the circle extends in a natural way to a quasi-isometry of the hyperbolic plane.*

Proof. Extending quasi-symmetry: Sketch.   It is easy to check that an earthquake is uniformly bounded if the homeomorphism of the circle at infinity is quasi-symmetric. Apply the preceding result.

> **Extending quasi-symmetry**

III.1.7.4. Corollary: Quasi-symmetry on surfaces.   *Let M and N be complete hyperbolic surfaces, and suppose that* $\phi: M \to N$ *is a quasi-isometry between them. Then for any quasi-symmetric homeomorphism at infinity from* ∂M *to* ∂N *which is isotopic to* $\partial\phi$, *there is a uniformly bounded earthquake map from M to N, and a quasi-isometry from M to N, with given boundary values at infinity.*

References

Kerckhoff, S.P. (1983), "The Nielsen realization problem," *Ann. Math.*, **117**, pp. 235–265.

Thurston, W.P. "Three-dimensional geometry and topology," http://msri.org/publications/books/gt3m/

PART IV

Lectures on Measures on Limit Sets of Kleinian Groups

S.J. Patterson, Mathematisches Institut, Universität Göttingen,
Bunsenstr. 3-5, D-37073 Göttingen Germany

Chapter IV.1

The Problems with which we shall be Concerned

This part represents a course of five talks given at the Warwick Symposium on Hyperbolic Geometry and 3-dimensional Manifolds. Although these notes are rather more detailed than the lectures could have been, I have retained the form of the lectures as far as possible. One addition is a fairly long bibliography which, while far from complete, is, I trust, representative. I would like to thank D.B.A. Epstein for the opportunity to lecture so expansively on this subject, and M. Denker, D.A. Kazhdan, and B. Stratmann for some very illuminating conversations about the material of the fifth lecture.

We shall begin by recalling some notations and definitions from hyperbolic geometry. The two models which we shall use are the Poincaré ball-model and the upper-half-space model.

On $\mathbb{R}^N$ denote by $|| \ldots ||$ the euclidean norm and by $\langle , \rangle$ the euclidean inner product. Let $B^{N+1} = \{x \in \mathbb{R}^{N+1} \mid ||x|| < 1\}$.

For the Poincaré ball-model we shall use B^{N-1}. For $x, x' \in B^{N-1}$ define

$$L(x, x') = \frac{1}{2} + \frac{||x - x'||^2}{(1 - ||x||^2)(1 - ||x'||^2)}$$

Let

$$Q = \{(x, y) \in \mathbb{R}^{N+1} \times \mathbb{R} : y^2 - ||x||^2 = 1, y > 0\}$$

and we introduce the stereographic projection

$$p: Q \to B^{N+1}; \quad (x, y) \mapsto x/(y + 1).$$

293

Then one verifies

$$L_B(p(x_1, y_1), p(x_2, y_2)) = \frac{1}{2}(y_1 y_2 - \langle x_1, x_2 \rangle).$$

From this it follows that the group of orientation preserving diffeomorphisms of B^{N+1} preserving L, denoted by $Con(N)$, is isomorphic to the connected component of the identity of $O(N + 1, 1)$; this condition means that for $g \in Con(N)$, $x, x' \in B^{N+1}$ one has

$$L_B(g(x), g(x')) = L_B(x, x').$$

The group $Con(N)$ acts transitively on B^{N+1} and the stabilizer of a point is isomorphic to $O(N + 1)$.

The infinitesimal analogue of L_B is the metric

$$ds^2 = \frac{||dx||^2}{(1 - ||x||^2)^2}.$$

The group $Con(N)$ will preserve this. Note that this metric is conformally equivalent to the euclidean metric and hence the elements of $Con(N)$ preserve euclidean angles, and are in this sense conformal maps. This explains the notation $Con(N)$. Corresponding to this metric one has the volume element σ so that

$$d\sigma(x) = \frac{dm_{N+1}(x)}{(1 - ||x||^2)_{N+1}}$$

where m_{N+1} is the $(N + 1)$-dimensional Lebesgue measure. We shall need later the corresponding Laplace operator, Δ. All these are, in the appropriate senses, $Con(N)$-invariant. If $[x_1, x_2]$ denotes the distance with respect to ds^2 then one has

$$2L_B(x_1, x_2) = \cosh 2[x_1, x_2].$$

The group $Con(N)$ also acts on the boundary S^N of B^{N+1}. For $x \in B^{N+1}$, $\zeta \in S^N$ we define the Poisson kernel

$$P_B(x, \zeta) = \frac{(1 - ||x||^2)}{(||x - \zeta||^2)}.$$

From the definition of the metric one sees that the (euclidean) conformal distortion of $g \in Con(N)$ at x is given by

$$j_B(g, x) = \frac{(1 - ||gx||^2)}{(1 - ||x||^2)}$$

and that this is also defined for $x \in S^N$. One has

$$P_B(gx, g\zeta) \cdot j_B(g, \zeta) = P_B(x, \zeta).$$

One can use this to prove that for $s \in \mathbb{C}$,

$$\Delta P(\cdot, \zeta)^s = -s(N - s) P_B(\cdot, \zeta)^s$$

by verifying it at 0. However the computations are easier in the upper-half-space model. This model will play a subordinate role but it is often useful for computations. We let $\mathbb{H}^{N+1} = \mathbb{R}^N \times \mathbb{R}_+^\times$; let Im: $\mathbb{H}^{N+1} \to \mathbb{R}_+^\times$ be the projection onto the second component. On $\mathbb{H}^{N+1}$ we introduce the metric

$$ds^2 = \frac{1}{4} \mathrm{Im}(z)^{-2} ||dz||^2$$

and the function

$$L_H(z_1, z_2) = \frac{1}{2} + \frac{||z_1 - z_2||^2}{4\mathrm{Im}(z_1)\,\mathrm{Im}(z_2)}.$$

One can verify that the map

$$a: B^{N+1} \rightarrow \mathbb{H}^{N+1};$$

$$x \mapsto (x + (0, \ldots, 0, 1)) \cdot ||x + (0, \ldots, 0, 1)||^{-2} - \left(0, \ldots, 0, \frac{1}{2}\right)$$

satisfies

$$L_H(a(x_1), a(x_2)) = L_B(x_1, x_2)$$

from which it follows that the given metric on $\mathbb{H}^{N+1}$ pulls back to that on B^{N+1} and in particular that a is conformal. It is clearly bijective and it allows us to let Con(N) act on $\mathbb{H}^{N+1}$ and to perform computations there. In this case conformal distortion is given by

$$j_H(g, x) = \frac{\mathrm{Im}(gx)}{\mathrm{Im}(x)}.$$

The Poisson kernel is given by

$$P_H(x, \zeta) = \frac{\mathrm{Im}(x)}{||x - \zeta||^2}$$

where $x \in \mathbb{H}^{N+1}$ and $\zeta \in \mathbb{R}^N$ considered as $\mathbb{R}^N \times \{0\}$. Again one has

$$P_H(g(z), g(\zeta)) j_H(g, \zeta) = P_H(z, \zeta)$$

as long as $g(\zeta)$ is defined (i.e. not "infinity"). One has

$$\Delta \cdot \mathrm{Im}^s = -s(N-s)\mathrm{Im}^s$$

from which one can easily deduce

$$\Delta P_H(\cdot, \zeta)^s = -s(N-s)P_H(\cdot, \zeta)^s$$

where Δ is the corresponding Laplace operator to the metric.

We shall be considering discrete subgroups G of Con(N). These act discontinuously on B^{N+1} or $\mathbb{H}^{N+1}$. One is really interested in two particular cases, that where $N = 1$ (Fuchsian groups) and $N = 2$ (Kleinian groups). It is these cases which play such an important rôle in function-theoretic and topological investigations. One should note that in these cases

$$\mathrm{Con}(1) = PSL_2(\mathbb{R})$$
$$\mathrm{Con}(2) = PSL_2(\mathbb{C}).$$

Examples of groups G as above are numerous. In both cases one can use the Klein combination theorem (Marden, 1977, Section 4.2) to construct examples. There are function-theoretic methods. Every hyperbolic Riemann surface can be represented conformally as $G\backslash B^2$ for some G and there is an even deeper representation of a union of Riemann surfaces by means of Kleinian groups (Marden, 1977, Section 6.6). More recently Thurston's work has brought to light an entirely unexpected "uniformization theory" for 3-manifolds from which further examples flow (Sullivan, 1981b; Thurston, 1982, 1979).

We shall consider G as acting on B^{N+1}. That G acts discontinuously on B^{N+1} does not mean that the same holds true of S^N. We define the *ordinary set*, $\Omega(G)$, of G in S^N to be the set of points on which G acts discontinuously, i.e. of points ζ which possess a neighbourhood U so that

$$\{g \in G \ : \ g(U) \cap U \neq \emptyset\}$$

is finite. The complement of $\Omega(G)$ in S^N is called the *limit set* of G, denoted by $L(G)$. If Card $L(G) > 2$ it is a minimal G-invariant closed set (in $\overline{B^{N+1}}$ or in S^N); if $u \in \overline{B^{N+1}}$ one can describe $L(G)$ as the derived set of $G\{u\}$.

We shall need a concept due to Beardon, that of the *exponent of convergence* of G. This is defined as

$$\delta(G) = \inf\left\{s > 0 \ : \ \sum_{g \in G} L(x_1, gx_2)^{-s} < \infty\right\}.$$

This does not depend on the choice of x_1 and x_2. By definition $\delta(G) \geq 0$; from the easily verified fact that

$$\int_{B^{N+1}} L(x,y)^{-s} \, d\sigma(y) < \infty$$

if $s > N$ one can easily deduce that $\delta(G) \leq N$. Thus

$$0 \leq \delta(G) \leq N.$$

During the conference, Gromov expressed the view that the following definition of $\delta(G)$ was preferable, as this inequality was immediate. One forms the manifold, or better, orbifold, $G \backslash B^{N+1}$ (orbifold (Thurston) = V-manifold (Satake)). Let $D_a(R)$ be a ball of hyperbolic radius R and centre $a \in B^{N+1}$. Let $V_a(R)$ be the volume of the projection of $D_a(R)$ in $G \backslash B^{N+1}$. Then

$$\delta(G) = N - \limsup_{R \to \infty} \frac{\log V_a(R)}{2R}.$$

Since the right hand has the value O for the case $G = I$ it is immediate that $\delta(G)$ lies in the region given above.

With a few further assumptions the range of $\delta(G)$ can be narrowed further (see Beardon, 1968, 1971; Beardon–Maskit, 1974).

We shall need some types of finiteness conditions on G. Let for any set $T \subset \overline{B^{N+1}}, H(T)$ be the hyperbolic convex cover of T; that is, the smallest hyperbolically convex set in B^{N+1} such that $T \subset \overline{H(T)}$, the closure in B^{N+1}. If there exists a compact set $K \subset B^{N+1}$ so that $G \cdot K \supset H(L(G))$ then G is called *convex cocompact*. Let $H_d(T)$ denote the d-neighbourhood of $H(T)$. We say that G is *geometrically finite* if for any $d > 0$ there exists $K \subset H_d(L(G))$ so that $G \cdot K = H_d(L(G))$ and $\sigma(K) < \infty$. This definition is due to Thurston (1979, Definition 8.4.1). In a generalization of a well-known theorem of Siegel's, Thurston proves that G is geometrically finite if and only if it possesses a fundamental polyhedron with finitely many faces (Thurston, 1979, Proposition 8.4.3). We will not discuss fundamental domains here, and refer the reader interested in further details to Beardon (1983) and Beardon–Maskit (1974).

Note that $B^{N+1} - \overline{H_d(L(G))}$ is a G-invariant open neighbourhood of $\Omega(G)$, which is the main fact that we will use.

One knows that if G contains no unipotent elements then these two concepts are equivalent (see Beardon, 1983; Thurston, 1979); on the other hand, if G contains parabolic elements it cannot be convex cocompact. If G is geometrically finite it is finitely presented; the converse is true if $N = 1$ but not otherwise.

The strongest result known for finitely generated groups is the remarkable theorem of Ahlfors (1964) (see also Sullivan, 1982–1983) that when $N = 2$, G is
finitely generated then $G \backslash \Omega(G)$ is a finite union of Riemann surfaces of finite
type.

The problems with which we shall be concerned in these lectures concern
the limit set, and so also the ordinary set of G and the asymptotic distribution
of a set of the form $G\{w\}$ $(w \in B^{B+1})$. The topological structure of the limit
set involves many interesting questions investigated by, among others, Abikoff,
Maskit and Sasaki, but we shall not go into these here. Rather we shall be
interested in the measure-theoretic description of $L(G)$. In the case of $N = 1$
the set $L(G)$ is a Cantor-like set. In the case $N = 2$ it can take on many forms.
These are also examples of "fractals" in the sense of Mandelbrot.

Typical of the type of problem with which we shall be involved are the
following:

(1) What is the Hausdorff dimension of $L(G)$, $\dim_H(L(G))$? We shall see that
    under certain circumstances it is equal to $\delta(G)$. On the other hand the
    Ahlfors conjecture – that if G is finitely generated and $L(G) \neq S^N$ with
    $N = 2$ then $m_N(L(G)) = 0$ – is still undecided. It is backed by relatively
    little evidence, but see Thurston, (1982).

(2) In analogy to the classical "circle-problem", as to the number of pairs
    of integers (x_1, x_2) with $x_1^2 + x_2^2 < r$ one can ask for the asymptotic
    behaviour of

$$\text{Card}\{g \in G \; : \; L(x_1, g(x_2)) < X\}$$

where x_1, $x_2 \in B^{N+1}$ are fixed and $X \to \infty$. Whereas the classical circle
problem can be solved in the first approximation by elementary methods,
this is not so in the case of hyperbolic geometry. We shall see that the
leading term above is, under good circumstances,

$$c(x_1, x_2)X^{\delta(G)}.$$

Note that both of these problems deal with asymptotic distribution of $G(\{x\})$
even though they look rather different.

The definitions of the ordinary set and limit set remind one at once of the
wandering and non-wandering sets of ergodic theory and there are indeed very
strong connections between these questions and the geodesic flow on $G \backslash B^{N+1}$
considered from an ergodic-theoretic point of view.

Moreover there are a number of strong analogies between the questions
discussed here and the theory of the behaviour of iterates of a rational map
$R: \mathbb{C}_\infty \to \mathbb{C}_\infty$. These have been recognized and used to produce some very

striking results by Sullivan and his co-workers. For a discussion of the analogies see Sullivan (1981a, 1982–1983); for further results see Mané–Sad–Sullivan (1983).

There is one aspect of the theory on which we shall not be able to touch. This is the relationship between the nature of the limit set and the Mostow rigidity theorem for subgroups of Con(2). For this see Sullivan's paper (1981a; see also Sullivan, 1978).

Chapter IV.2
A Measure on the Limit Set

Let G be a discrete subgroup of $\mathrm{Con}(N)$ and $L(G)$ its limit set. The objective of this lecture is to construct a "canonical" measure μ supported on $L(G)$. This will encode the asymptotic distribution of $G(\{x\})$. Before we state the main result recall that if we have a measurable map $\phi \colon S^N \to S^N$ and a measure m defined on S^N then one defines a new measure $\phi_* m$ by

$$(\phi_* m)(E) = m(\phi^{-1}(E)).$$

Here "measurable" means "Borel measurable" and all measures will be defined on the σ-algebra of Borel measurable sets.

Theorem. *If G is a discrete subgroup of $\mathrm{Con}(N)$ then there is a probability measure supported on $L(G)$ which satisfies*

$$g_* \mu = j(g)^{-\delta(G)} \cdot \mu.$$

In this statement $j(g)$ is the function $\zeta \mapsto j(g, \zeta)$.

Sullivan (1979) defines a *conformal density of dimension* α on S^N to be a measure M satisfying, for $g \in G$,

$$g_* m = j(g)^{-\alpha} m.$$

The special feature is that the change of scale of m depends only on the conformal distortion. Lebesgue measure is a conformal density of dimension N for any group G.

Proof. (*cf.* Patterson, 1976a; Sullivan, 1979): it is best to assume first that

$$\sum_{g \in G} L(w_1, g w_2)^{-\delta(G)}$$

diverges. Let δ_a be the Dirac measure at a. For $s > \delta(G)$ we define the measure

$$m_{s,w} = \sum_{g \in G} L(gw, 0)^{-s} \delta_{gw}.$$

This measure is supported on $\overline{G(\{w\})}$. We consider the probability measure $m_{s,w}/\|m_{s,w}\|$ as defined on $\overline{B^{N+1}}$. By Helly's theorem there exists a sequence $s_j > \delta(G)$, $s_j \to \delta(G)$ such that the $m_{s_j,w}/\|m_{s_j,w}\|$ converge weakly to some measure μ on $\overline{B^{N+1}}$. This is the measure we seek. Observe that since $\|m_{s_j,w}\| \to \infty$ one has $\mu(B^{N+1}) = 0$ from which it follows that the support of μ is $L(G)$.

Next we investigate $F(w') = \int P(w', \zeta)^{\delta(G)} \, d\mu(\zeta)$. From the definition of μ one sees that this is

$$\lim_{s_j \to \delta(G)} \left\{ \sum_{g \in G} L(gw, 0)^{-s_j} \right\}^{-1} \sum_{g \in G} L(gw, 0)^{-s_j} \left(\frac{1 - \|w'\|^2}{\|gw - w'\|^2} \right)^{s_j}.$$

It is elementary to verify that this is equal to

$$\lim_{s_j \to \delta(G)} \left\{ \sum_{g \in G} L(gw, 0)^{-s_j} \right\}^{-1} \sum_{g \in G} L(gw, w')^{-s_j}.$$

This shows, amongst other things, that

$$F(gw') = F(w')$$

for $g \in G$. It is a direct consequence of this (Sullivan, 1980, Theorem 2) that

$$g_* \mu = j(g)^{-\delta(G)} \mu.$$

One should also note that in view of the fact that $P(\,\cdot\,, \zeta)^{\delta(G)}$ is an eigenfunction of the Laplace operator, one also has

$$\Delta F = -\delta(G)(N - \delta(G))F.$$

This fact also underlines the significance of the measure μ.

Note that we have not proved that μ is unique; in general it will not be.

We now return to the proof of the theorem. At the outset we had assumed that $\sum_{g \in G} L(w_1, gw_2)^{-\delta(G)}$ diverged. To overcome this difficulty we consider a function $k: \mathbb{R}_+ \to \mathbb{R}$ which is increasing but which satisfies for each $\varepsilon > 0$ that there should exist $y_0 > 0$ so that for $x > 1$, $y > y_0$ one has

$$k(xy) \le x^\varepsilon k(y).$$

Moreover

$$\sum_{g \in G} k(L(w_1, gw_2)) \, L(w_1, gw_2)^{-\delta(G)}$$

should diverge. Then one forms

$$m_{s,w} = \sum k(L(gw, 0)) \, L(gw, 0)^{-s} \, \delta_{gw}$$

and the argument can be carried through as before without any essential change.

One problem that arises immediately in applications of this measure is the question of whether μ can have atoms. We shall next give some conditions which ensure that this does not occur. To do this we have to introduce some new concepts.

A *radial limit point* ζ of G is an element of $L(G)$ such that for any $w \in B^{N+1}$ there exists $c > 0$ and a sequence $g_j \in G, j \geq 1$ such that

$$\|g_j(w) - \zeta\| \leq c_j(1 - \|g_j(w)\|^2)$$

and

$$\|g_j(w)\|^2 \to 1.$$

This condition means that the $g_j(w)$ approach ζ inside a cone which lies inside B^{N+1}. The set of all radial limit points of G we denote by $L_r(G)$.

A *cusp* ζ of G is an element of S^N such that if we map B^{N+1} to $\mathbb{H}^{N+1}$ in such a fashion that ζ is sent to "∞" then the stabilizer in G of ∞ contains a subgroup of the form

$$\{z \mapsto z + \lambda \; : \; \lambda \in \Lambda\}$$

where $\Lambda \subset \mathbb{R}^N$ is a discrete subgroup. The *rank* of ζ is defined as the rank of Λ, when this is chosen to be maximal.

If G is geometrically finite then $L(G) - L_r(G)$ consists precisely of the cusps; see Beardon–Maskit, (1974) for the case $N = 2$, but Alan Beardon assures me that the general case is known.

Proposition. *Let μ be a conformal density of dimension $\delta(G)$ supported on $L(G)$ constructed as above. Suppose that $L(G)$ is infinite. Then if ζ is an atom of μ it can be neither a radial limit point nor a cusp with the following property.*

There exist an open set $F \subset S^N$ and a compact set $K \subset S^N - \{\zeta\}$ such that:

(a) $\bigcup_{g \in G_\zeta} gF = S^N - \{\zeta\}$, $G_\zeta = \mathrm{Stab}_G(\zeta)$
(b) *If* $(F - K) \cap g(F - K) \neq \emptyset$, $g \in G$, *then* $g = I$

Proof. (Patterson, 1976a; Sullivan, 1984): We deal with the case of a radial limit point first. Here one has

$$\mu \left(\bigcup g\{\varsigma\} \right) \le 1.$$

Let g_j be a sequence as in the definition of a radical limit point. One has

$$\sum_j j(g_j^{-1}, \varsigma)^{\delta(G)} = \mu \left(\bigcup g_j\{\varsigma\} \right) \le 1.$$

But we have

$$(1 - ||g_j w||^2) = j(g_j, w)(1 - ||w||^2)$$

and

$$||g_j w - \varsigma|| = j(g_j, w)^{1/2} j(g_j^{-1}, \varsigma)^{-1/2} ||w - g_j^{-1}\varsigma||$$
$$\ge j(g_j, w)^{1/2} j(g_j^{-1}, \varsigma)^{-1/2}(1 - ||w||).$$

From these and the definition of a radial limit point we see that there exists $c' > 0$ so that

$$j(g_j, w) j(g_j^{-1}, \varsigma) \ge c'$$

and

$$j(g_j, w) \to 0 \quad \text{as } j \to \infty.$$

Thus $j(g_j^{-1}, \varsigma) \to \infty$ as $j \to \infty$ and so the series above diverges.

To deal with the case of a cusp it is convenient to use the $\mathbb{H}^{N+1}$ model of hyperbolic geometry, indeed it was for this purpose that we introduced it at all. We shall assume that ς is an atom of μ. We have now that μ gives rise to a conformal density μ_0 on $\mathbb{R}^N$ but it has finite mass in the following sense:

$$\int (1 + ||\varsigma||^2)^{-\delta(G)} \, d\mu_0(\varsigma) < \infty.$$

The eigenfunction $F(w)$ now takes on the form

$$F(w) = m \operatorname{Im}(w)^{-\delta(G)} + \int P(w, \varsigma)^{\delta(G)} \, d\mu_0(\varsigma)$$

where $m > 0$ as ∞ corresponded to an atom. Suppose $w = (0, y)$; write the integral on the right-hand side as

$$\int \left\{ \frac{y(1 + ||\varsigma||^2)}{y^2 + ||\varsigma||^2} \right\}^{\delta(G)} \frac{d\mu(\varsigma)}{(1 + ||\varsigma||^2)^{\delta(G)}}.$$

Suppose now that y is large. The contribution to this integral from the region $||\zeta||^2 \le y^\varepsilon$ is $O(y^{-\delta(G)+\varepsilon})$ whereas that from the region $||\zeta||^2 > y^\varepsilon$ is $o(y^{\delta(G)})$. One easily sees from these considerations that locally uniformly in the $\mathbb{R}^N$-component of w one has

$$F(w) = m \operatorname{Im}(w)^{\delta(G)} + o(\operatorname{Im}(w)^{\delta(G)}).$$

Let k be a function as at the outset and define

$$\Phi(w, w'; s) = \sum k(L(w, gw')) \, L(w, gw')^{-s}.$$

One supposes w to be fixed and takes $\operatorname{Im}(w')$ to be large. Using the condition on the cusp formulated in the proposition one sees that one can assume that the $\mathbb{R}^N$-projections of the $g(w)$ lie in a set of the form $G_b K$ where $K \subset \mathbb{R}^N$ is compact. From this and a simple estimate one derives

$$\Phi(w, w'; s) \le c \operatorname{Im}(w')^{r-s} \cdot \Phi(w, w; s)$$

for some constant $c > 0$ and s such that $\delta(G) < s \le \delta(G) + 1$. Here r is the rank of ζ.

Since

$$F(w') = \lim_{j \to \infty} \frac{\Phi(w, w'; s_j)}{\Phi(w, 0; s_j)}$$

we see that

$$F(w') \le c' \operatorname{Im}(w')^{r-\delta(G)}.$$

This is only compatible with the previous result if $r - \delta(G) \ge \delta(G)$. However, one has the result of Beardon (1968), generalized a little, that

$$\delta(G) > \frac{r}{2}$$

and this leads to the contradiction. Hence ζ could not have been an atom of G.

We remark here that the condition of the proposition will be satisfied if

(a) ζ is of maximal rank, N, or
(b) if G is geometrically finite.

The condition fails for the degenerate groups in general.

The results which we have already obtained will now be complemented by a simple, beautiful lemma of Sullivan (1979, Section 2). First we introduce some notation. For $w^* \in B^{N+1}$ let

$$b(w^*, r) = \{w \in B^{N+1} : L(w, w^*) < r\}$$

and

$$I(w^*, r) = \{\zeta \in S^N : \overline{0\zeta} \cap b(w^*, r) \neq \varnothing\}$$

where $\overline{0\zeta}$ is the radius joining 0 to ζ. Thus $I(w^*, r)$ is the shadow cast by a candle at the origin by the sphere $b(w^*, r)$ on the boundary S^N. Denote by E the euclidean diameter of a set $E \subset S^N$.

Sullivan's Shadow Lemma. *Let μ be a finite conformal density of dimension α for a discrete group G. Suppose μ does not consist of a single atom. Then there exist constants $r, c, c' > 0$ so that for all but a finite set of $g \in G$ one has*

$$cI(g(0), r)^\alpha < \mu(I(g(0), r)) \leq c'I(g(0), r)^\alpha.$$

Proof. Consider $g^{-1}I(g(0), r)$. One has the "earthquake picture".[1]

By choosing r large enough and excluding a finite set of g one can ensure that there exists a $c_0 > 0$ so that

$$\mu(g^{-1}(I(g(0), r)) > c_0$$

since μ does not consist of a single atom. Of course one has

$$\mu(g^{-1}I(g(0), r)) \leq \mu(S^N).$$

What we shall show is that there exists $c_1, c_1' > 0$ so that for $\eta \in g^{-1}(I(g(0), r))$,

$$c_1(1 - ||g(0)||^2) \leq j(g, \eta) \leq c_1'(1 - ||g(0)||^2).$$

This will clearly suffice as the radius of $b(g(0), r)$ is comparable with $(1 - ||g(0)||^2)$ and we can apply the transformation formula for μ. The argument is simplest if one assumes g to be hyperbolic; the other cases can be dealt with similarly. We shall now show that there is a $d > 0$ so that the distance of the repulsive fixed point of g from $g^{-1}(I(g(0), r))$ exceeds d.

To see this consider spheres S_u, passing through 0 and $g^{-1}(0)$ and tangent to S^N at u. The set of possible u describes an $N - 1$ sphere in S^N which separates S^N into two pieces, T, T' say. Suppose L is a "cone"[2] passing through the two fixed points of g and through 0; then L is stable under the subgroup of $\text{Con}(N)$ fixing these two points. Moreover it is minimal with respect to this property, and $g^{-1}(0)$ must lie on this cone. The horospheres S_u represent the limiting case in which the "angle" of the cone has been kept constant but the

---

[1] This is not an earthquake in the sense of Thurston, but an analogue of the shadow of the molten centre of the earth for seismic waves – Ed.

[2] That is constant distance surface from a geodesic – Ed.

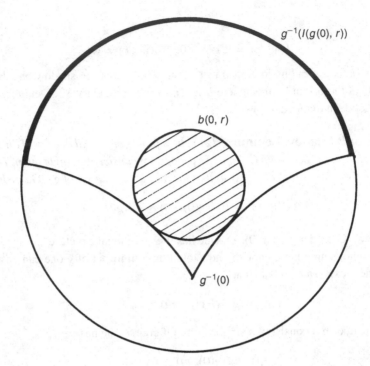

Figure IV.2.1

two endpoints have been brought together. Thus one endpoint lies in T and the other in T'. By construction, if $1 - ||g^{-1}(0)||^2$ is sufficiently small, one has that $g^{-1}I(g(0), r)$ is contained in one of the T or T', say T' since the sphere described by the possible u's has diameter comparable with $(1 - ||g(0)||^2)^{1/2}$ but $S^N - g^{-1}(I(g(0), r))$ contains a sphere of radius $d' > 0$.

Since $(1 - ||g(0)||^2)^{1/2} \to 0$, one has that the repulsive fixed point, which has to be that lying in T, is at a distance exceeding some $d > 0$ from $g^{-1}(I(g(0), r))$.

If this is so then one can deduce the required inequalities as follows. Let f be the attractive fixed point of g and f' the repulsive one. Let $\lambda = j(g, f') = j(g, f)^{-1}$. Then, noting that

$$j(g, 0) = (1 - ||g(0)||^2)$$

and for suitable constants $c_2, c_2' > 0$.

$$c_2(1 - ||g(0)||^2) \leq d(g(0), I(g(0), r)) \leq c_2'(1 - ||g(0)||^2)$$

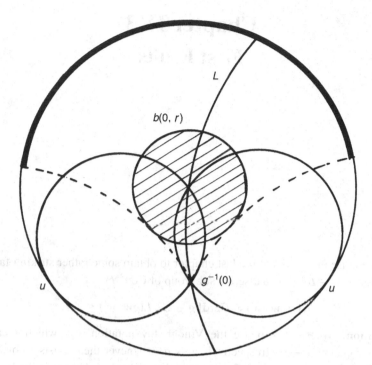

Figure IV.2.2

where d denotes the euclidean distance between two sets, we have

$$||g(0) - g(f)||^2 = \lambda^{-1} j(g, 0)1.$$

That is

$$||g(0) - f||^2 = \lambda^{-1} j(g, 0).$$

Likewise

$$||g(0) - f'|| = \lambda j(g, 0).$$

Since $||g(0) - f'||$ exceeds some fixed constant one can derive easily from these inequalities that $j(g, 0)$, λ^{-1} and $||g(0) - f||$ are all comparable with one another. It follows also that for $x \in I(g(0), r)$ one has $||x - f|| \le c_3 j(g, 0)$ for some $c_3 > 0$. Next from

$$||g(\eta) - f'|| = j(g, \eta) ||\lambda \eta - f'||^2$$

that $j(g, \eta)$ is comparable with λ^{-1}, as required.

Chapter IV.3

First Fruits

One can use the results of the last chapter to obtain some rather striking facts. Let for $w, w' \in B^{N+1}$, G a discrete subgroup of $\text{Con}(N)$,

$$N(w, w'; X) = \text{Card}\{\, g \in G : L(gw, w') \leq X \,\}$$

Notation: We shall also use the Vinogradov notation $\ll$ which means: $f(x) \ll g(x)$ as $x \to \infty$ (respectively $x \to a$) whenever there exists a constant $c > 0$ so that $|f(x)| \leq c\, g(x)$ for all sufficiently large x (respectively $|x - a|$ sufficiently small).

The first theorem gives *a priori* bounds on the function $N(w, w'; X)$, (Sullivan, 1979, Section 2).

Theorem 1.   *One has*

(i) $\delta(G) = \lim_{X \to \infty} \log N(w, w'; X) / \log X$.

(ii) $\sum_{g \in G} L(w, gw')^{-s} \ll (s - \delta(G))^{-1}$ *as* $s \to \delta(G), s > \delta(G)$.

(iii) $N(w, w'; X) \ll X^{\delta(G)}$.

Proof.   Let d denote the right-hand side of the equation in (i). If $d < \delta(G)$ then we would have that there existed a $d' < \delta(G)$ so that

$$N(w, w'; X) \ll X^d.$$

But

$$\sum_{g \in G} L(w, w')^{-s} = \int_0^\infty \xi^{-s} \, dN(w, w'; \xi)$$

$$= s \int_0^\infty \xi^{-(s+1)} N(w, w'; \xi) \, d\xi$$

308

for $s > \delta(G)$ by partial integration of Stiltjes integrals. The right-hand side of the above equation converges if $s > d'$ and moreover

$$\xi^{-s} N(w, w'; \xi) \to 0$$

for $s > d'$. It follows then that the left-hand expression converges for $s > d'$. But by definition of $\delta(G)$ one would then have $d' \geq \delta(G)$. As this is untenable we have now shown that $d \geq \delta(G)$.

Clearly (i) will now follow from (iii). The same argument as above yields (ii) from (iii). Thus it is only necessary to prove (iii).

For this purpose we fix Δ, $0 < \Delta < 1$. We consider $g \in G$ such that $L(gw, w') \in [X - \Delta X, X]$. This is a spherical shell of bounded hyperbolic thickness. Consider now r as in Sullivan's Shadow lemma. Then $b(gw, r)$ projects from 0 to $I(gw, r)$. Since all the $b(gw, r)$ have a given hyperbolic radius the number of g such that a point of some $b(gw, r)$ with $L(gw, w') \in [X - \Delta X, X]$ projects to a given fixed point $\zeta \in S^N$ is bounded independently of ζ, X. Thus we can conclude that for a conformal density of dimension α and finite total mass

$$\sum \mu(I(gw, r)) \leq c \, \mu \left(\bigcup I(gw, r) \right)$$
$$\leq c \, \mu(S^N)$$

for g in the set described above and a certain constant $c > 0$. But the right-hand side exceeds

$$c'(N(w, w'; X) - N(w, w'; X - \Delta X))X^{-\alpha}$$

since $\mu(I(gw, r))$ is comparable with $X^{-\alpha}$ by Sullivan's lemma. Thus we obtain

$$N(w, w'; X) - N(w, w'; X - \Delta X) \ll X^{\alpha}.$$

This yields, on summing,

$$N(w, w'; X) \ll X^{\alpha}.$$

Since we have a conformal density of dimension $\delta(G)$ we have now proved (iii). We have indeed proved somewhat more, namely Sullivan (1979, Section 2).

Theorem 2. *Let μ be a conformal density of dimension α for G. Then* $\alpha \geq \delta(G)$.

This leads to the following result, conjectured in Patterson (1976b; 1979) but proved independently of the conjectures by Sullivan (1979, Section 2).

Theorem 3.   Suppose $\Gamma_1 \subset \Gamma_2 \subset \cdots \subset G = \bigcup \Gamma_j$ *are discrete subgroups of* Con(N). *Then*

$$\delta(G) = \lim \delta(\Gamma_j).$$

Proof.   Let μ_j be the conformal density for Γ_j of dimension $\delta(\Gamma_j)$ constructed in the previous chapter. Then, at least on a subsequence, (μ_j) will have a weak limit on $\overline{B^{N+1}}$ which will be a conformal density of dimension $\lim \partial(\Gamma_j)$ for G. Thus, by Theorem 2

$$\lim \delta(\Gamma_j) \geq \delta(G).$$

The reverse inequality is however trivial, so that the theorem follows.

Up to this point we have dealt with results that are very general. To go further we have to put some sort of finiteness conditions on G. To begin with, we study the case of G convex cocompact (Patterson, 1976a; Sullivan, 1979, Section 3; Sullivan, 1980, Theorem 6).

Theorem 4.   *Suppose G is convex cocompact. Suppose μ is a conformal density of finite total mass supported on $L(G)$. Then the dimension of μ is $\delta(G)$ and μ is a multiple of the $\delta(G)$-dimensional Hausdorff measure. In particular, the Hausdorff dimension of μ is $\delta(G)$.*

One consequence of this is that μ is unique and hence if $E \subset L(G)$ is a measurable G-invariant subset then $\mu(E) = 0$ or $\mu(L(G) - E) = 0$. The earliest exact computations of the Hausdorff dimension of $L(G)$ for certain special classes of G were achieved by Akaza. For an account of the method, more akin to the usual computations of the Hausdorff dimension of Cantor sets than the one explained here (see Akaza–Furusawa, 1980).

Before we sketch the proof of this theorem we state another one.

Theorem 5.   *Suppose G is convex cocompact. Then there exists $c_1(w, w')$, $c_2(w, w') > 0$ for $w, w' \in B^{N+1}$ so that:*

(i) $N(w, w'; X) \geq c_1(w, w') X^{\delta(G)}$.

(ii) $\sum_{g \in G} L(w, w')^{-s} \geq c_2(w, w')/(s - \delta(G))$, $s > \delta(G), s \rightarrow \delta(G)$. *In particular* $\sum_{g \in G} L(w, w')^{-\delta(G)}$ *diverges.*

It is rather remarkable that under these hypotheses it is possible to prove the divergence of this series.

The proofs of these results are based on Hedlund's Lemma which I shall state in a form particularly suited to our applications.

Hedlund's Lemma. *Suppose G is convex compact. Let w, $w' \in B^{N+1}$ be given. There exists $r_0 > 0$ and Δ such that $0 < \Delta < 1$ with the following property:*

Let $\zeta \in L(G)$, $r > r_0$. Then there exists $g \in G$ with $L(gw, w') \in [X - \Delta X, X]$ such that

$$\zeta \in I(gw, r).$$

Proof. Let $H(L(G))$ be defined as in the first lecture. We can suppose that $0 \in H(L(G))$. Take $\rho: 0 < \rho < 1$ and note that since $\rho\zeta$ lies on the radius joining 0 and ζ, it lies also in $H(L(G))$. Thus there exists $g \in G$ such that $g^{-1}(\rho\zeta)$ lies in a fixed compact subset of B^{N+1}. Hence, for a suitable $c_1 > 0$ one will have $L(w, g^{-1}(\rho\zeta)) < c_1$, or

$$L(gw, \rho\zeta) < c_1.$$

Next, by the triangle inequality, there exists $c_2 > 1$ so that

$$c_2^{-1}L(\rho\zeta, 0) \leq L(gw, w') \leq c_2 L(\rho\zeta, 0).$$

We choose ρ so that

$$c_2\left(\frac{1}{2} + (1 - \rho^2)^{-1}\right) = X$$

and

$$\Delta = 1 - c_2^{-1}.$$

Then

$$L(gw, w') \in [X - \Delta X, X].$$

On the other hand, if $r_0 = c_1$ then by definition of $I(gw, r)$ one has that $\zeta \in I(gw, r_0)$.

It is immediate from combining Sullivan's lemma and Hedlund's lemma that for any spherical cap $C \subset S^N$ centred at $\zeta \in L(G)$ one has

$$c_3'|C|^\alpha \leq \mu(C) \leq c_3|C|^\alpha$$

where α is the dimension of μ. It is now an easy exercise in Hausdorff measures to conclude that μ is absolutely continuous with respect to Hausdorff measure on $L(G)$. Moreover, any two such densities are absolutely continuous with respect to one another, again by the result above. If the Radon–Nikodym derivative of μ with respect to Hausdorff measure were non-constant then there would exist a G-invariant set E such that $m_{\delta(G)}(E) > 0, m_{\delta(G)}(L(G) - E) > 0$. But then the conformal densities

$$S \mapsto \mu(S \cap E)$$

and

$$S \mapsto \mu(S - E)$$

would be conformal densities for G not absolutely continuous with respect to one another. This is impossible and so μ must be, up to a multiple, $m_{\delta(G)}$.

Theorem 5 can also be proved easily from Hedlund's lemma. It consists of noting now that the inequalities in the proof of Theorem 1 (iii) can all be reversed.

If G is now assumed to be geometrically finite then the situation becomes less satisfactory.

The argument above can be modified but it involves ingredients which will be introduced in the next two lectures. One of the difficulties is that the canonical measure is no longer Hausdorff measure. However, the Hausdorff dimension of $L(G)$ is still $\delta(G)$; this is proved for $N = 1$, 2 in Sullivan (1984, Section 5) but, as far as I can see, the arguments apply generally. Moreover the conformal density of dimension $\delta(G)$ supported on $L(G)$ with finite mass is unique. Sullivan was also able (Sullivan, 1984) to describe μ geometrically, at least in "good" cases. The analogues of Theorem 5 are not known in general. In the case that $\delta(G) > N/2$ then it is known by techniques to be discussed in the next lecture, but if $\delta(G) \le N/2$ then there only results available if $N = 1$. This will also be discussed in the next lecture.

There is one rather remarkable consequence of these results when $N = 1$. Any G can be exhausted by an increasing sequence of finitely generated groups Γ_j, and these are also geometrically finite. Thus one can conclude from Theorem 3 that in this case

$$\dim_H (L(G)) \ge \delta(G)$$

and indeed if $L_r(G)$ denotes the radial limit set then

$$\dim_H L_r(G) = \delta(G)$$

(see Sullivan, 1979, Section 6).

Analogues of these results are not known in higher dimensions.

It is worth noting that even when $N = 1$ one can have

$$\delta(G) \le \dim_H L(G);$$

indeed there exists for each $\varepsilon > 0$ a group G with $L(G) = S^N$ but $\delta(G) < \varepsilon$ (Patterson, 1979; 1983). It appears from computations shown to me by Mumford and Wright that the assertions of Theorem 5 fail for some finitely generated but not geometrically finite groups G when $N = 2$. One is led to ask whether the assertions of Theorem 5 characterize geometrically finite groups. For one result in this direction, Hayman's conjecture, see Patterson (1977, Section 3).

There are some cases in which one knows in advance something about the Hausdorff dimension of $L(G)$. If, for example, $\Omega(G)$ is not connected then $\dim_H L(G) \geq N - 1$ since the boundary of $\Omega(G)$ will be $(N - 1)$ dimensional, at least. Indeed in many cases one will have strict inequality. For example when $N = 2$ then a group is called quasi-Fuchsian if it is a quasi-conformally conjugate to a Fuchsian group, that is a discrete subgroup of Con(1). One can verify that if $L(G)$ is not a circle then it is non-rectifiable and so, if G is convex cocompact, one deduces from Theorem 4 that $\delta(G) > 1$, $\dim_H L(G) > 1$. These results were first derived by Bowen (1979); Bowen–Series (1979) and Sullivan (1982–1983).

These examples naturally raise the question as to the dependence of $\delta(G)$ on G, when G is deformed. The continuity of $\delta(G)$ on Teichmüller space is easy to verify (see, for example, Patterson, 1976b). More recently Phillips and Sarnak have shown in an interesting special case that $\delta(G)$ is analytic (Phillips–Sarnak, 1984a; *cf.* also Epstein, 1983, Chapter 3) on the Teichmüller space. The behaviour under singular deformations is less well understood but see Phillips–Sarnak (1984b) for some interesting results.

Chapter IV.4
Spectral Theory

In the second lecture, along with the measure μ we constructed a function F:

$$F(w) = \int P(w, \zeta)^{\delta(G)} d\mu(\zeta)$$

which satisfied

$$F(gw) = F(w) \quad (g \in G)$$

$$\Delta F = -\delta(G)(N - \delta(G))F$$

and

$$F \geq 0.$$

Recall that we had a hyperbolic volume element and a measure σ which yields a measure on $G \backslash B^{N+1}$. Likewise Δ can be defined as an essentially self-adjoint operator on a dense subspace of $L^2(G \backslash B^{N+1}, \sigma)$ (*cf.* Elstrodt, 1973a, I.1.3); we shall not go into the technical difficulties which arise at this point.

Theorem 1. *Suppose G is geometrically finite and $\delta(G) > N/2$. Then $F \in L^2(G \backslash B^{N+1}, \sigma)$. Moreover the spectrum of $-\Delta$ is contained in $[\delta(G)(N - \delta(G)), \infty]$ and $\delta(G)(N - \delta(G))$ is an isolated point of multiplicity 1 represented by F.*

Remark. The spectrum of $-\Delta$ is an object of intense investigation at present. In the special case $N = 1$ and G without parabolic elements the nature of the spectrum was determined in studies (Elstrodt, 1973a, b, 1974; Fay, 1977; Patterson, 1975, 1976c). Recently Lax and Phillips have shown that in the

314

region $[0, (N/2)^2]$ there is a discrete spectrum of finite multiplicity whereas in $[(N/2)^2, \infty]$ there is a continuous spectrum of infinite multiplicity (Lax–Phillips, 1982, 1984). This continuous spectrum can be described by means of the theory of Eisenstein series which has been developed by Mandouvalos, 1984).

One consequence of Theorem 1 is the following.

Theorem 2. *Suppose G is geometrically finite and $\delta(G) > N/2$. Let μ be a conformal density of finite total mass supported on $L(G)$. Then the dimension of μ is $\delta(G)$ and μ is unique up to a constant multiple. In particular, if $E \subset L(G)$ is G-invariant then either $\mu(E) = 0$ or $\mu(L(G) - E) = 0$.*

This theorem complements Theorem 4 of the previous chapter. Actually the restriction $\delta(G) > N/2$ can be removed, as was done by Sullivan (1984). His proof involves a much more detailed geometric study of μ.

A further consequence of the same line of thought is given in Theorem 3.

Theorem 3. *Suppose G is geometrically finite and $\delta(G) > N/2$. Then there exists $c > 0$ so that as X*

$$N(w, w'; X) < c \, F(w) \, F(w') \, X^{\delta(G)}$$

and the function, defined in $\mathrm{Re}(s) > \delta(G)$,

$$\sum_{g \in G} L(w, gw')^{-s} - c' \, F(w) \, F(w')/(s - \delta(G))$$

has an analytic continuation as an analytic function into a half plane of the form $\mathrm{Re}(s) > \delta(G) - \eta$ for some $\eta > 0$.

Observe that these results sharpen those of the last lecture in an essential fashion. The first assertion here can be improved by giving an estimate for the error of the form X^η for some $\eta < \delta(G)$; see Lax–Phillips (1982) for this, or, for an alternative method, the discussion in Patterson (1977, Section 3).

The first of these assertions provides a very convenient method for computing $\delta(G)$ practically. One has that

$$\log \left(\frac{N(w, w'; 2X)}{N(w, w'; X)} \right) \to \delta(G) \log 2$$

and this is, in computer experiments, quite rapid. It would be very reassuring to have given $\varepsilon > 0$ *a priori* estimates for suitable $X_0(\varepsilon)$ in terms, say, of a fundamental domain for G, w and w' so that for $X > X_0(\varepsilon)$

$$\left| \log \left(\frac{N(w, w'; 2X)}{N(w, w'; X)} \right) - \delta(G) \log 2 \right| < \varepsilon.$$

Such estimates do not exist at present, although it should be feasible to find them.

It should be the case that the statements of Theorem 3 hold also when $\delta(G) \leq N/2$ but we shall discuss this question after having given some idea of the proofs. It does seem plausible that the validity of the statements characterizes geometrically finite groups out of the collection of all discrete subgroups of $\mathrm{Con}(N)$ as was discussed in the third chapter.

Indications of proofs. That $F \in L^2(G \backslash B^{N+1}, \sigma)$ is a matter of simple estimations. For everything else we need the concept, due to Selberg, of a *point-pair invariant*. This is a function

$$k: B^{N+1} \times B^{N+1} \to \mathbb{C}$$

which satisfies, for all $g \in \mathrm{Con}(N)$, w, $w' \in B^{N+1}$,

$$k(gw, gw') = k(w, w').$$

If k is a point-pair invariant then there exists $k_0: [1, \infty] \to \mathbb{C}$ so that

$$k(w, w') = k_0(L(w, w'))$$

and each such function is clearly a point-pair invariant. We shall suppose that the point-pair invariants satisfy certain growth and continuity conditions about which it is better not to be specific. To each point-pair invariant we ascribe the integral operator

$$f \mapsto \left(w \mapsto \int_{B^{N+1}} k(w, w') f(w') \, d\sigma(w') =: k * f(w) \right)$$

if f is G-invariant and we let

$$k_G(w, w') = \sum_{g \in G} k(w, gw');$$

then

$$k * f(w) = \int_{G \backslash B^{N+1}} k_G(w, w') f(w') \, d\sigma(w').$$

For suitable k the k_G map $L^2(G \backslash B^{N+1}, \sigma)$ into itself. The operators $f \mapsto k * f$ commute amongst themselves and also with Δ. Moreover there is a function

$\hat{k}(s)$, analytic in s, so that if $\Delta f = -s(N-s)f$ then $k*f = \hat{k}(s)f$. Thus the spectral theory of the differential operator Δ can be investigated through the algebra of integral operators $f \mapsto k*f$. These results are due to Selberg and can be found in this article (Selberg, 1956).

In particular the resolvent kernel of $-\Delta$ can be expressed in this form although there is a singularity along the diagonal. This is a point-pair invariant r_t with the property that $\hat{r}_t(s) = (t(N-t) - s(N-s))^{-1}$, (see Elstrodt, 1973a, I. 4; Mandouvalos, 1984, Theorem 3.3). The property that we shall need of this is that for $L(w, w')$ large

$$r_t(w, w') = \sum_{0 \le r \le R} a_r(t) L(w, w')^{-t-r} + O(L(w, w')^{-t-R-1})$$

for any integral $R \ge 0$. In particular the analytic properties of $t \mapsto (r_t)_G$ are determined by those of

$$\sum_{g \in G} L(w, gw')^{-t}.$$

We shall regard the spectrum of Δ on $L^2(G \backslash B^{N+1}, \sigma)$ as being parametrized by a subset Σ_G of $\mathbb{C}$ so that $s \in \Sigma_G \Leftrightarrow s(N-s) \in$ Spectrum of $-\Delta$. Since $-\Delta$ is positive one has

$$\Sigma_G \subset [0, N] \cup \left\{ s: \mathrm{Re}(s) = \frac{N}{2} \right\}.$$

Note that s and $N-s$ correspond to the same point of the spectrum. Note here the analogy with the theory of the hydrogen atom. The parallels with quantum mechanics and the Schrödinger operator are very strong; especially striking are the parallels between the theory of Eisenstein series and scattering theory.

From the general spectral theory one has a representation of the operator $f \mapsto k_G * f$ through

$$\int_{\mathbb{C}} p(w, w'; s) \hat{k}(s) \, d\rho(s),$$

where ρ is a spectral measure supported on Σ_G and $p(w, w'; s)$ is the projection on to the $s(N-s)$-eigenspace of $-\Delta$. It is given in terms of generalized eigenfunctions. See (Patterson 1977, Section 2) for a discussion, or alternatively (Fay, 1977; Elstrodt, 1973a, Section 4).

This representation has only a meaning in L^2-terms but one can give it a meaning in the point-wise sense.

We can now simplify the argument if we make use of the results of Lax and Phillips referred to above. If we apply this spectral representation to the resolvent kernel and know that the spectrum in $[N - \delta(G), \delta(G)]$ is finite we see that if $F_1, \ldots, F_1$ is an orthogonal basis of the $\delta(G)(N - \delta(G))$-eigenspace of $-\Delta$ then,

$$\sum_{g \in G} L(w, gw')^{-s} - c \sum_i F_i(w)\overline{(F_i(w'))}/(s - \delta(G))$$

is analytic in a half plane $\mathrm{Re}(s) > \delta(G) - \eta$ for some $\eta > 0$ and $c > 0$. The same argument also shows that $\Sigma_G \cap [0, N] \subset [N - \delta(G), \delta(G)]$ (if $\delta(G) \geq N/2$) and $\Sigma_G \cap [0, N] \subset \{N/2\}$ (if $\delta(G) < N/2$).

We return to the argument of the second chapter. From this we obtain for each $w' \in B^{N+1}$ a measure μ_w, so that

$$\lim_{s \searrow \delta(G)} (s - \delta(G)) \sum_{g \in G} L(w, gw')^{-s}$$

is equal to

$$\int P(w, \zeta)^{\delta(G)} d\mu_{w'}(\zeta).$$

Moreover the various $\mu_{w'}$ can be verified to be absolutely continuous with respect to each other. Since this is also of the form

$$c \sum F_i(w) \overline{F}_i(w')$$

we deduce that F_i can be taken to be real and $F_1 = F$, $\mu_1 = \mu_0 = \mu$. If we represent F_i as

$$F_i(w) \int P(w, \zeta)^{\delta(G)} d\mu_i(\zeta)$$

then

$$\mu_{w'} = \sum F_i(w') \mu_i.$$

Let ϕ_i be the Radon-Nikodym derivative of μ_i with respect to μ_0. Since $\mu_{w'}$ is absolutely continuous with respect to μ_0 these exist and are G-invariant. Conversely any bounded G-invariant function ψ on $L(G)$ gives rise to a square-integrable eigenfunction by

$$\int P(w, \zeta)^{\delta(G)} \psi(\zeta) d\mu(\zeta).$$

This space is however finite-dimensional and one can easily derive from this that there exists M subsets $B_1, \ldots, B_M$ of $L(G)$, disjoint from one another with positive μ-measure so that each ϕ_i is a linear combination of the characteristic functions of the B_i. Let

$$\tilde{F}_i(w) = \int_{B_i} P(w, \zeta)^{\delta(G)} \, \mathrm{d}\mu_0(\zeta).$$

We can write $\sum F_i(w) \, F_i(w')$ as

$$\sum A_i(w') \, \tilde{F}_i(w)$$

where, since $\mu_{w'}(B_i) > 0$, we have $A_i(w') > 0$. Also,

$$\int \left\{ \sum_i A_i(w') \, \tilde{F}_i(w) \right\} \tilde{F}_j(w') \, \mathrm{d}\sigma(w') = \tilde{F}_j(w)$$

since $\sum F_i(w) \, F_i(w')$ has the reproducing property on the eigenspace under consideration. This gives rise to the relation

$$\sum c_{ij} \tilde{F}_i(w) = \tilde{F}_j(w)$$

where the c_{ij} are all positive. This is a non-trivial linear relation between the linearly independent $\tilde{F}_i$. This is only possible if $M = 1$. Thus $F = F_1$ is the unique eigenfunction. The assertions of Theorems 1, 2, and 3 now follow easily from this, the Wiener–Ikehara theorem being used for the first part of Theorem 3.

This argument is the one used in Patterson (1976a). It would have been possible to replace it by that of Sullivan (1984, Theorem 1). This theorem is based on ergodic-theoretic considerations and so is of rather wider applicability. The argument just given is rather more special but a little shorter.

The disadvantage of this sort of argument is that the assumption $\delta(G) > N/2$ is unavoidable. This is a serious restriction although, as was pointed out at the end of the previous chapter, one can easily give a condition that $\delta(G) > N - 1$.

If one wants to study the function $N(w, w'; X)$ or $\sum_{g \in G} L(w, gw')^{-s}$ in more depth then one has to introduce the "Eisenstein series" and undertake their analytic continuation. In the simplest case – $N = 1$, G without parabolic elements – this was done in Patterson (1975, 1976c, d) and the necessary machinery for the general problem has been developed by Mandouvalos in Mandouvalos (1984), when it is supplemented by the results of Lax-Phillips (1982, 1984). In the case that $\delta(G) < N/2$ this type of analytic continuation can be proved by a relatively simple type of argument first used in Patterson (1976c, Section 4) and generalized in Mandouvalos (1984, Chapter 4(v)). One can then obtain the analytic continuation of certain classes of $G \subset \mathrm{Con}(N)$ for which one has $\delta(G) \geq N/2$ by embedding $\mathrm{Con}(N)$ in $\mathrm{Con}(M)$ for $M > 2\delta(G)$ and using the previously developed theory for G as a discrete subgroup of $\mathrm{Con}(M)$.

From Patterson (1976d, Section 4), Theorem 1 of Chapter IV. 3 and a well-known lemma of Landau one has that ($N = 1$, $\delta(G) > 1/2$)

$$\sum_{g \in G} L(w, gw')^{-s} - c(w, w')/(s - \delta(G))$$

has an analytic continuation for some $c(w, w')$ to a neighbourhood of $\delta(G)$. Moreover by Theorem 4 of Chapter IV. 3 one has that $c(w, w') = c_0 \, F(w) \, F(w')$ where

$$F(w) = \int_{L(G)} P(w, \zeta)^{\delta(G)} \, dm_{\delta(G)}(\zeta)$$

and c_0 is a suitable constant. Unfortunately, to prove that

$$N(w, w'; X) \sim c_1 \, F(w) \, F(w') \, X^{\delta(G)}$$

which is what one expects, one would have to show that $\sum_{g \in G} L(w, gw')^{-s}$ or, equivalently, the Eisenstein series have no poles along $\mathrm{Re}(s) = \delta(G)$ other than $s = \delta(G)$. That this is unlikely is underlined by Theorem 5 of Chapter IV. 3. It is related to the theory of the Selberg zeta function in Patterson (1976d, Section 6).

In this case, $\delta(G) \leq N/2$, the functions F are no longer square-integrable. However, if G is geometrically finite then F is square-integrable on $G \backslash V_R$ where V_R is the R-neighbourhood of the set $H(L(G))$, see Sullivan (1984). This fact plays an important rôle in Patterson (1975, 1976c, d; Sullivan, 1984) and it is tempting to speculate on a more general approach to the problems of this section involving this fact. It is an interesting question, posed by Sullivan (1983, pp. 182–183), as to whether F being square-integrable on $G \backslash V_R$ characterizes geometrically finite groups, F being an eigenfunction of the type constructed in the second lecture.

Chapter IV.5
Geodesic Flows

In this chapter we shall look at the consequences of the ideas developed above for the analysis of the geodesic flow on $G \backslash B^{N+1}$. The same idea can on the other hand be used in certain cases to give an entirely different proof of the existence of the measure of the second chapter, as was shown by Rufus Bowen and Caroline Series. This approach leads to a quite different intuitive understanding and thereby to some striking results not presently accessible to the methods described above. In particular the wonderful theorem of Mary Rees will be discussed at the end of the chapter.

It will help to set up some group theoretic machinery to describe the problem. We let

$$K_0 = \text{Stab}_{\text{Con}(N)}(0)$$

which is isomorphic to $SO(N+1)$. Consider the next group,

$$A = \text{Stab}_{\text{Con}(N)}(\{(1, 0, \ldots, 0), (-1, 0, \ldots, 0)\}).$$

If $A_0 = A \cap K_0$ then $A \cong A_0 \times A_1$ where $A_1 \cong \mathbb{R}_+^\times$: it is, essentially, a split torus. If $r \colon \mathbb{R}_+^\times \to A_1$ is suitably normalized then

$$r(t)(0) = \left(\frac{1-t^2}{1+t^2}, 0, \ldots, 0\right)$$

which describes the geodesic joining $(1, 0, \ldots, 0)$ and $(-1, 0, \ldots, 0)$ at unit speed.

If, finally, we let $P = \text{Stab}_{\text{Con}(N)}(\{1, 0, \ldots, 0\})$ then $P = AU$ where U is a unipotent group, here isomorphic to $\mathbb{R}^N$. Clearly $\text{Con}(N)/K_0$ can be identified with B^{N+1} and $\text{Con}(N)/P$ with S^N. The group $\text{Con}(N)$ also operates on $S^N \times S^N \backslash \{\text{diag}\}$ and this can be identified with $\text{Con}(N)/A$. Geometrically this means that we can look at the set of geodesics where a geodesic is to be described by giving its two endpoints on S^N. These geodesics are directed in that the first endpoint can be regarded as a beginning and the second as the end.

Next we can regard the unit sphere of the tangent space at 0 as K_0/A_0. Consequently we can regard the sphere bundle as

$$\pi \colon \operatorname{Con}(N)/A_0 \to \operatorname{Con}(N)/K_0.$$

The group of operators induced by right multiplication by A_1 on $\operatorname{Con}(N)/A_0$ gives the geodesic flow.

Let now G be a discrete subgroup of $\operatorname{Con}(N)$. Then we can describe the sphere bundle over $G\backslash B^{N+1}$ as

$$G\backslash\operatorname{Con}(N)/A_0 \to G\backslash\operatorname{Con}(N)/K_0 \cong G\backslash B^{N+1}$$

and the geodesic flow on $G\backslash\operatorname{Con}(N)/A_0$ is given by right multiplication by A_1. This point of view was first pointed out by Gelfand and Fomin (1952) who then used their knowledge of the decomposition of $L^2(G\backslash\operatorname{Con}(N))$ into irreducible $\operatorname{Con}(N)$-representations and of the structure of these to prove the L^2 version of the Hopf ergodic theorem.

On $S^N \times S^N\backslash\{\mathrm{diag}\}$ we can construct a G-invariant measure $\tilde\mu$ by

$$d\tilde\mu(\zeta, \zeta') = \frac{d\mu(\zeta) \cdot d\mu(\zeta')}{||\zeta - \zeta'||^{2\delta(G)}}$$

where μ is a conformal density of dimension $\delta(G)$. From what has been said above it can be regarded as a measure on $\operatorname{Con}(N)/A_0 A_1$, invariant under the left G-action. Since A_1 has an obvious invariant measure we obtain a G-invariant measure on $\operatorname{Con}(N)/A_0$ invariant under the right A_1 action. That is, we have a measure on the sphere bundle of $G\backslash B^{N+1}$ invariant under the geodesic flow. We shall, to avoid too many symbols, write $\tilde\mu$ for any of these measures.

This measure only measures those geodesics both of whose endpoints are in $L(G)$. One might have expected that a suitable measure on the geodesic flow was one for which one measured those elements which, for $t \to +\infty$ converged to an element of $L(G)$. However, one has the following phenomenon first noted by Hopf (1939, 1971, 1937) in his proof of the ergodicity for groups with finite co-volume, that if $a_1, a_2 \in \operatorname{Con}(N)/A_0$ and if the positive endpoints of a_1 and a_2 in $\operatorname{Con}(N)/A_0 A_1 \cong S^N \times S^N\backslash\{\mathrm{diag}\}$ are the same, then there exists $t_0 > 0$ so that the distance between $\pi(a_1 r(t))$ and $\pi(a_2 r(tt_0))$ in $G/K_0 \cong B^{N+1}$ tends *exponentially* to 0. This means that the negative endpoint, where the element of the geodesic came from in the long distant past, is almost irrelevant.

One should also note that μ, as a measure on $S^N \cong \operatorname{Con}(N)/P$ induces a measure on $G\backslash\operatorname{Con}(N)/A_0$ but the measure (which is U-invariant) is mapped into a certain multiple of itself by the right action of $a \in A_1$. This measure, the one which we have principally investigated up to the present, will be of much less relevance here.

The following theorem describes Sullivan's results on the geodesic flow specialized to the case of geometrically finite G; these are taken from Sullivan (1979, 1984, 1983). The actual results proved are rather more general but are conditional and it would complicate matters to quote them in detail here.

Theorem 1. *Suppose G is geometrically finite. Then $\tilde{\mu}$ is a finite measure and the geodesic flow, with invariant measure $\tilde{\mu}$, is ergodic. The measure-theoretic entropy of the geodesic flow is $\delta(G)$ (with respect to $\tilde{\mu}$) and $\tilde{\mu}$ is the measure maximizing entropy. Moreover, if G is convex-cocompact then the topological entropy of the geodesic flow is also $\delta(G)$.*

For the concepts involved here see (Bowen, 1975) and (Denker–Grillenberger–Sigmund, 1976).

This theorem can be regarded as a very extensive generalization of Hopf's ergodic theorem (Hopf, 1939, 1971, 1937). Apropos of this theorem and how it was regarded in the years following its appearance see Weil (1947, p. 316). The proof of the theorem above is based on the same basic ideas as Hopf's but there are additional considerations needed here. It will not be possible here to even attempt a sketch of this.

From the proof it also follows that $\sum_{g \in G} L(w, gw')^{-\delta(G)}$ diverges. From the considerations adumbrated in the previous chapter one would expect the stronger result, that

$$\liminf_{s \searrow \delta(G)} (s - \delta(G)) \sum_{g \in G} L(w, gw')^{-s} > 0$$

for G geometrically finite, should hold, but this remains, in this generality, a conjecture.

The statement can be sharpened somewhat. Rudolph (*cf.* Sullivan, 1983, p. 183) has shown that the geodesic flow is Bernoulli, and so, mixing.

Before we move on to our major theme here, it is worthwhile to look briefly at the ideas of Gelfand and Fomin again. As we have seen, $\tilde{\mu}$ can be regarded as a positive linear functional on the space of bounded continuous functions $C_b(G\backslash\mathrm{Con}(N)/A_0 A_1)$, or, equivalently, as an A-invariant functional on $C_b(G\backslash\mathrm{Con}(N))$. The space $C_b(G\backslash\mathrm{Con}(N))$ is a dense subspace of $L^2(G\backslash\mathrm{Con}(N))$ and one might hope that one could understand the nature of the average temporal development of $t \mapsto \phi(gr(t))$. This is indeed the case but the information obtained concerns rather the dissipative part. The Hopf ergodic theorem shows that

$$\frac{1}{|x|} \int_X \phi(ga) \, da \rightarrow M(\phi) \quad (\tilde{\mu} - a.e.)$$

as X becomes a longer and longer interval in A_1 with I as 'lower' endpoint. On the other hand the Gelfand–Fomin approach shows that

$$\int_X \phi(ga) \, da$$

converges weakly in $L^2(G\backslash\mathrm{Con}(N))$ and even with an error term $O(\exp(-\mathrm{Min} \ (N - \delta(G), N/2)|X|))$; if $\delta(G) < N$ this limit is zero.

The linear functional M can be described explicitly in terms of $\tilde{\mu}$. Moreover, if F denotes, as usual the eigenfunction which we regard as a function on $G\backslash\mathrm{Con}(N)$ by letting $F(g)$ be $F(g(0))$, then

$$F(g_1) F(g_2) = \int \phi(g_3^{-1}, g_3^{-1}g_2) \, d\tilde{\mu}(g_3)$$

where ϕ is so chosen that

$$\int_{A_1} \phi(ag_1, ag_2) \, da = 2^{\delta(G)} \, P(g_1(0), (1, 0, \ldots, 0))^{\delta(G)}$$

$$\times P(g_2(0), (-1, 0, \ldots, 0))^{\delta(G)}.$$

This formula is the point of Sullivan (1984, Section 5). Moreover $F(g_1) F(g_2)$ appears as the fundamental expression in the considerations of the previous lecture.

One might hope that in view of this one could understand the linear form M better. One question which one would like to answer is the following. Let V be an irreducible smooth representation of $\mathrm{Con}(N)$ and $\Lambda: V \rightarrow \mathbb{C}$ a G-invariant linear form such that for each $v \in V$ the function $g \mapsto \Lambda(gv)$ lies in the domain of M. Then M induces an A_1-invariant linear form on V. For which V is this form defined, for which is it 0?

After this digression we can now return to our main theme. The properties of μ recall those of Gibbs measures, for which see Bowen (1975) and Denker–Grillenberger–Sigmund (1976). This has been made much more precise by Bowen (1979) and Series (1981a, b). The basis of this approach is the Hedlund–Morse idea of symbolic dynamics. In this we shall model the geodesic flow by means of a shift operator acting on a sequence space.

To explain this let D be a fundamental domain for G, finite faced if G is geometrically finite. Let $\mathcal{A}$ (the alphabet) be the set of $g \in G$ such that gD and

D have a face in common. Thus if G is geometrically finite $\mathcal{A}$ will be a finite set; in all cases $\mathcal{A}$ is a set of generators of G. The set $\mathcal{A}$ is, of course, by no means canonical. Suppose that $0 \in D$.

Let $\zeta \in L(G)$. We join 0 to ζ by a radius, which is also a hyperbolic geodesic. Then 0ζ passes through the fundamental domains

$$D, \; g_1 D, \; g_1 g_2 D, \; g_1 g_2 g_3 D, \ldots$$

where $g_1, g_2, g_3, \ldots \in \mathcal{A}$. We assume that the geodesic 0ζ only passes through N-dimensional faces of D and not the lower dimensional boundary components for simplicity of exposition. To ζ we associate the function

$$s_\zeta \colon \mathbb{N} \to \mathcal{A}; \; j \to g_j.$$

Let Σ_+ be the set of all these functions. The map $\zeta \mapsto s_\zeta$ is clearly injective and hence we can define the inverse map

$$p \colon \Sigma_+ \to L(G).$$

One could enlarge Σ_+ to included all geodesics from 0 but then some elements of Σ_+ would break off. This will not be important to us. It would also be sensible to introduce the space of sequences corresponding to a geodesic joining two different limit points but this will not be relevant to our needs.

Series (1981a) shows how to describe Σ_+ a priori (following old ideas of Dehn) and also how to overcome the fudge made above when $N = 1$. It is also possible to treat the case $N > 1$ by methods that are more ergodic-theoretical; for this see Rees (1981a, §1).

Denote by

$$[a_1, \ldots, a_m] = \{\sigma \in \Sigma_+ \; : \; \sigma(j) = a_j \; (1 \le j \le m)\}$$

for $a_1, \ldots, a_M \in \mathcal{A}$. Let,

$$T \colon \Sigma_+ \to \Sigma_+; \; T(\sigma)(j) = \sigma(j+1) \quad (j \ge 1)$$

be the shift operator. It is clearly defined and one has that for all but a countable set of pairs $(\sigma, \sigma') \in \Sigma_+^2$ the following two statements are equivalent:

(i) *there exists $g \in G$ such that $p(\sigma') = g(p(\sigma))$,*
(ii) *there exist $m, n > 0$ such that $T^m(\sigma) = T^n(\sigma')$.*

In this sense the shift operator T encodes the action of G on $L(G)$ now modelled by Σ_+.

In order to deal with problems of the kind which we have been discussing it is necessary to introduce some metrical structure onto Σ_+ and this is the objective of the following definitions.

The shift operator $T: \Sigma_+ \to \Sigma_+$ induces an operator $\tilde{T}: L(G) \to L(G)$. Now let us choose open sets $O_a \supset p([a])$ $(a \in \mathcal{A})$ which are pair-wise disjoint. One can extend the action of T to O_a by $\tilde{T}|O_a = a^{-1}|O_a$. We can define $\tilde{T}$ thus on $\bigcup_{a \in \mathcal{A}} O_a$. We define also the conformal distortion of $\tilde{T}$ to be

$$j(\tilde{T}, \zeta) = j(a^{-1}, \zeta) \quad \text{for } \zeta \in O_a.$$

It is intuitively fairly clear that $\tilde{T}$ should be *expanding* so that the theory of Bowen (1975) will be applicable. Moreover one has

$$L(G) = \bigcap_{j \geq 0} \tilde{T}^{-j}(O)$$

Now that we have defined $j(\tilde{T}, \zeta)$ we can restrict it to $L(G)$; we form from it the following *negative* function of $x \in \Sigma_+$:

$$\phi(x) = -\log j(\tilde{T}, p(x)).$$

We have now to recall some basic notions from the theory of Gibbs measures. Given a distribution function f on a sequence space Σ with shift operator $T: \Sigma \to \Sigma$ one forms a certain functional $P_T(f)$ called the *pressure* of f. This functional is positive on positive functions, convex, continuous and subadditive. One can also define the *topological entropy* of T, $h_{\text{top}}(T)$, to be $P_T(0)$.

One can also define for any T-invariant probability measure m on Σ a measure-theoretical measure $h_m(T)$. Then one has fundamental inequality

$$h_m(T) + \int f \, dm \leq P_T(f).$$

There is a T-invariant probability measure $\mu(f, T)$ on Σ, a so-called *Gibbs measure*, such that

$$h_{\mu(f,T)}(T) + \int f \, d\mu(f, T) = P_T(f).$$

For details as to the classes of functions involved and definitions see Bowen (1975) and Denker–Grillenberger–Sigmund (1976).

If $\mu(f, T)$ is a Gibbs measure then there exist constants $c_1, c_2 > 0$ so that if Z is an m-cylinder then the ratio of $\mu(f, T)(Z)$ and

$$\exp \left\{ P_r(f) + \sup_{x \in Z} \sum_{k=0}^{m-1} f(T^k(x)) \right\}$$

lies between c_1 and c_2; see Bowen (1975, p. 32).

The functional $P_r(f)$ has its origins in the thermodynamic pressure, although this might not be very evident; the reader might find the discussion in Bowen (1975) or Ruelle (1978) helpful.

One has

$$h_m(T) \leq h_{\text{top}}(T)$$

and if m is a Gibbs measure $\mu(0, T)$ then equality is attained. The measure-theoretic entropy measures the amount of disorder introduced by T, either in a thermodynamic context or in the context of information theory and coding (Billingsley, 1965; Khinchin, 1957).

The point behind Bowen and Series' construction is that one chooses f to be of the form $\delta\phi$, where ϕ is as above. We are now back with our original Σ_+, T. As ϕ is negative, $P_T(\phi)$ is not necessarily positive; indeed $P_T(0) \geq 0$ but $P_T(\phi) \leq 0$. One defines $\delta_1(G)$ to be the value of δ for which

$$P_T(\delta\phi) = 0.$$

In view of the inequality above, $\mu = \mu(\delta_1(G)\phi T)$, is a measure on Σ_+, or, if we like, on $L(G)$ via p. It has the property that $\mu(Z)$ is comparable with

$$\exp\left\{ \operatorname*{Sup}_{x \in Z} \sum_{k=0}^{m-1} \delta_1(G)\phi(T^k(x)) \right\}.$$

On examining what this means we see that it gives an estimate for sets of the form $I(g(0), r)$ of the same type as discussed in the third chapter. More precisely $p(Z)$ is, if Z is $[a_1, \ldots, a_m]$, essentially,

$$L(G) \cap (a_1, \ldots, a_m)(O_{a_m}).$$

By construction and the chain rule,

$$\exp\left\{ \operatorname*{Sup}_{x \in Z} \sum_{0 \leq k < m} \delta_1(G)\phi(T^k(x)) \right\}$$

is

$$\operatorname*{Sup}_{\xi \in O_{a_m} \cap L(G)} j((a_1, \ldots, a_m)^{-1}, \xi)^{-\delta_1(G)}.$$

This is, again by the chain rule

$$\operatorname*{Sup}_{\xi \in (a_1, \ldots, a_m)(O_{a_m}) \cap L(G)} j(a_1, \ldots, a_m, \xi)^{\delta(G)}.$$

Since μ is translation invariant one obtains the translation property of the theorem of the second lecture. It can now be identified as a conformal density of dimension $\delta(G)$, since it follows that $\delta_1(G) = \delta(G)$. The information given above is actually rather more precise and yields estimates of the kind for which we had previously to apply Sullivan's Shadow lemma. Note here that the technique involves splitting G up into shells of a given length which is reminiscent of the method used by Akaza, cf. (Akaza–Furusawa, 1980).

The success of this method depends on the choice of $\delta \cdot \phi$ in the construction of the Gibbs measure. Although this many seem unmotivated and it does to the writer, this type of choice is usual in the theory of Axiom A flows; see (Bowen, 1975, p. 95ff). In particular one can regard $j(g, \zeta)^{\delta(G)}$ as the "jacobian" of g on the "$\delta(G)$-dimensional space" $L(G)$.

There are strong relations between the definitions of the topological entropy and Hausdorff dimension which makes this relation rather more plausible (see Billingsley, 1965, Chapter 4, Section 14; Bowen, 1973).

There are other aspects of the theory of Gibbs measures which have not been used in the discussion above. These are, in general the case,

(a) *for "good" functions, for a suitable norm $\| \ \|$ one has*
$$\int f_1 \cdot (T^n) * f_2 \, d\mu = \int f_1 \, d\mu \int f_2 \, d\mu + O(\|f_1\| \cdot \|f_2\| \gamma^n)$$
as $n \to \infty$ for some $\gamma < 1$,
(b) *for some $\sigma \geq 0$*

$$\mu \left\{ x \in \Sigma : \sum_{0 \leq j \leq n} (T^j) * (f)(x) - n \int f \, d\mu \leq r \cdot \sqrt{n} \right\} \to$$

$$\frac{1}{\sigma\sqrt{2\pi}} \int_{-\infty}^{r} \exp(-r^2/2\sigma^2) dr, \text{ as } n \to \infty.$$

The probabilistic meaning of these is clear.

A Gibbs measure need not be unique; in the case in question Rees used the measure constructed in Chapter IV. 2 to give a particularly well-behaved Gibbs measure with the following property (see Rees, 1981a, p. 119):

There exist $c > 0$, $\eta > 0$ with $\eta < 1$ so that, if Z is an m-cylinder and $W \subset Z$, then

$$|\mu(Z)/\mu(T(Z)) - \mu(W)/\mu(T(W))| < c\eta^m \mu(Z)/\mu(T(Z)).$$

She used the properties of this to prove the following particularly beautiful theorem (Rees, 1981a, b):

Theorem 2. *Let $G \subset \mathrm{Con}(1)$ be a finitely generated discrete subgroup. Let A be an abelian group and $\alpha: G \to A$ be a homomorphism. Let $G_1 = \mathrm{Ker}(\alpha)$.*

Then

$$\delta(G_1) = \delta(G).$$

In fact she has proved a much more precise theorem, determining when $\sum L(w_1, gw_2)^{-\delta(G_1)}$ converges or diverges. There are examples of homomorphisms $\alpha: G \to A$, where A is no longer abelian where $\delta(\mathrm{Ker}(\alpha)) < \delta(G)$ (see Patterson, 1979) and one suspects that $\delta(\mathrm{Ker}(\alpha)) = \delta(G)$ precisely when $\mathrm{Im}(\alpha)$ is amenable (or, perhaps, what would imply this, when $\mathrm{Im}(\alpha)$ is of subexponential growth) (for the notions of the order of growth of a finitely generated group and of an amenable group see Gromov (1981) and Brooks (1981). That this is plausible follows from the discussion below, where we indicate an asymptotic estimation (3) which reflects the order of growth of $\mathrm{Im}(\alpha)$ in an analytic expression. The examples of Patterson (1979) are certainly consistent with this expectation. (I am indebted to Kazhdan for pointing out that it was more natural to expect the correct condition on $\mathrm{Im}(\alpha)$ to be "amenable" rather than "of polynomial growth" which is what I had first expected.) The results of Brooks (1981) and Phillips–Sullivan (1981) are also significant here.[1]

The idea behind the proof of Theorem 2 can be sketched as follows. We suppose that α is surjective. Let $\hat{A}$ be the dual group of A. For $\chi \in \hat{A}$ define

$$G(w_1, w_2; s, \chi) = \sum_{g \in G} L(w_1, gw_2)^{-s} \chi(g).$$

Then

$$\int_{\hat{A}} G(w_1, w_2; s, \chi)\, d\chi = \sum_{g \in G_1} L(w_1, gw_2)^{-s}.$$

The group $\hat{A}$ is compact. We have to show that the left-hand side has no analytic continuation to a neighbourhood of $\delta(G)$. It is possible that when $\delta(G) > N/2$ one could use perturbation theoretic methods in conjunction with the ideas described in the previous lecture; then one could produce rather detailed information on $G(w_1, w_2; s, \chi)$ from which the assertion above could be derived. This remains merely a hope but for some interesting results in this direction see Epstein (1983).

Next we define

$$T_m(w_1, w_2; \chi) = \sum L(w_1, gw_2)^{-\delta(G)} \chi(g)$$

---

[1] After these notes were written, Brooks confirmed this suggestion when $\delta(G) > N/2$; see Brooks, (1985). The question of the divergence or convergence type of G_1 has also been considerably clarified, see Varopoulos (1987).

where the sum is over those g of length m in $\mathcal{A}$. This is comparable with

$$T_m^0(\chi) = \sum \mu(Z)\chi^0(Z)$$

where Z runs over all m-cylinders and,

$$\chi^0(a_1, \ldots, a_m) = \chi(a_1, \ldots, a_m).$$

Thus

$$T_m^0(1) = 1$$

and moreover $|T_m^0(\chi)| \le 1$ with $|T_m^0(\chi)| = 1$ if and only if $\chi = 1$ or $\chi = \chi_0$ where

$$\chi_0(a_j) = -1$$

for all j, since $a_j \in \mathcal{A}$ implies that $a_j^{-1} \in \mathcal{A}$. It may or may not happen that χ_0 exists.

If G has no parabolic elements then there exist $c_1, c_2 > 0$ so that

$$c_1 m \le \log \mu(Z)^{-1} \le c_2 m \tag{1}$$

where Z is an m-cylinder.

Now

$$\int_{\hat{A}} T_m^0(\chi)\,d\chi$$

is comparable with

$$\sum_{\substack{g \in G \\ \text{length}(g) = m}} \{L(w_1, gw_2)\}^{-\delta(G)}$$

and in view of (1) it suffices to show that there exist $c > 0$, $l \ge 0$ so that

$$\int_{\hat{A}} T_m^0(\chi)\,d\chi \ge c \cdot m^{-1}. \tag{2}$$

From this it would follow that $\sum_{g \in G} L(w_1, gw_2)^{-s}$ diverges for $s \le \delta(G)$ so that

$$\delta(G_1) = \delta(G).$$

Moreover $\sum_{g \in G_1} L(w_1, gw_2)^{-\delta(G)}$ diverges if $l \le 1$. The converse inequality to (2) would imply that this condition is also sufficient.

The proof of (2) consists of splitting $\hat{A}$ into U_m (a suitable neighbourhood of 1), V_m (a suitable neighbourhood of χ_0, if this exists), and $\hat{A} - (U_m \cup V_m)$. One then shows that the integral

$$\int_{\hat{A}-(U_m\cup V_m)} T_m^0(\chi)\,d\chi$$

is relatively insignificant since $T_m^0(\chi)$ itself is small. On the other hand one can estimate $T_m(\chi)$ near 1 or χ_0. These give asymptotic formulae for

$$\int_{U_m} T_m(\chi)\,d\chi \text{ and } \int_{V_m} T_m(\chi)\,d\chi.$$

Of these the second is oscillating and one obtains that (2) holds in the slightly weaker form

$$\int_{\hat{A}} T_m^0(\chi) + T_{m+1}^0(\chi)\,d\chi \geq m^{-1} \tag{2'}$$

which is sufficient for our purposes.

The derivation of (2) from (a) and (c) involves a great deal of delicate analysis which we cannot describe here. To make it plausible we shall assume that (a) holds without an error term. Thus

$$\mu(Z \cup T^{-m}Z') = \mu(Z)\mu(Z')$$

if Z is an s-cylinder with $s \leq m$. From this follows at once

$$\mu(a_1,\ldots,a_m) = \mu(a_1)\cdots\mu(a_m)$$

and consequently

$$T_m(\chi) = T_1(\chi)^m.$$

Suppose now that $\hat{A} \cong (\mathbb{R}/\mathbb{Z})^K$. Considering $T_1(\chi)$ as a function on $\mathbb{R}^K$ one has a Taylor expansion of the form $1 - Q(\xi) + O(\|\xi\|^4)$ where Q is a positive definite quadratic form on $\mathbb{R}^K$ and $\|\ \|$ is the usual euclidean norm. This follows from the definition of $T_m^0(\chi)$ when one writes $\chi(Z) = \exp(2\pi i\, l_Z(\xi))$, with l_Z a linear form, since the quadratic term is

$$-(2\pi)^2 \sum_Z \mu(Z)\, l_Z(\xi)^2$$

the sum being over all 1-cylinders. Writing $\hat{A}$ additively one also has

$$T_m^0(\chi + \chi_0) = (-1)^m\, T_m^0(\chi)$$

so that we also have an expansion about χ_0.

In this case we define $U_m = \{\xi : \|\xi\| \leq 1/m^{1/2}\}$, $V_m = \chi_0 + U_m$. Then one can easily verify that

$$\int_{\hat{A}-(U_m \cup V_m)} T_1(\chi)^m \, d\chi = O(\beta^m)$$

for some $\beta < 1$. On the other hand,

$$\int_{U_m} T_1(\chi)^m \, d\chi \sim c \cdot m^{-K/2} \tag{3}$$

for some $c > 0$, and

$$\int_{V_m} T_1(\chi)^m \, d\chi \sim c \, (-1)^m m^{-K/2}. \tag{3'}$$

Note here that $\mathrm{Im}\,(\alpha)$ *is of polynomial growth of order K. The proofs of these statements is straightforward from the Taylor expansion above.*

It is clear that (3) and (3') given a result of the type (2').

This sketch may give some insight into the rather difficult technical arguments of Rees (1981a, b) where the simplifying assumptions that G has no parabolic elements and that (a) holds without an error term are dispensed with.

There is a final point which should be emphasized here. The property (a) of Gibbs measures has no analogue in the language which we have used in the earlier lectures since there we would not have spoken of T. Thus this approach does lead to insights which are inaccessible to a more "classical" approach and conversely.

References

Ahlfors, L. (1964), "Finitely generated Kleinian groups," *Amer. J. Math.*, **86**, 413–429.

Akaza, T. and Furusawa, H. (1980), "The exponent of convergence of Poincaré series of some Kleinian groups," *Tôhoko Math. J.*, **32**, 447–452.

Beardon, A.F. (1968), "The exponent of convergence of Poincaré series," *Proc. London Math. Soc.*, **18**, 461–483.

Beardon, A.F. (1971), "Inequalities for certain Fuchsian groups," *Acta Math.*, **127**, 221–258.

Beardon, A.F. (1983), *The Geometry of Discrete Groups*, Springer, New York.

Beardon, A.F. and Maskit, B. (1974), "Limit points of Kleinian groups and finite sided fundamental polyhedra," *Acta Math.*, **132**, 1–12.

Billingsley, P. (1965), *Ergodic theory and information*, John Wiley.

Bowen, R. (1973), "Topological entropy for non-compact sets," *Trans. Amer. Math. Soc.*, **184**, 125–136.

Bowen, R. (1975), "Equilibrium states and the ergodic theory of Anosov diffeomorphisms," *Springer Lecture Notes*, **470**.

Bowen, R. (1979), "Hausdorff dimension of quasi-circles," *Publ. Math. IHES*, **50**, 11–25.

Bowen, R. and Series, C. (1979), "Markov maps associated with Fuchsian groups," *Publ. Math. IHES*, **50**, 153–170.

Brooks, R. (1981), "The fundamental group and the spectrum of the Laplacian," *Comm. Math. Helv.*, **56**, 581–598.

Brooks, R. (1985), "The bottom of the spectrum of a Riemannian covering," *J.f.d.r.u.a. Math.*, **357**, 101–114.

Denker, M., Grillenberger, C. and Sigmund, K. (1976), "Ergodic theory on compact spaces," *Springer Lecture Notes*, **527**.

Elstrodt, J. (1973a), "Die Resolvente zum Eigenwertproblem der automorphen Formen in der hyperbolischen Ebene I," *Math. Ann.*, **203**, 295–300.

Elstrodt, J. (1973b), "Die Resolvente zum Eigenwertproblem der automorphen Formen in der hyperbolischen Ebene II," *Math. Z.*, **132**, 99–134.

Elstrodt, J. (1974), "Die Resolvente zum Eigenwertproblem der automorphen Formen in der hyperbolischen Ebene III," *Math. Ann.*, **203**, 99–132.

Epstein, C.L. (1983), *The spectral theory of geometrically periodic hyperbolic 3-manifolds*, Ph. D. thesis, Courant Institute, N.Y. University.

Fay, J.D. (1977), "Fourier coefficients of the resolvent for a Fuchsian group," *J. d. reine u. angew. Math.*, **293**, 142–203.

Gelfand, I.M. and Fomin, S.V. "Geodesic flows on manifolds of constant negative curvature," *Uspekhi Mat. Nauk*, **7**, 118–137, 1952. *Amer. Math. Soc. Translations*, **1**, 1965, 49–65.

Gromov, M. (1981), "Groups of polynomial growth and expanding maps," *Publ. Math. I.H.E.S.*, **53**, 53–78.

Hopf, E. (1937), *Ergodentheorie, Ergebnisse der Math.*, Vol. 5, Springer-Verlag.

Hopf, E. (1939), "Statistik der geodä Linien in Mannigfaltigkeiten negativer Krümmung," *Ber. Verh. Sächs. Akad. Wiss. Leipzig*, **91**, 261–304.

Hopf, E. (1971), "Ergodic theory and the geodesic flow on surfaces of constant negative curvature," *Bull. Amer. Math. Soc.*, **77**, 863–877.

Khinchin, A.I. (1957), *Mathematical Foundations of Information Theory*, Dover.

Lax, P.D. and Phillips, R.S. (1982), "The asymptotic distribution of lattice points in Euclidean and non-Euclidean space," *J. Fnl. Ahal.*, **46**, pp. 280–350.

Lax, P.D. and Phillips, R.S. (1984), "Translation representations for automorphic solutions of the wave equation in non-Euclidean spaces I," *Commun. Pure Appl. Math.*, **XXXVII**, 303–28.

Mané, R., Sad, P. and Sullivan, D. (1983), "On the dynamics of rational maps," *Ann scient Ec Norm Sup.*, **16**, 193–217.

Mandouvalos, N. (1984), *The theory of Eisenstein Series and Spectral Theory for Kleinian Groups*, Ph. D. thesis, Cambridge.

Marden, A. (1977), "Geometrically finite Kleinian groups and their deformation spaces," In *Discrete Groups and Automorphic Functions* (ed. W.J. Harvey), Academic Press, pp. 259–293.

Patterson, S.J. (1975), "The Laplacian operator on the Riemann surface I," *Comp. Math.*, **31**, 83–107.

Patterson, S.J. (1976a), "The limit set of a Fuchsian group," *Acta Math.*, **136**, 241–273.

Patterson, S.J. (1976b), "The exponent of convergence of Poincaré series," *Monatsh. f. Math.*, **82**, 297–315.

Patterson, S.J. (1976c), "The Laplacian operator on the Riemann surface II," *Comp. Math.*, **32**, 71–112.

Patterson, S.J. (1976d) "The Laplacian operator on the Riemann surface III," *Comp. Math.*, **33**, 227–259.

Patterson, S.J. (1977), "Spectral theory and Fuchsian groups," *Math. Proc. Camb. Phil. Soc.*, **81**, 59–75.

Patterson, S.J. (1979), "Some examples of Fuchsian groups," *Proc. Lond. Math. Soc.* **39**, 276–298.

Patterson, S.J. (1983), "Further remarks on the exponent of convergence of Poincaré series," *Tohoku Math. J.*, **35**, 357–373.

Phillips, R.S. and Sarnak, P. (1984a), *On the Spectrum of the Hecke Groups, Duke Math. J.* **52** (1985), 211–221.

Phillips, R.S. and Sarnak, P. (1984b), "The Laplacian for domains in hyperbolic space and limit sets of Kleinian groups," *Acta Math.* **155** (1985), 173–241.

Phillips, A. and Sullivan, D. (1981), "Geometry of leaves," *Topology*, **20**, 209–218.

Rees, M. (1981a), "Checking ergodicity of some geodesic flows with infinite Gibbs measure," *Ergod. Thy Dyn. Systs.*, **1**, 107–133.

Rees, M. (1981b), "Divergence type of some subgroups of finitely generated Fuchsian groups," *Ergod. Thy Dyn. Systs.*, **1**, 209–221.

Ruelle, D. (1978), *Thermodynamical Formalism*, Addison-Wesley/CUP.

Selberg, A. (1956), "Harmonic analysis and discontinuous groups in weakly symmetric Riemannian spaces with applications to Dirichlet series," *J. Indian Math. Soc.* **20**, 47–87.

Series, C. (1981a), "Symbolic dynamics for geodesic flows," *Acta Math.*, **146**, 103–128.

Series, C. (1981b), "The infinite word problem and limit sets in Fuchsian groups," *Ergod. Thy Dyn. Systs.*, **1**, 337–360.

Sullivan, D. (1978), *On the ergodic theory at infinity of an arbitrary discrete group of hyperbolic motions, Ann. of Math. Studies*, **91**, 465–496.

Sullivan, D. (1979), "The density at infinity of a discrete group of hyperbolic motions," *Publ. Math. IHES*, **50**, 171–202.

Sullivan, D. (1980), λ-potential theory on manifolds, preprint.

Sullivan, D. (1981a), "Conformal dynamical systems," *Proc. Intl. Symp. Dyn. Systs. Rio de Janeiro*, 725–752.

Sullivan, D.P. (1981b), *Travaux de Thurston sur les groupes quasi-fuchsiens et les variétés hyperboliques de dimension 3 fibrés sur S^1*, Séminaire Bourbaki 554, Lecture Notes 842, Springer Verlag.

Sullivan, D. (1982–1983), *Seminar on conformal and Hyperbolic Geometry*, 1982–1983. Notes by M. Baker and J. Seade: Mimeographed notes from IHES.

Sullivan, D. (1983), "Discrete conformal groups and measurable dynamics," *Proc. Symp. Pure Maths*, **39**, 169–185.

Sullivan, D. (1984), "Entropy, Hausdorff measures old and new, and limit sets of geometrically finite Kleinian groups," *Acta Math.*, **153**, 259–77.

Thurston, W.P. (1979), *The Geometry and Topology of 3-Manifolds*, http://msri.org/publications/books/gt3m/

Thurston, W.P. (1982), "Three-dimensional manifolds, Kleinian groups and hyperbolic geometry," *Bull. AMS*, **6**, 357–381.

Varopoulos, N. Th. (1987), "Finitely generated Fuchsian groups," *J. Reine Angew. Math.* **375/376**, 394–405.

Weil, A. (1947a), "L'avenir des mathématiques," in *La pensée mathématique*, ed. F. Le Lionnais, Cahiers du sud, 307–320, 1947. Oeuvres Scientifiques.

Printed in the United States
by Baker & Taylor Publisher Services